Course 2 Core-Plus Mathematics

Contemporary Mathematics in Context

2nd Edition

Christian R. Hirsch • James T. Fey • Eric W. Hart
Harold L. Schoen • Ann E. Watkins
with
Beth E. Ritsema • Rebecca K. Walker • Sabrina Keller
Robin Marcus • Arthur F. Coxford • Gail Burrill

 **Glencoe**

New York, New York   Columbus, Ohio   Chicago, Illinois   Woodland Hills, California

Glencoe

The *McGraw-Hill* Companies

This material is based upon work supported, in part, by the National Science Foundation under grant no. ESI 0137718. Opinions expressed are those of the authors and not necessarily those of the Foundation.

Send all inquiries to:
Glencoe/McGraw-Hill
8787 Orion Place
Columbus, OH 43240-4027

ISBN: 978-0-07-877258-0 (Student Edition)
MHID: 0-07-877258-3 (Student Edition)

Core-Plus Mathematics
Contemporary Mathematics in Context
Course 2 Student Edition

Printed in the United States of America.

3 4 5 6 7 8 9 10 079/043 16 15 14 13 12 11 10 09 08

Core-Plus Mathematics 2
Development Team

Senior Curriculum Developers
Christian R. Hirsch (Director)
Western Michigan University

James T. Fey
University of Maryland

Eric W. Hart
Maharishi University of Management

Harold L. Schoen
University of Iowa

Ann E. Watkins
California State University, Northridge

Contributing Curriculum Developers
Beth E. Ritsema
Western Michigan University

Rebecca K. Walker
Grand Valley State University

Sabrina Keller
Michigan State University

Robin Marcus
University of Maryland

Arthur F. Coxford (deceased)
University of Michigan

Gail Burrill
Michigan State University
(First edition only)

Principal Evaluator
Steven W. Ziebarth
Western Michigan University

Advisory Board
Diane Briars
Pittsburgh Public Schools

Jeremy Kilpatrick
University of Georgia

Robert E. Megginson
University of Michigan

Kenneth Ruthven
University of Cambridge

David A. Smith
Duke University

Mathematical Consultants
Deborah Hughes-Hallett
University of Arizona / Harvard University

Stephen B. Maurer
Swarthmore College

William McCallum
University of Arizona

Doris Schattschneider
Moravian College

Richard Scheaffer
University of Florida

Evaluation Consultant
Norman L. Webb
University of Wisconsin-Madison

Technical Coordinator
James Laser
Western Michigan University

Collaborating Teachers
Mary Jo Messenger
Howard County Public Schools, Maryland

Jacqueline Stewart
Okemos, Michigan

Graduate Assistants
Allison BrckaLorenz
Christopher Hlas
University of Iowa

Michael Conklin
University of Maryland

Dana Grosser
Anna Kruizenga
Nicole Lanie
Diane Moore
Western Michigan University

Undergraduate Assistants
Cassie Durgin
University of Maryland

Rachael Kaluzny
Jessica Tucker
Ashley Wiersma
Western Michigan University

Core-Plus Mathematics 2 Field-Test Sites

Core-Plus Mathematics 2 builds on the strengths of the 1st edition which was shaped by multi-year field tests in 36 high schools in Alaska, California, Colorado, Georgia, Idaho, Iowa, Kentucky, Michigan, Ohio, South Carolina, and Texas. Each revised text is the product of a three-year cycle of research and development, pilot testing and refinement, and field testing and further refinement. Special thanks are extended to the following teachers and their students who participated in the testing and evaluation of 2nd Edition Course 2.

Hickman High School
Columbia, Missouri
  Melissa Hundley
  Stephanie Krawczyk
  Cheryl Lightner
  Amy McCann
  Tiffany McCracken
  Ryan Pingrey
  Michael Westcott

Holland Christian High School
Holland, Michigan
  Tim Laverell
  Brian Lemmen
  Betsi Roelofs
  John Timmer
  Mike Verkaik

Jefferson Junior High School
Columbia, Missouri
  Lori Kilfoil

Malcolm Price Lab School
Cedar Falls, Iowa
  Megan Balong
  Dennis Kettner
  James Maltas

Oakland Junior High School
Columbia, Missouri
  Dana Sleeth

Riverside University High School
Milwaukee, Wisconsin
  Cheryl Brenner
  Dave Cusma
  Alice Lanphier
  Ela Kiblawi
  Ulices Sepulveda

Rock Bridge High School
Columbia, Missouri
  Nancy Hanson
  Emily Hawn
  Lisa Holt
  Betsy Launder
  Linda Shumate

Sauk Prairie High School
Prairie du Sac, Wisconsin
  Joel Amidon
  Shane Been
  Kent Jensen
  Joan Quenan
  Scott Schutt
  Mary Walz

Washington High School
Milwaukee, Wisconsin
  Anthony Amoroso

West Junior High School
Columbia, Missouri
  Katie Bihr

Overview of Course 2

1 FUNCTIONS, EQUATIONS, AND SYSTEMS

Functions, Equations, and Systems reviews and extends student ability to recognize, describe, and use functional relationships among quantitative variables, with special emphasis on relationships that involve two or more independent variables.

Topics include direct and inverse variation and joint variation; power functions; linear equations in standard form; and systems of two linear equations with two variables, including solution by graphing, substitution, and elimination.

2 MATRIX METHODS

Matrix Methods develops student understanding of matrices and ability to use matrices to represent and solve problems in a variety of real-world and mathematical settings.

Topics include constructing and interpreting matrices, row and column sums, matrix addition, scalar multiplication, matrix multiplication, powers of matrices, inverse matrices, properties of matrices, and using matrices to solve systems of equations.

3 COORDINATE METHODS

Coordinate Methods develops student understanding of coordinate methods for representing and analyzing properties of geometric shapes, for describing geometric change, and for producing animations.

Topics include representing two-dimensional figures and modeling situations with coordinates, including computer-generated graphics; distance in the coordinate plane, midpoint of a segment, and slope; coordinate and matrix models of rigid transformations (translations, rotations, and line reflections), of size transformations, and of similarity transformations; animation effects.

Overview of Course 2

REGRESSION AND CORRELATION

Regression and Correlation develops student understanding of the characteristics and interpretation of the least squares regression equation and of the use of correlation to measure the strength of the linear association between two variables.

Topics include interpreting scatterplots; least squares regression, residuals and errors in prediction, sum of squared errors; Pearson's correlation coefficient, lurking variables, and cause and effect.

NONLINEAR FUNCTIONS AND EQUATIONS

Nonlinear Functions and Equations introduces function notation, reviews and extends student ability to construct and reason with functions that model parabolic shapes and other quadratic relationships in science and economics, with special emphasis on formal symbolic reasoning methods, and introduces common logarithms and algebraic methods for solving exponential equations.

Topics include formalization of function concept, notation, domain and range; factoring and expanding quadratic expressions, solving quadratic equations by factoring and the quadratic formula, applications to supply and demand, break-even analysis; common logarithms and solving exponential equations using base 10 logarithms.

NETWORK OPTIMIZATION

Network Optimization develops student understanding of vertex-edge graphs and ability to use these graphs to solve network optimization problems.

Topics include optimization, mathematical modeling, algorithmic problem solving, digraphs, trees, minimum spanning trees, distance matrices, Hamilton circuits and paths, the Traveling Salesperson Problem, critical paths, and the PERT technique.

Overview of Course 2

7 TRIGONOMETRIC METHODS

Trigonometric Methods develops student understanding of trigonometric functions and the ability to use trigonometric methods to solve triangulation and indirect measurement problems.

Topics include sine, cosine, and tangent functions of measures of angles in standard position in a coordinate plane and in a right triangle; indirect measurement; analysis of variable-sided triangle mechanisms; Law of Sines and Law of Cosines.

8 PROBABILITY DISTRIBUTIONS

Probability Distributions further develops student ability to understand and visualize situations involving chance by using simulation and mathematical analysis to construct probability distributions.

Topics include Multiplication Rule, independent events, conditional probability, probability distributions and their graphs, waiting-time distributions, expected value, and rare events.

Contents

Contents

Contents

Contents

Preface

The first three courses in *Core-Plus Mathematics* provide a significant common core of broadly useful mathematics for all students. They were developed to prepare students for success in college, in careers, and in daily life in contemporary society. Course 4 continues the preparation of students for success in college mathematics and statistics courses. The program builds upon the theme of mathematics as sense-making. Through investigations of real-life contexts, students develop a rich understanding of important mathematics that makes sense to them and which, in turn, enables them to make sense out of new situations and problems.

Each course in *Core-Plus Mathematics* shares the following mathematical and instructional features.

- **Integrated Content** Each year, the curriculum advances students' understanding of mathematics along interwoven strands of algebra and functions, statistics and probability, geometry and trigonometry, and discrete mathematics. These strands are unified by fundamental themes, by common topics, and by mathematical habits of mind or ways of thinking. Developing mathematics each year along multiple strands helps students develop diverse mathematical insights and nurtures their differing strengths and talents.
- **Mathematical Modeling** The curriculum emphasizes mathematical modeling including the processes of data collection, representation, interpretation, prediction, and simulation. The modeling perspective permits students to experience mathematics as a means of making sense of data and problems that arise in diverse contexts within and across cultures.
- **Access and Challenge** The curriculum is designed to make mathematics accessible to more students while at the same time challenging the most able students. Differences in student performance and interest can be accommodated by the depth and level of abstraction to which core topics are pursued, by the nature and degree of difficulty of applications, and by providing opportunities for student choice on homework tasks and projects.

- **Technology** Numeric, graphic, and symbolic manipulation capabilities such as those found on many graphing calculators are assumed and appropriately used throughout the curriculum. The curriculum materials also include a suite of computer software called *CPMP-Tools* that provide powerful aids to learning mathematics and solving mathematical problems. (See page xvii for further details.) This use of technology permits the curriculum and instruction to emphasize multiple representations (verbal, numerical, graphical, and symbolic) and to focus on goals in which mathematical thinking and problem solving are central.
- **Active Learning** Instructional materials promote active learning and teaching centered around collaborative investigations of problem situations followed by teacher-led, whole-class summarizing activities that lead to analysis, abstraction, and further application of underlying mathematical ideas and principles. Students are actively engaged in exploring, conjecturing, verifying, generalizing, applying, proving, evaluating, and communicating mathematical ideas.
- **Multi-dimensional Assessment** Comprehensive assessment of student understanding and progress through both curriculum-embedded assessment opportunities and supplementary assessment tasks supports instruction and enables monitoring and evaluation of each student's performance in terms of mathematical processes, content, and dispositions.

Integrated Mathematics

Core-Plus Mathematics replaces the traditional Algebra-Geometry-Advanced Algebra/Trigonometry-Precalculus sequence of high

Preface

school mathematics courses with a sequence of courses that features concurrent and connected development of important mathematics drawn from four strands.

The Algebra and Functions strand develops student ability to recognize, represent, and solve problems involving relations among quantitative variables. Central to the development is the use of functions as mathematical models. The key algebraic models in the curriculum are linear, exponential, power, polynomial, logarithmic, rational, and trigonometric functions. Modeling with systems of equations, both linear and nonlinear, is developed. Attention is also given to symbolic reasoning and manipulation.

The primary goal of the Geometry and Trigonometry strand is to develop visual thinking and ability to construct, reason with, interpret, and apply mathematical models of patterns in visual and physical contexts. The focus is on describing patterns in shape, size, and location; representing patterns with drawings, coordinates, or vectors; predicting changes and invariants in shapes under transformations; and organizing geometric facts and relationships through deductive reasoning.

The primary role of the Statistics and Probability strand is to develop student ability to analyze data intelligently, to recognize and measure variation, and to understand the patterns that underlie probabilistic situations. The ultimate goal is for students to understand how inferences can be made about a population by looking at a sample from that population. Graphical methods of data analysis, simulations, sampling, and experience with the collection and interpretation of real data are featured.

The Discrete Mathematics strand develops student ability to solve problems using vertex-edge graphs, recursion, matrices, systematic counting methods (combinatorics), and voting methods. Key themes are discrete mathematical modeling, optimization, and algorithmic problem-solving.

Each of these strands of mathematics is developed within focused units connected by fundamental ideas such as symmetry, matrices, functions, data analysis, and curve-fitting. The strands also are connected across units by mathematical habits of mind such as visual thinking, recursive thinking, searching for and explaining patterns, making and checking conjectures, reasoning with multiple representations, inventing mathematics, and providing convincing arguments and proofs.

The strands are unified further by the fundamental themes of data, representation, shape, and change. Important mathematical ideas are frequently revisited through this attention to connections within and across strands, enabling students to develop a robust and connected understanding of mathematics.

Active Learning and Teaching

The manner in which students encounter mathematical ideas can contribute significantly to the quality of their learning and the depth of their understanding. *Core-Plus Mathematics* units are designed around multi-day lessons centered on big ideas. Each lesson includes 2–5 mathematical investigations that engage students in a four-phase cycle of classroom activities, described in the following paragraph—Launch, Explore, Share and Summarize, and Check Your Understanding. This cycle is designed to engage students in investigating and making sense of problem situations, in constructing important mathematical concepts and methods, in generalizing and proving mathematical relationships, and in communicating, both orally and in writing, their thinking and the results of their efforts. Most classroom activities are designed to be completed by students working collaboratively in groups of two to four students.

The launch phase of a lesson promotes a teacher-led class discussion of a problem situation and of related questions to think about, setting the context for the student work to follow. In the second or explore phase, students investigate more focused problems and questions related to the launch situation. This investigative work is followed by a teacher-led class discussion in which students summarize mathematical ideas developed in their groups, providing an opportunity to construct a shared understanding of important concepts, methods, and approaches. Finally, students are given tasks to complete on their own to check their understanding of the concepts and methods.

Each lesson also includes homework tasks to engage students in applying, connecting, reflecting on, extending, and reviewing their mathematical understanding. These On Your Own tasks are central to the learning goals of each lesson and are intended primarily for individual work outside of class. Selection of tasks should be based on student performance and the availability of time and technology. Students can exercise some choice of tasks to pursue; and at times, they should be given the opportunity to pose their own problems and questions to investigate.

Multiple Approaches to Assessment

Assessing what students know and are able to do is an integral part of *Core-Plus Mathematics*. There are opportunities for assessment in each phase of the instructional cycle. Initially, as students pursue the investigations that comprise the curriculum, the teacher is able to informally assess student understanding of mathematical processes and content and their disposition toward mathematics. At the end of each investigation, a class discussion to Summarize the Mathematics provides an opportunity for the teacher to assess levels of understanding that various groups of students have reached as they share and explain their findings. Finally, the Check Your Understanding tasks and the tasks in the On Your Own sets provide further opportunities to assess the level of understanding of each individual student. Quizzes, in-class tests, take-home assessment tasks, and extended projects are included in the teacher resource materials.

Acknowledgments

Development and evaluation of the student text materials, teacher materials, assessments, and computer software for *Core-Plus Mathematics 2nd Edition* was funded through a grant from the National Science Foundation to the Core-Plus Mathematics Project (CPMP). We express our appreciation to NSF and, in particular, to our program officer John Bradley for his long-term trust, support, and input.

We are also grateful to Texas Instruments and, in particular, Dave Santucci for collaborating with us by providing classroom sets of graphing calculators to field-test schools.

As seen on page iii, CPMP has been a collaborative effort that has drawn on the talents and energies of teams of mathematics educators at several institutions. This diversity of experiences and ideas has been a particular strength of the project. Special thanks is owed to the exceptionally capable support staff at these institutions, particularly to Angela Reiter, Matthew Tuley, and Teresa Ziebarth at Western Michigan University.

We are grateful to our Advisory Board, Diane Briars (Pittsburgh Public Schools), Jeremy Kilpatrick (University of Georgia), Robert E. Megginson (University of Michigan), Kenneth Ruthven (University of Cambridge), and David A. Smith (Duke University) for their ongoing guidance and advice. We also acknowledge and thank Norman L. Webb (University of Wisconsin-Madison) for his advice on the design and conduct of our field-test evaluations.

Special thanks are owed to the following mathematicians: Deborah Hughes-Hallett (University of Arizona/Harvard University), Stephen B. Maurer (Swarthmore College), William McCallum (University of Arizona), Doris Schattschneider (Moravian College), and to statistician Richard Scheaffer (University of Florida) who reviewed and commented on units as they were being developed, tested, and refined.

Our gratitude is expressed to the teachers and students in our 10 evaluation sites listed on page iv. Their experiences using the revised *Core-Plus Mathematics* units provided constructive feedback and suggested improvements that were immensely helpful.

Finally, we want to acknowledge Lisa Carmona, James Matthews, Carrie Mollette, Gary Walker, Karen Vujnovic, and their colleagues at Glencoe/McGraw-Hill who contributed to the publication of this program.

To the Student

Have you ever wondered ...

- How computer animations are created?
- How athletic footwear stores predict footwear inventory needs?
- How a complex construction project is scheduled to be completed in the shortest amount of time?
- How the intensity of sound at a concert depends on your distance from the speakers?
- How detailed maps of rugged mountain ranges are made?
- How unusual it is for a wife to be taller than her husband?
- How insurance companies determine what to charge for a policy?

The mathematics you will learn in *Core-Plus Mathematics* Course 2 will help you answer questions like these.

Because real-world situations and problems often involve data, shape, quantity, change, or chance, you will study concepts and methods from several interwoven strands of mathematics. In particular, you will develop an understanding of broadly useful ideas from algebra and functions, geometry and trigonometry, statistics and probability, and discrete mathematics. In the process, you will also see many connections among these strands.

In this course, you will learn important mathematics as you investigate and solve interesting problems. You will develop the ability to reason and communicate about mathematics as you are actively engaged in understanding and applying mathematics. You will often be learning mathematics in the same way that many people work in their jobs—by working in teams and using technology to solve problems.

In the 21st century, anyone who faces the challenge of learning mathematics or using mathematics to solve problems can draw on the resources of powerful information technology tools. Calculators and computers can help with calculations, drawing, and data analysis in mathematical explorations and solving mathematical problems.

Graphing calculators and computer software tools will be useful in work on many of the investigations in *Core-Plus Mathematics*.

To the Student

The curriculum materials include computer software called *CPMP-Tools* that will be of great help in learning and using the mathematical topics of each CPMP course.

The software toolkit includes four families of programs:

- Algebra—The software for work on algebra problems includes an electronic spreadsheet and a computer algebra system (CAS) that produces tables and graphs of functions, manipulates algebraic expressions, and solves equations and inequalities.
- Geometry—The software for work on geometry problems includes an interactive drawing program for constructing, measuring, and manipulating geometric figures and a set of custom tools for studying computer animation and geometric models of physical mechanisms.
- Statistics—The software for work on data analysis and probability problems provides tools for graphic display and analysis of data, simulation of probabilistic situations, and mathematical modeling of quantitative relationships.

- Discrete Mathematics—The software for work on graph theory problems provides tools for constructing, manipulating, and analyzing vertex-edge graphs.

In addition to the general purpose tools provided for work on tasks in each strand of the curriculum, *CPMP-Tools* includes files of most data sets essential for work on problems in *Core-Plus Mathematics* Course 2. When you see an opportunity to use computer tools for work on a particular investigation, select the *CPMP-Tools* menu corresponding to the content involved in the problem. Then select the submenu items corresponding to the required mathematical operations and data sets.

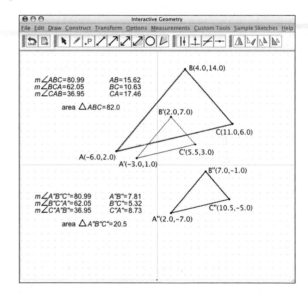

In Course 2, you're going to learn a lot of useful mathematics. It will make sense to you and you can use it to make sense of your world. You're going to learn a lot about working collaboratively on problems and communicating with others as well. You're also going to learn how to use technological tools intelligently and effectively. Finally, you'll have plenty of opportunities to be creative and inventive. Enjoy!

FUNCTIONS, EQUATIONS, AND SYSTEMS

Mathematical problems that arise in science, government, business, sports, and the arts usually involve combinations of several variables and several conditions relating those variables.

For example, karate students often try to demonstrate their skill by breaking boards. Success depends on finding the right combination of speed, mass, and aim for the karate chop and on the length, width, thickness, and strength of the target board.

Through work on the investigations in this unit, you will develop the understanding and skill needed to solve problems that involve several variables and relationships among those variables. Key ideas will be developed through your work on three lessons.

Lessons

1 Direct and Inverse Variation

Use algebraic ideas and symbols to express direct and inverse variation relationships among variables, to write those relationships in useful equivalent forms, and to solve problems where direct and inverse variation are involved.

2 Multivariable Functions

Use functions of two variables to represent quantitative relationships that involve combinations of direct and inverse variation and sums of direct variation expressions. Use graphs to study solutions for linear equations with two variables.

3 Systems of Linear Equations

Use estimation and symbolic reasoning to solve systems of two linear equations with two variables by methods involving inspection of graphs, substitution of variables, and elimination of variables.

Direct and Inverse Variation

The Winter Olympic Games include downhill, Super G, slalom, and giant slalom ski-race events. The downhill has the longest runs (over 3,000 meters) and the greatest vertical drops (over 800 meters) with the greatest distance between turns and the highest speeds—Olympic downhill racers achieve speeds of up to 85 miles per hour! The slalom, most technical of the four events, has the shortest runs and the smallest vertical drops with the least distance between turns. However, the goal in all four events is the same—to reach the finish line in the shortest time.

If you have ever raced down a hill—on foot, skis, a sled, a bike, inline skates, or a skateboard—then experience tells you that the time it takes depends on many variables.

Consider various sports that involve downhill racing. Think about the factors that decrease or increase the time it takes to travel from top to bottom.

a For downhill or slalom skiing, how will changes in the length of the course affect race time? How will changes in vertical drop of the course affect race time? How will changes in the distance between turns affect race time?

b What other factors will affect downhill or slalom ski-race times? How will changes in each of those variables increase or decrease race time?

c Pick another downhill race sport that interests you and think about the variables that affect race time in that sport event. What changes in those variables will increase the time to travel from top to bottom? What changes will decrease the time?

In previous courses, you used tables of values, graphs, and symbolic rules to represent functions relating independent and dependent variables. You recognized and described a variety of common patterns in those relationships. In this lesson, you will develop your understanding and skill in dealing with two special types of relationships—direct and inverse variation.

Investigation 1 — On a Roll

In downhill racing on skis, sleds, bikes, or skateboards, changes in two variables may have opposite effects on time from top to bottom. This makes it difficult to predict the combined effect if two variables change at the same time. However, through experimentation, you can examine the effects of change in each variable separately and then build a model of the multivariable relation.

As you work on the questions of this investigation, look for answers to these questions:

How do course length and steepness affect run time for a downhill race?

How can the relationship between those variables be expressed in symbolic form?

Platform Height, Ramp Length, and Ride Time Ramps are often used in skateboarding, not just for "getting air," but also for starting into terrain parks or street races. Many ramps are attached to a raised platform as pictured below.

1. The height of the platform and length of the ramp affect the time it takes to roll down the ramp.

   a. For a fixed ramp length, how do you think the time it takes to ride down the ramp will change as platform height increases?

   b. For a fixed platform height, how do you think the time it takes to ride down the ramp will change as ramp length increases?

   c. Suppose that one skateboard ramp is twice as long as another ramp. What relationship between platform heights for those ramps do you think will allow skateboarders starting at the top of each ramp at the same time to reach the bottom at the same time?

To explore the effects of platform height and ramp length on the time it takes to ride downhill, you can conduct an experiment designed by Galileo over 500 years ago. To get ideas about the effects of gravity, he timed trips of a ball rolling down ramps of various heights and lengths. You can build a ramp with a piece of V-shaped wood or metal. You will find it easier to do accurate timing if the ramps are fairly long and gently sloped.

Carry out the ramp experiments described in Problems 2 and 3 below. Divide the experimental work among the members of your class, with each team taking one fixed ramp length or platform height to study.

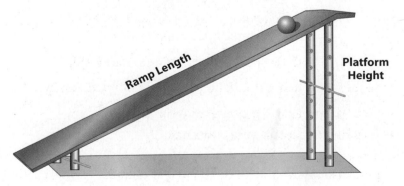

Ramp Length

Platform Height

2. To see how the time T it takes a ball to roll down a ramp changes as the ramp length L increases, experiment with ramps that have lengths varying from 3 feet to 8 feet but a platform height of 0.5 feet in each case. Then rerun the experiment using a platform height of 0.25 feet in each case.

   a. Record the results of each experiment in a table like this:

Ramp Length (in feet)	3	4	5	6	7	8
Roll Time (in seconds) at 0.5-ft Height						
Roll Time (in seconds) at 0.25-ft Height						

   b. To study the patterns relating roll time to ramp length, make plots of the (L, T) data—one for each platform height.

   c. Examine the data patterns from the two experiments. For each platform height, describe the relationship between roll time and ramp length.

3. To see how the time T it takes a ball to roll down a ramp changes as the platform height H increases, experiment with platform heights varying from 0.25 feet to 1.5 feet but fixed ramp length of 8 feet in each case. Then change the fixed ramp length to 4 feet and see how roll time is related to platform height in that case.

   a. Record the results of each experiment in a table like this:

Platform Height (in feet)	0.25	0.5	0.75	1.0	1.25	1.5
Roll Time (in seconds) for 8-ft Ramp						
Roll Time (in seconds) for 4-ft Ramp						

   b. To study the patterns relating platform height and roll time, make plots of the (H, T) data—one for each ramp length.

   c. Examine the data patterns from the two experiments. For a fixed ramp length, describe the relationship between roll time and platform height.

4. Compare the results from your experiments in Problems 2 and 3 to your responses to the questions of Problem 1. Discuss any surprises and try to explain why the results make sense.

Families of Variation Patterns The data patterns and graphs that show how roll time depends on ramp length and platform height may remind you of other relationships between variables that you have seen in prior mathematical studies.

5. Decide which of the following functions have table and graph patterns that:

   • are similar to the (*ramp length, time*) relationship.

   • are similar to the (*platform height, time*) relationship.

   • are different from those relationships.

   Be prepared to explain your decisions.

   a. The sales tax on store purchases in Michigan is a function of the purchase price and can be calculated with the formula $T = 0.06p$.

b. When a club was planning its Halloween party at the Fun House, the planners figured the cost per person using $C = \frac{225}{n}$, where n is the number of club members who would attend.

c. The number of tickets sold to a charity basketball game is a function of the price charged with rule $N = 4{,}000 - 50p$.

d. When a doctor or nurse gives an injection of medicine like penicillin, the amount of active medicine t hours later can be estimated by a function like $M = 300(0.6^t)$.

e. The circumference of a circle is related to the radius by the formula $C = 2\pi r$.

f. When a football punt leaves the kicker's foot, its height above the ground at any time in its flight is given by a function like $h = -16t^2 + 80t + 4$.

g. The average speed for the Daytona 500 race is a function of the time it takes to complete the race with rule $s = \frac{500}{t}$.

6 Using your data plots from Problems 2 and 3, experiment with function graphs to find function rules that seem to be good models for the relationships between:

a. roll time T and ramp length L for each platform height you tested in Problem 2.

b. roll time T and platform height H for each ramp length you tested in Problem 3.

Basic Variation Patterns The situations in this investigation involved a variety of functions relating dependent and independent variables. Several examples involved special patterns called *direct* and *inverse variation*.

Direct Variation: If the relationship of variables y and x can be expressed in the form:

$$y = kx \text{ for some constant } k,$$

then we say that **y varies directly with x** or that **y is directly proportional to x**. The number k is called the *constant of proportionality* for the relationship.

The close connection between multiplication and division of numbers implies that if y is directly proportional to x, then $\frac{y}{x} = k$. The symbolic form $\frac{y}{x} = k$ shows that the ratio of y to x is constant, for any corresponding values of y and x.

7 Explain why the perimeter P of a square is directly proportional to the length s of a side.

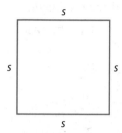

a. What equation shows this direct proportionality relationship?

b. What is the constant of proportionality?

c. How does the value of P change as the value of s increases steadily? How is this pattern of change related to the constant of proportionality?

8 Identify the direct variation relationships in Problem 5. For each:

a. Explain how the value of the dependent variable changes as the value of the independent variable steadily increases.

b. Describe the relationship of the variables involved by completing a sentence like this: "The variable _____ is directly proportional to _____, with constant of proportionality _____."

c. Express the relationship between the variables in an equivalent symbolic form that shows the constant ratio of the two variables.

Inverse Variation: If the relationship of variables y and x can be expressed in the form:

$$y = \frac{k}{x} \text{ for some constant } k,$$

then we say that **y varies inversely with x** or that **y is inversely proportional to x**. The number k is called the *constant of proportionality* for the relationship.

Once again, the close connection between multiplication and division of numbers implies that if y is inversely proportional to x, then $xy = k$. The symbolic form $xy = k$ shows that the product of y and x is constant, for any corresponding values of x and y.

9 The time t required to download a 4-megabyte music file from an Internet music seller is inversely proportional to the rate r at which data is transferred to the receiving computer.

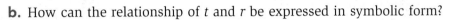

a. How long will it take to download a 4-megabyte file if the transmission occurs at a rate of 2.5 megabytes per minute? How long if the transmission rate is 0.8 megabytes per minute?

b. How can the relationship of t and r be expressed in symbolic form?

c. How does the value of t change as the value of r increases steadily? How is this pattern of change related to the constant of proportionality?

10 Identify the inverse variation relationships in Problem 5. For each:

a. Explain how the value of the dependent variable changes as the value of the independent variable increases steadily.

b. Express the relationship between the variables in two different but equivalent symbolic forms.

c. Describe the relationship of the variables involved by completing a sentence like this: "The variable _____ is inversely proportional to _____, with constant of proportionality _____."

11 Examine the tables below, each of which describes a relation between x and y.

Table I

x	25	50	60	100	150
y	8	4	3.33	2	1.33

Table II

x	10	15	25	40	100
y	6	9	15	24	60

Table III

x	5	15	30	64	80
y	9.6	3.2	1.6	0.75	0.6

a. Which relations involve direct variation? What is the constant of proportionality in each case?

b. Which relations involve inverse variation? What is the constant of proportionality in each case?

Summarize
the Mathematics

Functions that model direct and inverse variation relationships have tables, graphs, and rules that are related in ways that make reasoning about them easy.

a Suppose y is directly proportional to x with constant of proportionality $k > 0$.

    i. If the value of x increases by 1, how will the value of y change?

    ii. If the value of x doubles, how will the value of y change?

    iii. What will the graph of the function look like?

b A function with rule $y = mx + b$ ($b \neq 0$) is *not* a model of direct variation.

    i. How is the graph of such a linear function different from that of the related direct variation function $y = mx$? How are graphs of the two types of functions similar?

    ii. How is the table of (x, y) values for such a linear function different from that of the related direct variation function $y = mx$? How are tables of the two types of functions similar?

c Suppose y is inversely proportional to x with constant of proportionality $k > 0$.

    i. How will the value of y change as the value of x increases?

    ii. If the value of x doubles, how will the value of y change?

    iii. What will the graph of the function look like?

Be prepared to share your ideas and reasoning with the class.

✔ Check Your Understanding

For each of the following functions, indicate whether it involves direct variation, inverse variation, or neither of those special relationships. For those that do involve direct or inverse variation, identify the constant of proportionality and write a sentence like those in Problems 8 and 10 that describes the relationship.

a. The number of sheets in a stack of copier paper is related to the height of the stack in centimeters by $N = 100h$.

b. If you step on a dirty nail, you might get bacteria in the wound. An initial number of 50 bacteria could grow to $B = 50(2^t)$ bacteria at a time t hours later.

c. If a car uses g gallons of gasoline in a 200-mile test run, its fuel efficiency is calculated by the formula $E = \dfrac{200}{g}$.

d. The stretch of a bungee cord (in feet) is related to the jumper's weight (in pounds) by $C = 0.2w$.

e. The upward velocity v (in feet per second) of a high volleyball serve is related to time in flight t (in seconds) by $v = 64 - 32t$.

The most common forms of direct and inverse variation can be expressed by the equations $y = kx$ and $y = \frac{k}{x}$. But there are other important examples of direct and inverse variation—expressed with rules in the form $y = kx^r$ and $y = \frac{k}{x^r}$. Because the pattern relating x and y is determined by the exponent or power r in each case, these relationships are called **power functions**.

As you work on the problems of this investigation, look for answers to these questions:

What are the patterns of variation that can be modeled well by power functions?

What practical and scientific problems can be solved by use of power functions?

Modeling Sound and Light Intensity

The intensity of sound from a voice or of light from a flashlight is related to the distance from the source to the receiving ear or surface. The more distant the source, the lower the sound or light intensity at the receiving end.

1. The following graphs show three possible patterns for functions relating sound or light intensity to distance. The graphs show distance from source to receiver as the independent variable and sound or light intensity as the dependent variable.

Graph I   **Graph II**   **Graph III**

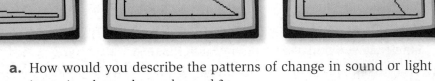

a. How would you describe the patterns of change in sound or light intensity shown by each graph?

b. Which graph do you believe is most likely to model accurately the relationship between sound or light intensity and distance from the source? Be prepared to explain your thinking.

② You could test your ideas about the (*distance, intensity*) relationship by collecting data from an experiment. But you can also get good ideas by mathematical reasoning alone. Consider what would happen if you were to enter a dark room and shine a small flashlight directly at a flat surface like a wall. The flashlight will create a circle of light on the wall.

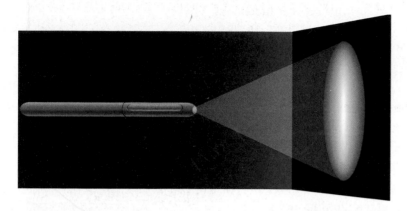

a. Complete entries in the following table that contains measurements of light circle diameter for one flashlight that has been held at several distances from a wall. Distance and diameter measurements are in feet. Express the area in terms of π.

Light Circle Measurements

Distance from Light Source, *x*	1	2	3	4	5	6
Diameter of Light Circle, *d*	2	4	6	8	10	12
Radius of Light Circle, *r*						
Area of Light Circle, *A*						

b. Write rules that show:

   i. diameter of light circle as a function of distance from the light source.

   ii. radius of light circle as a function of distance from the light source.

   iii. area of light circle as a function of distance from the light source.

c. Describe the relationships of the geometric variables diameter, radius, and area by completing sentences like this: "The variable _____ is _____ proportional to _____, with constant of proportionality ____."

d. Light energy is measured in a unit called *lumens*. The intensity of light is measured in lumens per unit of area. As the light circle of a flashlight or lamp increases in size, the intensity of light decreases.

To explore how that decrease in light intensity is related to distance from source to target, suppose that the flashlight that gave (*distance, diameter*) values in Part a produces 160 lumens of light energy. Use the area data from Part a to complete this table relating light intensity *I* to distance *x*.

Light Intensity Measurements

Distance from Light, *x*	1	2	3	4	5
Area of Light Circle, *A*	π	4π			
Light Intensity, *I*	$\dfrac{160}{\pi}$	$\dfrac{160}{4\pi}$			

e. Write a rule that shows light intensity I as a function of distance x from source to receiving surface.

f. Study the graph of the light intensity function in Part e.

    i. Which of the graph shapes in Problem 1 seems to best model the pattern of change in light intensity as distance from source to receiver increases?

    ii. Explain in words what the pattern of change shown by the light intensity function and its graph tells about the effective range of a flashlight or lamp.

The Power Function Family The functions describing dependence of light circle area and light intensity on distance are only two of many direct and inverse variation patterns that can be modeled by rules in the form $y = kx^r$ and $y = \dfrac{k}{x^r}$.

Use your calculator or a computer graphing tool to explore the relationship between the power r, the constant of proportionality k, and the numerical patterns relating x and y. To get a good picture of each graph, be sure to set your graphing window so you see both positive and negative values of x and y. You can use the zoom feature of your graphing tool to see more of each graph.

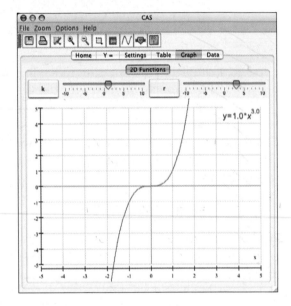

3 How is shape of the graph for a direct variation function $y = kx^r$ related to the values of r and k?

a. To see patterns that help answer this question, you might begin by studying examples in which $k = 1$ and $r = 1, 2, 3, 4, 5,$ and 6. Then explore what happens for different positive and negative values of the proportionality constant k. Describe the patterns you observe.

b. Based on the calculations involved in the different rules used in Part a, explain why the different observed patterns make sense.

4 How is the shape of the graph for an inverse variation function $y = \dfrac{k}{x^r}$ related to the values of r and k?

a. To see patterns that help answer this question, you might begin by studying examples in which $k = 1$ and $r = 1, 2, 3, 4, 5,$ and 6. Then explore what happens for different positive and negative values of the proportionality constant k. Describe the patterns that you observe.

b. Based on the calculations involved in the different rules that you used in Part a, explain why those patterns make sense.

Modeling Roll-Time Data Patterns Scientists often get ideas for theories by studying patterns in experimental data. They usually test their theories by comparing predictions of theory to results of further experiments.

   In searching for functions that are accurate models for the relationship between roll time T and platform height H, you probably found that the patterns of data from your experiments looked a lot like those of inverse variation functions. The following display shows a plot of data from experiments with an 8-foot ramp at various platform heights. A graph of $T = \frac{2}{H}$ is plotted on top of the data.

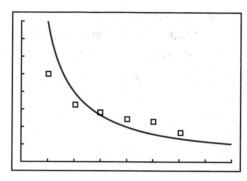

The match of function and data plot is not bad. However, scientific theories predict that (for an 8-foot ramp) roll time T and platform height H will be related by the function

$$T = \frac{2}{\sqrt{H}} \text{ which is equivalent to } T = \frac{2}{H^{0.5}} \text{ and } T = 2H^{-0.5}.$$

Scientists describe this relationship by saying, "Time is inversely proportional to the square root of platform height, with constant of proportionality 2."

5 Use the data from your own experiments of Investigation 1 and the function $T = \frac{2}{\sqrt{H}}$ to complete the following table.

Platform Height (in feet)	0.25	0.5	0.75	1.0	1.25	1.5
Experimental Roll Time (in sec) for 8-ft Ramp						
Theoretical Roll Time (in sec) for 8-ft Ramp						

a. Plot the experimental (H, T) values and the function $T = \frac{2}{\sqrt{H}}$ on a graph.

b. Why will the theoretical (predicted) and experimental (actual) results be somewhat different?

6 When you suspect that the relationship between two variables will be modeled well by a power function, you can use a calculator or computer curve-fitting tool to find a power regression rule for the function.

a. Use the (*distance, area*) values from your table in Problem 2 Part a and a curve-fitting tool to find the power model that fits the data pattern. Explain why the rule derived in this way is similar to or different from what you developed by reasoning alone.

b. Use the (*distance, intensity*) values from your table in Problem 2 Part d and a curve-fitting tool to find the power model that fits the data pattern. Explain why the rule derived in this way is similar to or different from what you developed by reasoning alone.

c. Use the (*platform height, experimental roll time*) values from your table in Problem 5 and a curve-fitting tool to find the power model that fits the data pattern.

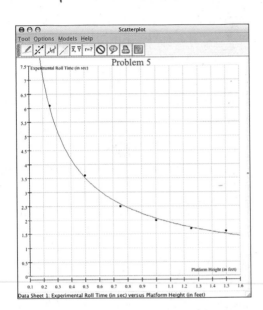

Summarize
the Mathematics

The situations in this investigation involved extensions of the basic direct and inverse variation patterns you explored in Investigation 1.

a What types of graphs occur for direct variation power functions $y = kx^r$:

  i. when r is a positive even integer and k is positive?
  ii. when r is a positive odd integer and k is positive?

b How do the answers to Part a change if k is negative?

c What types of graphs occur for inverse variation power functions $y = \dfrac{k}{x^r}$:

  i. when r is a positive even integer and k is positive?
  ii. when r is a positive odd integer and k is positive?

d How do the answers to Part c change if k is negative?

e What types of graphs occur for direct and inverse variation when r is 0.5 or −0.5?

Be prepared to share your ideas and reasoning with the class.

✔Check Your Understanding

Without using your calculator or computer graphing tool, match each function rule with its graph. In each case, write a sentence explaining the type of variation that relates x and y. For example, "y is (inversely) proportional to … with constant of proportionality … ."

a. $y = 0.5x^2$

b. $y = \dfrac{5}{x^2}$

c. $y = x^3$

d. $y = \dfrac{1}{x^3}$

e. $y = 3\sqrt{x}$

f. $y = (0.5^x)$

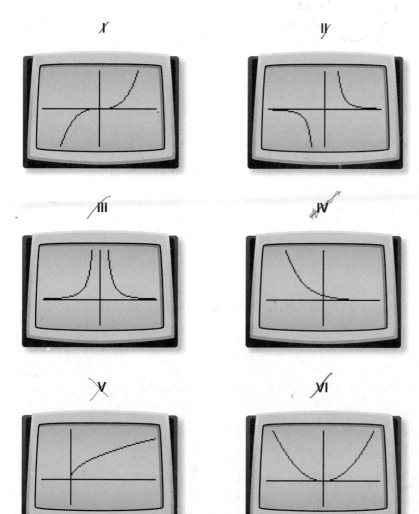

I II

III IV

V VI

On Your Own

These tasks provide opportunities for you to use and strengthen your understanding of the ideas you have learned in the lesson.

1 For each of the following relations, indicate whether it is an example of direct variation, inverse variation, or neither.

For those relations that are examples of direct or inverse variation, identify the constant of proportionality and write a sentence describing the relationship in the language of direct and inverse variation or proportionality.

a. The rule $E = 7.50h$ gives wages earned at a job as a function of number of hours worked.

b. When members of an agriculture cooperative pool their money to buy a \$40,000 hay baler, the cost per member depends on the number of members, and is calculated using $c = \dfrac{40,000}{n}$.

c. Daily profit at the Wheel Away roller rink is related to the number of paying customers by the rule $P = 8n - 350$.

d. Because, on average, Latisha makes 60% of the free throws she shoots in a basketball game, the number of points she scores on free throws depends on the number of shots she gets.

e. If a runner can average 5 miles per hour for quite a long run, the distance covered is a function of the total running time.

f. If a marathon runner covers 26 miles, the average running speed for that race is a function of the time the runner takes to complete it.

g. In any square, the side length s and the diagonal length d are related by $\dfrac{d}{s} = \sqrt{2}$.

h. In any rectangle with area 40 square inches, the length L and width W are related by $LW = 40$.

2 In many European countries, shoppers pay a *value added tax* (VAT) on most purchases. The VAT rate can be as high as 25%.

a. Suppose that the VAT rate in a country is 25%. Express the direct relationship between the value added tax T and item price P in two equivalent forms.

b. Use either of the direct variation forms that you developed in Part a to write and solve equations to answer these questions about VAT amounts:

  i. What is the value added tax on an item priced at 40 Euros?

  ii. What is the price of an item on which the VAT is 15 Euros?

  iii. What is the total purchase price (including VAT) of an item selling for 95 Euros?

c. Which direct variation form was most helpful for answering the questions in Part b? Why?

3 The time required for a race car to complete a 400-mile race is inversely proportional to the average speed that the car maintains.

a. Express the relationship between race time and average speed in two equivalent forms.

b. Use either of the inverse variation forms you developed in Part a to write and solve equations to answer these questions about race speed and time.

  i. What average speed is required to complete the race in 2.5 hours?

  ii. How long will it take to complete the race if the average speed is 140 miles per hour?

c. Which inverse variation form was most helpful for answering the questions in Part b? Why?

4 Simple pendulums are interesting and important physical devices. You may have studied pendulum motion in science experiments by attaching a weight to a string and looking for patterns relating the weight, the length of the string, and the motion of the weight as it swings from side to side.

It turns out that the frequency of a pendulum (in swings per time unit) depends only on the length of the pendulum arm (not the weight of the bob or the initial starting point of the swings). The function $F = \dfrac{30}{\sqrt{L}}$ is a good model for the relationship between pendulum arm length L and frequency of swing F when length is measured in meters and frequency in swings per minute.

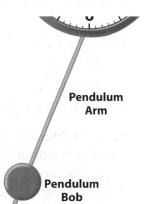

Pendulum Arm

Pendulum Bob

a. Write the rule for pendulum frequency in the form of a power function.

b. Sketch a graph of the frequency function and explain what its shape tells about the way that pendulum frequency changes as the pendulum arm length increases.

c. Estimate the frequency of pendulums with arm length 1 meter and 0.5 meters.

d. Estimate the pendulum length required for a frequency of 1 swing per *second*.

5 For each of the following functions, indicate whether it is an example of direct variation, inverse variation, or neither.

For those functions that are examples of direct or inverse variation, identify the constant of proportionality and write a sentence describing the relationship in the language of direct and inverse variation or proportionality.

a. The surface area of a cube with edge length e is given by $A = 6e^2$.

b. When a basketball player attempts a shot from mid-court, the height of the ball h is a function of time in flight t. Those variables might be related by a rule like $h = -16t^2 + 32t + 7$.

c. The volume V of a sphere is related to the radius r of the sphere by $V = \frac{4}{3}\pi r^3$.

d. The diameter d of a large tree is related to the circumference C of that tree by $d = \frac{C}{\pi}$.

e. The radius r of a circle is related to the area A of that circle by
$$r = \frac{\sqrt{A}}{\sqrt{\pi}}.$$

f. The balance B of a savings account after n years with initial investment of \$500 earning 6% annual interest is given by $B = 500(1.06^n)$.

6 Suppose that a spotlight is used for lighting objects at many different distances and that the area A of the light circle produced is related to the distance x to the lighted object by $A = 0.1x^2$.

a. If the spotlight produces 250 lumens of light energy, what function gives the intensity of the light I (in lumens per square foot) as a function of the distance x in feet from the spotlight to its target?

b. Write and solve equations that match these questions about the light intensity from the spotlight.

   i. What is the intensity of light on an object that is 20 feet from the spotlight source?

   ii. How far from the spotlight is an object that receives 100 lumens of light energy per square foot?

7 Without using a graphing tool, match each of the following functions with its graph in the collection that follows. The scales on the *x*-axis and *y*-axis are 1 on all graphs.

Function List: $y = 1.5x$ $y = 0.5x^2$ $y = -0.5x^2$

$y = 2x^3$ $y = \sqrt{x}$ $y = \frac{1}{x}$

$y = -\frac{1}{x}$ $y = \frac{1}{x^2}$ $y = \frac{1}{\sqrt{x}}$

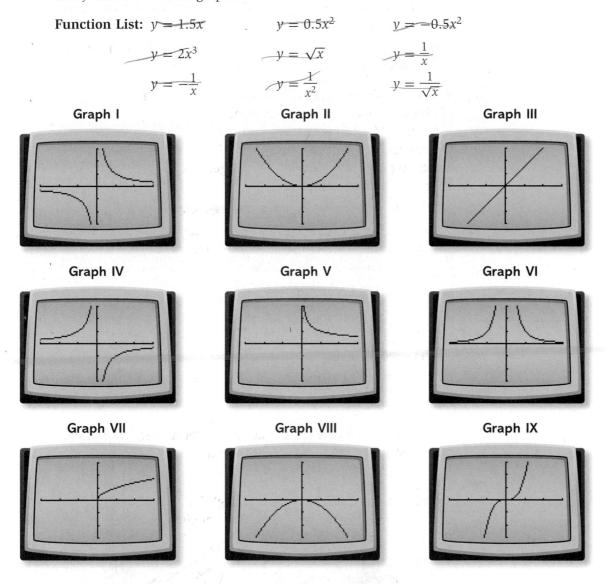

Graph I   Graph II   Graph III

Graph IV   Graph V   Graph VI

Graph VII   Graph VIII   Graph IX

Connections

These tasks will help you to build links between mathematical topics you have studied in the lesson and to connect those topics with other mathematics that you know.

8 In one trial of the *On a Roll* experiment, some students in Maryland came up with the following data relating roll time to ramp length for a ramp height of 0.5 ft.

Ramp Length (in feet)	3	4	5	6	7	8
Roll Time (in sec) at 0.5-ft Height	2.1	3.1	4.1	4.6	5.5	6.1

a. On grid paper, make a plot of the (*ramp length, roll time*) data and draw a line that seems to match the pattern of points in that plot.

b. Find the equation of the line that you drew to model the data pattern and explain what the numbers in that equation tell about the relationship between ramp length and roll time.

c. Use a calculator or computer linear regression tool to find the equation of a "best-fit" linear model for the (*ramp length, roll time*) pattern.

d. Compare the actual data to values of length and time predicted by the best-fit regression line and locate the largest error of prediction for the best-fit equation.

9 In one class that did the *On a Roll* experiment, students decided to test each combination of ramp length and platform height 3 times. Which of the following strategies would you recommend for a study aimed at finding an algebraic model for the relationship between roll time and the 2 independent variables, and why?

a. Use the mean of each set of 3 test-result roll times as the representative data.

b. Use the median of each set of 3 test-result roll times as the representative data.

c. Make a scatterplot using all data points.

d. Use results from the test roll that seemed to be done best.

e. Use some other strategy to find models of the relationships involved. Explain.

10 In testing the effect of platform height on roll time for the *On a Roll* experiment, it makes sense to use a single ramp length in all rolls. Suppose that an 8-foot ramp length was the choice.

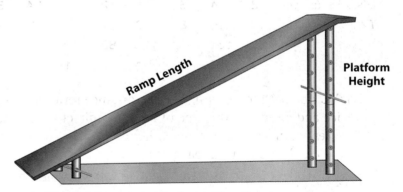

a. If the platform is 1 foot high, how can you calculate the distance from the base of the platform to the end of the ramp? What is the slope of the ramp in that case?

b. Complete a copy of a table like that shown below.

Platform Height (in feet)	Distance from Base to End of Ramp (in feet)	Slope of Ramp
1		
2		
0		
8		

11 Describe the symmetries of the graphs of the basic types of power functions.

 a. Functions of the form $y = x^2$, $y = x^3$, $y = x^4$, $y = x^5$

 b. Functions of the form $y = \frac{1}{x}$, $y = \frac{1}{x^2}$, $y = \frac{1}{x^3}$, $y = \frac{1}{x^4}$

 c. Based on your work in Parts a and b, make a conjecture about the symmetries of the graphs of functions of the form $y = x^n$, where n is an *even* integer.

 d. Make a conjecture about the symmetries of the graphs of functions of the form $y = x^n$, where n is an *odd* integer.

12 From television sets and radios to X-rays, microwave, and wireless telephone signals, we are surrounded by the energy of *electro-magnetic fields* (EMF). As sources of EMF have become more common in our everyday lives, scientists have become concerned about possible health hazards caused by exposure to EMF.

 Data in the following table show patterns of EMF measurements (in a unit called *milligauss*) at various distances from the front and back of a television set and from a VCR.

EMF Measurements

Distance (in cm)	2	4	6	8	12	16	24	32	48
TV Front	12	10	8	7	5	3	3	—	—
TV Back	—	184	—	126	82	49	20	8	2
VCR	23	13	6	4	2	1	—	—	—

 a. Look at the patterns relating EMF to distance for each electronic device and decide on the type of function (direct variation, exponential, power, or inverse variation) that you think would best model the data pattern in each case.

 b. Make a scatterplot of the data relating EMF from the front of a television set to distance from the set. Then use your calculator or computer curve-fitting tool to find two types of best-fit function models for the data pattern—a linear function and one other promising type. Decide which of the two seems to be the best model and explain your reasoning.

 c. Make a scatterplot of the data relating EMF from the VCR to distance from the VCR. Then use your calculator or computer curve-fitting tool to find two types of best-fit function models for the data pattern—linear and one other promising type. Decide which of the two seems to be the best model and explain your reasoning.

Reflections

These tasks provide opportunities for you to re-examine your thinking about ideas in the lesson.

13 When examining a table of (x, y) data, what clues would suggest that the relation between x and y is an inverse variation? A direct variation?

14 When Wanda and Troy were asked to explain how they recognized graphs of direct variation functions, they said, "The graph has to go through the origin $(0, 0)$." Give explanations or counterexamples to justify your answers to these questions.

  a. Does every direct variation graph contain $(0, 0)$?

  b. Is every function with a graph containing $(0, 0)$ an example of direct variation?

15 For each of the following rules, indicate whether it is an increasing linear function, decreasing linear function, exponential growth function, exponential decay function, quadratic function, or none of those types. Try to answer first without using a graphing tool and then check your ideas and resolve differences between your first ideas and what the calculator suggests.

  a. $y = 65 - 2.5x$ **b.** $y = 7 - 2x^2$

  c. $y = \frac{x}{3} + 42$ **d.** $y = 5(3^x)$

  e. $y = 2^x + x^2$ **f.** $y = 5(0.3^x)$

  g. $y = 0.5(1.5^x)$ **h.** $y = 4.5$

16 Of the various types of functions that you've worked with in this lesson:

  a. Which seem to occur most often in solving interesting or important problems?

  b. Which are usually easy to match with problem conditions?

  c. Which lead to equations that can be solved easily by reasoning and symbol manipulation alone?

  d. Which lead to equations that are most easily solved by use of calculator or computer tools?

17 Which approach to finding models for relationships between variables do you prefer? Why?

  • Collecting some data and looking for a pattern in that data

  • Thinking about how variables in the situation are related to each other and figuring out a function rule by logical reasoning alone

  • Some combination of data analysis and logical deduction

18 Directions on flash cameras suggest that the focus object of a flash picture should not be very far from the camera. How could your discoveries about the relationship between distance and light intensity explain that advice?

Extensions

These tasks provide opportunities for you to explore further or more deeply the mathematics you are learning.

19 In one trial of the *On a Roll* experiment, some students in Iowa came up with the following data relating roll time to platform height, using an 8-foot ramp.

Platform Height (in feet)	0.25	0.5	0.75	1.0	1.25	1.5
Roll Time (in sec) at 0.5-ft Height	7.5	3.6	2.9	2.5	2.3	1.7

  a. Use calculator or computer plotting and curve-fitting tools to find best-fitting linear, exponential, quadratic, and power regression models for the (*platform height, roll time*) data pattern.

  b. Compare graphs of the various best-fit modeling functions to a plot of the actual test data and see which function type seems the best fit.

  c. What difference between the graphs of an inverse variation model and an exponential decay model (near a platform height of 0) suggests that the inverse variation function is more likely to be appropriate in this context?

20 Suppose that the intensity I of light energy (in lumens per square foot) shining on an object x feet from a spotlight is given by $I = \dfrac{2{,}500}{x^2}$.
Compare the light intensity on objects at 10, 20, 40, and 80 feet from the spotlight to complete this conjecture: "If the distance from light source to light target doubles, then the light intensity is reduced by a factor of ____." Explain why the pattern that you observed makes sense in terms of the function rule.

Review

These tasks provide opportunities for you to review previously learned mathematics and to refine your skills in using that mathematics.

21 Answer each of the following arithmetic questions. Then support each claim with a check using a related arithmetic operation.

  Example: Does $15 - 12 = 3$?
       Yes, because $15 = 3 + 12$.

  a. Does $7.4 \times 5 = 37$?

  b. Does $24 \div 6 = 4$?

  c. Does $-13 = -39 \div 3$?

  d. Does $12 - 15 = -3$?

  e. Does $7.5 + 4.6 = 11.1$?

  f. Does $225 = 9 \times 25$?

22 Find values of x that make the following statements true.

a. $3^x \times 3^4 = 3^{11}$

b. $5^7 \times 5^3 = 5^x$

c. $5^0 = x$

d. $\sqrt{5} = 5^x$

e. $\dfrac{5}{7^3} = 5(7^x)$

23 Solve each of the following equations for x.

a. $3x - 8 = 29$

b. $3(x - 8) = 29$

c. $3(x - 8) + 17 = 29$

d. $7x + 12 = 3x - 8$

24 Find equations for the lines that contain these pairs of points.

a. $(0, 0)$ and $(2, 6)$

b. $(0, 0)$ and $(5, 8)$

c. $(1, 2)$ and $(7, 11)$

25 Use what you know about exponents and radicals to write each of the following function rules in equivalent form, $y = kx^r$.

a. $y = \dfrac{4}{x}$

b. $y = -7\sqrt{x}$

c. $y = \dfrac{4}{\sqrt{x}}$

d. $y = \dfrac{5}{x^3}$

e. $y = \dfrac{3}{4x^2}$

f. $y = (3x)^2$

26 In the diagram below, points A, C, and E are all on the same line. What is the value of x?

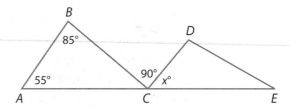

27 The sketch below shows a cylindrical oil storage tank with a diameter of 50 feet and a height of 20 feet.

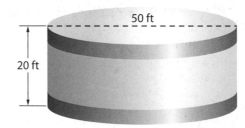

a. How long is each of the reinforcing bands around the tank?

b. What is the surface area of the tank—the vertical wall and the circular bases?

c. What is the volume of the tank?

d. One cubic foot is equivalent to about 7.5 gallons. What is the capacity of the oil tank in gallons?

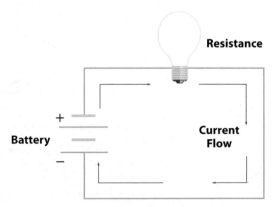

Multivariable Functions

Relations among several variables are often used to express scientific principles. For example, in the electrical circuits that provide power for flashlights, appliances, and tools, the key variables—*voltage*, *current*, and *resistance*—are related by a principle called Ohm's Law.

In a flashlight, the battery supplies electrical pressure (measured in *volts*) that creates a flow of electrons called current (measured in *amperes*) through the wire and bulb circuit. The wire and the bulb offer resistance (measured in *ohms*) to the flow of electrons.

Resistance

Battery

+

−

Current Flow

One way to think about the relationship among current, voltage, and resistance in an electrical circuit is to imagine the flow of water generated by pumps like those in circulating fountains or fish tanks. The pump (like a battery) pushes a current of water (like electrons) through a hose (like a wire), and squeezing the hose creates resistance to that flow.

Think About This Situation

Suppose that you were to run some experiments by changing the batteries and bulbs in a working flashlight circuit. Think about what would happen to the current flow and thus the light bulb.

a How would you expect the current to change if the original battery was replaced by one of higher voltage? What if the new battery had lower voltage?

b How would you expect the current to change if the original bulb was replaced by one of higher resistance? What if the new bulb had lower resistance?

c How would you expect the current to change if the original battery was replaced by one of higher voltage and the bulb by one of lower resistance? What if the voltage decreased and the resistance increased?

d How would you expect the current to change if both the voltage and resistance were increased? What if both the voltage and the resistance were decreased?

In this lesson, you will develop your skill in reasoning about multivariable relationships and functions.

Investigation 1 — Combining Direct and Inverse Variation

The flashlights we use for various lighting tasks come with many different combinations of batteries and bulbs. The relationship between current I, voltage V, and resistance R is important. If the current is too great, the bulb may break; if the current is too low, the bulb may barely light. As you work on the problems of this investigation, look for an answer to these questions:

What symbolic rules represent the relationship of current, voltage, and resistance in a simple electrical circuit?

How can relationships among several variables be written in useful equivalent forms?

1 Based on your reasoning about how changes in voltage and resistance affect current in a flashlight circuit, which of the following formulas might possibly express the numerical relationship among those variables? Be prepared to explain your reasoning.

$$I = V - R \qquad I = V + R \qquad I = V \cdot R \qquad I = \frac{V}{R} \qquad I = \frac{R}{V}$$

2 Engineers designing electrical circuits have tools to measure voltage, current, and resistance of designs they want to test. You could use those instruments to run some experiments and collect data on combinations of volts, amps, and ohms in flashlight circuits that function properly. Computer software like the "Light it Up!" custom tool can be used to simulate experiments with batteries and bulbs in search of the relationship among I, V, and R.

a. Use the simulation to find at least 5 combinations of values for these variables that will produce working battery and bulb circuits.

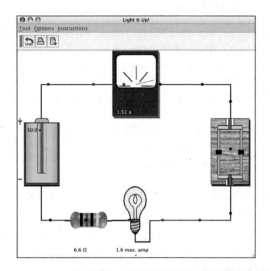

b. Write a formula showing how I depends on V and R in all cases. Compare your formula with those of your classmates. Resolve any differences.

c. Use the language of direct and inverse variation to explain in words how current varies as voltage and resistance change.

3 Design of a working battery and bulb circuit requires finding values for 3 related variables. In some cases, you might know values of V and R and need to calculate the corresponding value of I. In other cases, you might know values of I and R or values of I and V.

a. What is the resistance in a circuit that includes a 9-volt battery and has current flow of 2 amps? What if the circuit includes a 12-volt battery and has current flow of 5 amps? What formula expresses R as a function of V and I?

b. What size battery (in volts) is required to produce a current flow of 2.5 amps when the resistance of the circuit is 4 ohms? What if the circuit must have a current flow of 4 amps and the resistance is 7 ohms? What formula expresses V as a function of R and I?

c. How do Ohm's Law $I = \dfrac{V}{R}$ and the natural connection between multiplication and division provide algebraic justification for the formula in Part b that expresses V in terms of I and R?

4 The questions of Problems 2 and 3 asked you to write Ohm's Law in three different ways. You can use the formula $I = \dfrac{V}{R}$ to reason about how and why changes in R and V will cause changes in I. For example,

If V stays the same and R is increased, then I *will decrease* because *when the denominator of a positive fraction increases and the numerator stays the same, the value of the fraction will decrease.*

Use the formula and similar reasoning to complete each of the following sentences about how the three variables are related.

 a. If R stays the same and V is increased, then I _____ because _____.

 b. If R and V are each doubled, then I _____ because _____.

 c. If R is cut in half and V is doubled, then I _____ because _____.

⑤ **Revisiting Roll Time, Ramp Length, and Platform Height** Scientists can use theories about the effects of gravity on falling objects to deduce the relationship of roll time T, ramp length L, and platform height H, for an experiment like the one you did in Lesson 1. Ignoring possible effects of friction, theory predicts that these variables will be related by the function $T = \dfrac{L}{4\sqrt{H}}$.

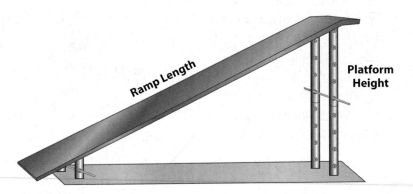

 a. Use the language of direct and inverse variation to describe the theoretical relationship of T to L and H. What is the constant of proportionality?

 b. Use the function $T = \dfrac{L}{4\sqrt{H}}$ to complete the following tables of predicted roll time values for various combinations of ramp length and platform height. Then plot the sample (L, T) and (H, T) values on separate graphs.

Ramp Length (in feet)	4	5	6	7	8
Roll Time (in sec) at 0.25-ft Height					

Platform Height (in feet)	0.25	0.5	0.75	1.0	1.25
Roll Time (in sec) for 8-ft Ramp					

 c. Compare patterns in the tables and plots in Part b to your experimental results from Lesson 1. Explain any differences.

Summarize
the Mathematics

The situations in this investigation involved relations among several variables. In each case, those relations could be expressed with a single rule showing one variable as a function of the others.

a Suppose that three variables x, y, and z are related by the rule $z = kxy$ with $k > 0$. Considering $x > 0$ and $y > 0$, how will the value of z change if:

    i. x is held constant and y increases?   **ii.** x is held constant and y decreases?

b Suppose that three variables x, y, and z are related by the rule $z = k \cdot \frac{x}{y}$ with $k > 0$. Considering $x > 0$ and $y > 0$, how will the value of z change if:

    i. x is held constant and y increases?   **ii.** x is held constant and y decreases?

c If three variables x, y, and z are related by $z = \frac{x}{y}$, what equivalent rules show:

    i. x as a function of y and z?   **ii.** y as a function of x and z?

d Use the language of direct and inverse variation to explain how z is related to x and y in the rule $z = \frac{x}{y}$.

Be prepared to share your ideas and reasoning with the class.

✓ Check Your Understanding

The volume V of an elastic container for air—like a tire or a balloon—is related to the temperature T and to the pressure P of the air inside the container. The formula $V = k \cdot \frac{T}{P}$ is one way of expressing that relationship. The number $k > 0$ is a constant depending on the units of measurement for volume, temperature, and pressure.

a. How will volume of a balloon change if:

  i. pressure remains constant, but temperature increases?

  ii. pressure remains constant, but temperature decreases?

  iii. temperature remains constant, but pressure increases?

  iv. temperature remains constant, but pressure decreases?

b. How can the relationship among P, T, and V be expressed to show:

  i. P as a function of T, V, and k?

  ii. T as a function of V, P, and k?

Problems that occur in business situations often require expressing income as a linear function of one variable like time worked or number of sales. For example, if an employee earns $7.25 per hour, then income I earned for working h hours is given by $I = 7.25h$. If a movie theater charges $8.50 for admission, then the income R from selling n admission tickets is given by $R = 8.50n$.

Questions about those situations can be expressed as equations or inequalities. For example, the question "How many hours does the employee need to work to earn $290?" is represented by the equation $290 = 7.25h$. The question "How many admission tickets does the movie theater need to sell to collect at least $1,000?" is represented by the inequality $8.50n \geq 1,000$.

In many business situations, income is a function of several variables. The income function is built by combining two or more simple linear functions. As you work on the problems of this investigation, look for answers to these questions:

How can you use linear functions of two independent variables to represent problem situations?

How can you graph and find solutions for linear equations in two variables?

1. Many middle- and high-school students work on a service-for-hire basis until they reach the minimum age required to apply for a "real" job. Bret mows lawns and washes cars in his neighborhood to earn spending money. He charges $20 per lawn and $10 per car wash.

   a. How much money would Bret earn for mowing 15 lawns? For washing 20 cars? What would be Bret's total income for 15 lawns and 20 car washes?

   b. Bret's total income I is a function of the numbers of lawns mowed L and cars washed C. Write a rule that expresses I as a function of L and C.

   c. Suppose that Bret has a goal to earn $1,200 and has scheduled 50 lawn-mowing jobs for the summer. How many cars must Bret wash to reach the income goal? What if Bret only schedules 40 lawn-mowing jobs?

2 Suppose that Bret sets an income goal of $2,000.

    a. Write an equation that represents the question "How many lawn mowings and how many car washes will it take to achieve Bret's income goal?"

    b. If Bret only mows lawns, how many would it take to reach the income goal?

    c. How many cars would Bret have to wash to meet the income goal by car washing only?

    d. Find 2 more combinations of number of lawn mowings and number of car washes that would achieve Bret's income goal. Be prepared to explain why these (L, C) combinations are *solutions* of the equation from Part a.

3 Plot the (L, C) pairs that you found to be solutions of $20L + 10C = 2,000$ on a coordinate grid like that pictured below. Identify coordinates of two other points that seem to fit the pattern of the plotted points. Check to see if those pairs of numbers are also solutions of the equation.

Combinations Giving $2,000 Income

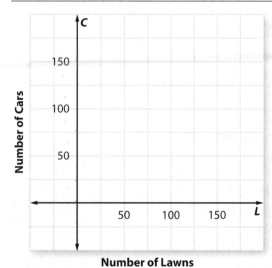

Number of Lawns

4 The graph of (L, C) pairs that will achieve the $2,000 income goal suggests that C is a linear function of L.

    a. Rewrite the equation $20L + 10C = 2,000$ to express C as a linear function of L.

    b. Write a rule expressing L as a function of C. Is L a linear function of C? Explain why or why not.

5 Many entertainment events do not charge one fixed price for admission. Discounted tickets are usually offered for matinee showings, children, and seniors. The income I, in dollars, for the Kingstown Playhouse is given by $I = 12r + 8d$, where r is the number of regular admission tickets and d is the number of discounted admission tickets.

a. How much does the playhouse charge for each regular admission ticket? How much for each discounted admission ticket?

b. How does I change as r increases? As d increases?

6 *A Mother's Dream* is currently showing at the Kingstown Playhouse. The production costs of the play total $24,000, and ticket prices are the same as in Problem 5.

a. Write an equation using variables r and d that represents the question "How many regular and discounted admissions are needed in order for the producers of the play to break even?"

b. Find 5 solutions to your equation from Part a. Record the solutions in a table like the one below.

r					
d					

c. Draw a graph showing all solutions to the equation from Part a for which values of both r and d make sense in this situation.

d. Write a rule that shows how to calculate the number of regular admission tickets r as a function of the number of discounted admissions d, if the play is to break even.

7 The Freedom High School booster club sponsors a fund-raising carnival each fall. In the basketball shooting game, participants pay $2 for a chance to make three shots in a row and win a basketball as a prize. A local sporting goods store sells the basketballs to the booster club for a special price of $5 each. The club can return any remaining balls when the carnival is over.

a. Write a rule that shows how fund-raiser profit P depends on the number of customers c and the number of winners w.

b. How does profit change as the number of customers increases? As the number of wins increases?

c. If the basketball shooting game has 25 customers and 8 winners, it will earn a profit of $10.

   i. Find 4 other pairs of number of customers and number of winners that will result in profit for the game that is greater than 0.

   ii. Write an algebraic condition that describes all pairs (c, w) that will guarantee a positive profit for the game.

   iii. Draw a graph that shows all (c, w) pairs that guarantee positive profit.

d. Suppose that the free-throw booth attracts 40 customers. How many customers were winners if the game generated profit of $50 for the booster club?

Summarize the Mathematics

For variables related by functions such as $I = 20x + 10y$, we say that I is a *linear function* of x and y. Equations of the form $2{,}000 = 20x + 10y$ are called *linear equations* in x and y. Linear equations in two variables can be rewritten to express one of the variables as a linear function of the other.

a Why does it make sense to call equations of the form $2{,}000 = 20x + 10y$ *linear equations*?

b Based on what you have learned previously about linear functions, why does it make sense to say that I is a linear function of x and y when these variables are related by a rule like $I = 20x + 10y$?

c Is (4, 10) a solution of $5x + 10y = 120$? How could you check using the equation? Using a graph?   (0, 12)   (24, 0)

d What strategies can be used to find solutions for equations such as $2{,}000 = 20x + 10y$?

e How can you rewrite equations like $2{,}000 = 20x + 10y$ to express x as a function of y? To express y as a function of x?

f What is the slope of the graph of solutions for the equation $2{,}000 = 20x + 10y$?

Be prepared to share your ideas and reasoning with the class.

✔ Check Your Understanding

When the event planners at the Cellar decided to sponsor a fall costume party, they set two ticket prices: $10 for advance tickets and $15 for tickets at the door.

a. What total ticket income will the party generate for the following combinations of ticket sales?

   i. 60 advance tickets and 85 tickets at the door

   ii. 98 advance tickets and 32 tickets at the door

b. Write a rule for income as a function of the numbers of tickets sold in advance and at the door.

c. Suppose that the event planners have a goal of $1,500 in ticket income and have sold 75 tickets in advance. How many tickets must be sold at the door to reach the income goal?

d. If the income goal is set at $2,000, what equation represents the question of how many tickets must be sold in advance and at the door to reach the income goal? Make a table and draw a graph showing non-negative solutions for this equation.

e. Rewrite the equation from Part d to express the number of tickets that need to be sold at the door as a function of the number of tickets sold in advance.

On Your Own

Applications

1 For each of the following functions, write a sentence describing the relationship among the variables in the language of direct and inverse variation.

   a. Earned wages E at a job are a function of hours worked h and hourly pay rate r, given by the rule $E = rh$.

   b. When members of a sailing club pool their money to buy a boat, the cost per member c depends on the number of members n and the cost of the boat B according to $c = \dfrac{B}{n}$.

   c. When a flashlight shines on a flat surface at night, the brightness B of the light on that surface is a function of the distance d from the flashlight to the surface and the lumen strength L of the flashlight beam; the rule that relates them is $B = \dfrac{L}{d^2}$.

   d. If a runner can average m miles per hour for quite a long run, the distance covered d is a function of the total running time t and average speed m. $D = t \cdot m$

2 When a car, van, or small truck is involved in a traffic accident, the likelihood that a passenger will be fatally injured depends on many conditions. Two key variables are speed and mass of the vehicle in which the passenger is riding.

   a. What general relationship would you expect between the rate of fatalities in auto accidents and the speed and mass of the vehicle in which a passenger is riding?

   b. What data on highway accidents would help you develop a function relating the rate of passenger fatalities, vehicle speed, and vehicle mass?

   c. If A represents the fatality rate in auto accidents, s represents vehicle speed, and m represents vehicle mass, which of the following functions would you expect to best express the relation among those variables? Explain your choice.

      Function I $A = 200(s + m)$ **Function II** $A = 200(s - m)$

      Function III $A = 200\dfrac{s}{m}$ **Function IV** $A = 200sm$

(3) An important consideration in construction is the weight a steel or wooden beam can hold without breaking. Some beam materials are stronger than others. But for any particular material, two important variables that influence the breaking weight are the length and thickness of the beam.

1   2   3   4

a. Which beam do you think would support the greatest weight? The least weight?

b. The beams differ in length and thickness. How would you expect those two variables to affect the breaking weight of a beam?

c. Breaking weight W depends on beam length L and thickness T. What sort of rule might be used to express W as a function of L and T?

d. The table that follows shows data collected by a class that used strands of raw spaghetti spanning gaps of various lengths to investigate the breaking weight of "spaghetti bridges":

Breaking Weight in grams

		Number of Strands			
		1	2	3	4
	2	92.5	145.1	188.1	261.6
	3	47.8	109.9	128.4	185.8
Gap Length (in inches)	4	38.6	69.9	98.5	124.7
	5	29.6	43.7	79.1	95.9
	6	23.8	28.3	66.5	78.4

Do the patterns of change shown in the data table support or change your thinking about the sort of rule that might be used to express W as a function of L and T?

e. Based on the data in the table above, the class developed the rule $W = 137\frac{T}{L}$ to express W as a function of T and L.

  i. Rewrite the rule to express T as a function of L and W.

  ii. Rewrite the rule to express L as a function of W and T.

(4) If a variable z is directly proportional to x and inversely proportional to y, the relationship of those variables can be expressed in the form $z = \dfrac{kx}{y}$, where k is the constant of proportionality. Write similar rules to represent the relationships described in these situations.

a. The volume V of a cylindrical container is directly proportional to its height h and the square of its radius r. $V =$

b. The force F required to lift some object with a lever is directly proportional to the mass m of the object and inversely proportional to the length L of the lever. $F = \dfrac{km}{L}$

C. ↓

c. The attraction F between two objects in space varies directly with the product of their masses m_1 and m_2 and inversely with the square of the distance d between them.

5 A group of 13 machine shop workers in Ohio regularly pooled their money to buy lottery tickets. On July 29, 1998, they won a lump sum payment of $161.5 million.

a. How much did each of the 13 winners receive, assuming they shared the winnings equally?

b. How would the amount received by each winner change if more workers had participated in the lottery pool? If fewer workers had participated?

c. How would the amount received by each winner change if the lottery jackpot had been larger? What if the lottery jackpot had been smaller?

d. Write a rule that gives the share of the winnings S for each person as a function of the lottery jackpot L and the number of people N in the pool.

e. Rewrite the rule in Part d to express N as a function of L and S.

6 The population density of any country or region is usually measured by calculating the number of people per unit of area. For example, in 1790, the United States had a population density of 4.5 people per square mile; in 2000, that figure had risen to 79.6 people per square mile. In that time period, both the number of people and the land area of the country grew.

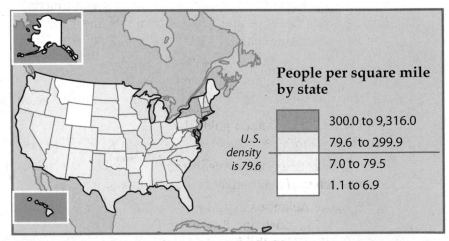

People per square mile by state

	300.0 to 9,316.0
	79.6 to 299.9
	7.0 to 79.5
	1.1 to 6.9

U.S. density is 79.6

Source: U.S. Census Bureau

a. What are some land areas that were added to the United States after 1790?

b. What function rule shows how to calculate population density D for a region from the area A and population P of that region?

c. How do increases in the population of a region affect the population density? How about decreases in the population?

d. What change in population density of a region will result if both population and land area increase? What if both decrease?

e. How would you complete the following sentence to describe the dependence of population density on both total population and land area?

> "The population density of a country is (directly/inversely) related to the total population and (directly/inversely) related to the area of the country."

7 A local radio station sponsors a T-shirt toss and floppy hat drop during home pro basketball games. The T-shirts cost the radio station $8 each, and the hats cost $12 each.

a. The promotional cost C for the radio station depends on the numbers of shirts s and hats h given away at the game. Write a rule expressing C as a function of s and h.

b. How will the radio station's cost change as the number of shirts given away increases? How will cost change as the number of hats given away increases?

c. Suppose the radio station has budgeted $2,400 per game for giveaways. Write an equation that represents the question "How many shirts and hats can the radio station give away for $2,400?"

d. List 4 solutions to your equation from Part c. Draw a graph that shows all of the possible solutions.

e. Rewrite your equation from Part c to express s as a function of h. Explain what the slope and s-intercept of this linear function tell you about the situation.

8 Aidan is planning to build a small goldfish pond in his back yard. He plans to keep fantail goldfish, which can grow to be 4 inches, and common goldfish, which can grow to be 8 inches.

a. The *fish load* of a pond is measured in total inches of fish lengths.

   i. Suppose Aidan wants to keep 5 fantail goldfish and 8 common goldfish. When the fish reach full size, what will the fish load of the pond be?

   ii. Write an algebraic rule that shows how pond fish load L depends on the number f of full-size fantail goldfish and the number c of full-size common goldfish in the pond.

   iii. How will the fish load of the pond change as the number of full-size fantail goldfish increases? How will it change as the number of full-size common goldfish increases?

b. Aidan is considering a pond kit that is rated for 140 inches of goldfish. How many full-size fantail and common goldfish can be supported in this pond? Be sure to include the following in your response:

  • an equation that represents the question;

  • a table showing several combinations of numbers of full-size fantail and common goldfish that meet the pond rating limit;

  • a graph of all possible solutions.

c. As the number of common goldfish in a pond increases, how will the number of fantail goldfish that can be supported in the pond change?

(9) Find three specific solutions of the linear equation $3x - 2y = 12$. Then graph the full solution set of this equation.

(10) Rewrite each of the following linear equations to express y as a function of x. Determine the slope and y-intercept for the graph of the solution set of each equation.

  a. $2x + y = 6$ **b.** $8x - 5y = 20$ **c.** $-4x - 3y = 15$

Connections

(11) Ms. Williams gives her students a 10-point quiz every week.

  a. Suppose a student has an average quiz score of 8 after 5 quizzes.

  i. How many total quiz points has the student earned?

  ii. If the student scores 2 points on the sixth quiz, what is the student's average quiz score?

  b. Suppose the student still has an average quiz score of 8 after 15 quizzes.

  i. How many total quiz points has the student earned?

  ii. If the student scores 2 points on the sixteenth quiz, what is the student's average quiz score?

  c. How does a single quiz score's impact on the average score change as the number of quizzes increases?

(12) The three key variables in racing are distance, speed, and time.

  a. If a runner covers 400 meters in 50 seconds, what is the runner's average speed? What if it takes the runner 60 seconds to cover the same distance? What formula expresses average speed s as a function of distance d and time t?

  b. If a NASCAR racer covers 240 miles at an average speed of 150 miles per hour, how long will the race take? What if the average speed is 180 miles per hour? What formula expresses race time t as a function of distance d and average speed s?

c. If a participant in a triathlon swims at an average speed of 1.2 meters per second for 40 minutes, how much distance will be covered? What if the average speed drops to 0.9 meters per second and the time increases to 50 minutes? What formula expresses distance d as a function of average speed s and time t?

13 For each of the following geometric formulas, solve for the stated variable and answer the related questions.

a. Solve $C = \pi d$ for d.

  i. If a circular swimming pool is 200 feet around, what is the diameter of the pool?

  ii. A circular garden is enclosed with 50 feet of fencing. What is the diameter of the garden?

b. Solve $A = \ell w$ for w.

  i. If the length of a rectangular sandbox is set at 16 feet, what width is required to obtain an area of 200 square feet?

  ii. If the length of the sandbox was to increase but the area was to remain 200 square feet, how would the width have to change?

c. Solve $P = 2\ell + 2w$ for w.

  i. If you have 52 feet of lumber to construct the sides of a rectangular sandbox, and the length is set at 16 feet, how wide can the sandbox be?

  ii. If the length of the sandbox was to increase but the perimeter was to remain 52 feet, how would the width have to change?

d. Solve $V = \ell w h$ for h.

  i. In designing a box to have volume 1,000 cm³, length 20 cm, and width 10 cm, what is the height?

  ii. If the volume of the box was to increase but the length and width were to remain unchanged, how would the height have to change?

14 In Course 1, you may have discovered *Euler's formula* $V + F = E + 2$, which relates the number of vertices V, the number of faces F, and the number of edges E of simple polyhedra.

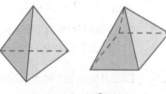

a. Write an equivalent form of Euler's formula that expresses E as a function of V and F. Check your answer using the polyhedra shown at the right.

b. Write an equivalent form of Euler's formula that expresses F as a function of V and E. Check your answer.

15 Consider the rules $z = 2x + 3y$ and $z = 5x - 4y$ expressing z as linear functions of x and y.

a. For each function, complete a table like that below showing the value of z for different values of x and y:

z = 2x + 3y

		x			
		0	1	2	3
y	0				
	1				
	2				
	3				

b. For $z = 2x + 3y$ and any fixed value of y, how does z change for each unit increase in x? For any fixed value of x, how does z change as y increases?

c. For $z = 5x - 4y$ and any fixed value of y, how does z change for each unit increase in x? For any fixed value of x, how does z change as y increases?

d. How do the patterns of change in Parts b and c relate to what you have learned previously about patterns of change in linear functions?

e. If $z = ax + by$, how will z change as x increases? As y increases?

16 Look back at your work in Applications Task 10. Each equation is a specific case of the general form of a linear equation $ax + by = c$, where a, b, and c are constants.

a. What is the slope of the graph of $ax + by = c$? Are there any conditions on a, b, or c?

b. What is the y-intercept of $ax + by = c$? Are there any conditions on a, b, or c?

Be prepared to provide algebraic reasoning that justifies your answers to these questions.

17 When radio or television transmitters are placed on tall towers, it is common to support those towers with wires anchored at various points on the tower and the ground.

There are three variables involved in placement of those support wires.

h = height of the point where the wire is anchored on the tower

b = distance of the ground anchor from the base of the tower

w = length of the support wire itself

Suppose that there is an anchor in the ground at a point 40 feet from the base of the tower.

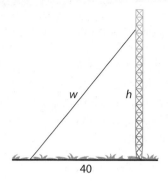

40

a. If $h = 60$ feet, what is w? What if $h = 90$ feet?

b. What function shows how the length w of the support wire from that anchor point depends on the height h of the point where it is to be connected to the tower?

c. Make a table that shows how the length of the support wire w changes as the anchor point on the tower moves up from 0 to 30 to 60 to 90 to 120 feet above the ground. Plot these (h, w) pairs on a graph and use your calculator for help in sketching a graph of the function in Part b for $0 \leq h \leq 150$.

d. How are the function, the table of values, and the graph for the situation similar to or different from those that you have worked with in earlier investigations?

18 Now suppose that there is a support wire, like that in Connections Task 17, that is 130 feet long.

a. What function shows how to calculate b if the 130-foot wire is to be attached to the tower at a point h feet up from the ground?

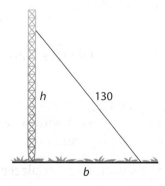

b. Make a table showing how the distance b from the ground anchor to the base of the tower changes as the height of the anchor point on the tower h moves up from 0 to 30 to 60 to 90 to 120 feet. Plot these (h, b) pairs on a graph and use your calculator or computer software for help in sketching a graph of the function in Part a for $0 \leq h \leq 130$.

Reflections

19 Three different kinds of functions you have studied can be used to model situations in which increase in the independent variable leads to decrease in the dependent variable—linear functions with negative slope, exponential decay functions, and inverse variation functions (for positive constant of proportionality). What clues do you use to decide which of those functions might be useful in modeling a particular data pattern?

20 Studying patterns of change in a function with two independent variables, like Ohm's Law, is a bit more complicated than work with single-variable functions. How do you go about understanding the pattern of change in a function like $c = \frac{B}{n}$? How about a function like $z = 5x + 7y$?

21 Mathematicians generally use the words *function*, *expression*, and *equation* with different meanings in mind.

  a. Which terms would you use to label each of the following algebraic items?

   i. $7x^2 - 5x$

   ii. $y = 7x^2 - 5x$

   iii. $7x^2 - 5x = 0$

  b. What do you think mathematicians have in mind as a rule for deciding when to use *function*, when to use *expression*, and when to use *equation*?

22 To quickly sketch the graph of the solution set of a linear equation like $6x - 4y = 12$, Vera described the following method.

> I find the coordinates of the *x*-intercept by mentally substituting O for *y* and then solve for *x*. To find the coordinates of the *y*-intercept, I substitute O for *x* and then solve for *y*. Then I draw the line through the two intercept points.

Try Vera's method for the equation $6x - 4y = 12$ and explain why it works.

Extensions

23 The time T it takes a ball to roll down a ramp of length L that is supported at one end by a platform of height H is described by the formula

$$T = k \cdot \frac{L}{\sqrt{H}},$$

where k is some positive constant.

a. Solve $T = k \cdot \dfrac{L}{\sqrt{H}}$ for k.

b. Refer back to the data you collected for the first ramp experiment in Lesson 1 (page 4). Calculate k from T, L, and H for each data point.

c. Describe the shape, center, and spread of the values you obtained for k.

d. Determine an estimate for the value of k and substitute this value for k in the original formula. How could you determine how well this function models the data from the ramp experiment?

24 When Leon learned Ohm's Law, he wrote a simple spreadsheet program to help with calculations in various homework tasks.

a. One part of the spreadsheet calculates current, when voltage and resistance are known. What formula would be used in cell **B3** if the spreadsheet user enters numerical values of voltage and resistance in cells **B1** and **B2**, respectively?

Ohm's Law.xls

◇	A	B	
1	voltage =		
2	resistance =		
3	current =		
4			

b. How could you modify the spreadsheet of Part a so that it calculates voltage when numerical values of current and resistance are entered?

c. How could you modify the spreadsheet of Part a so that it calculates resistance when numerical values of current and voltage are entered?

d. The following table starts to show values of current for every combination of voltage and resistance from 1 to 10 in steps of 1. Create a spreadsheet that produces entries in all of the cells by entering only 2 numbers and 3 formulas and then applying several fill down or fill right commands.

Ohm's Law.xls

◇	A	B	C	D	E	F	G	H	I	J	K
1		1	2	3	4	5	6	7	8	9	10
2	1	1	0.5	0.33	0.25	0.2	0.17	...			
3	2	2	1	0.67	0.5	0.4	...				
4	3	3	1.5	1	0.75	...					
5	4	4	2	1.33							
6	5	5	2.5								
7	6	6									
8	7	...									
9	8										
10	9										
11	10										

25 The resistance in an electrical wire depends on the length and the thickness of the wire as measured by its cross-sectional diameter.

Diameter, d

Length, L

a. Suppose you are experimenting with battery and bulb circuits of different designs. How would you expect the resistance of the circuit to change as the length of the wire used in the design is increased? As the diameter of the wire used is increased?

b. Suppose that the formula $R = 5 \cdot \dfrac{L}{d^2}$ was proposed for the relationship among resistance R, length L, and diameter d of the wire in an electrical circuit. How does this formula fit with your expectations from Part a?

c. Rewrite the formula from Part b to express L as a function of R and d. You can check your reasoning with the computer algebra system command **solve(R=5*L/d^2,L)**.

d. Rewrite the formula from Part b to express d as a function of R and L. You can check your reasoning with the computer algebra system command **solve(R=5*L/d^2,d)**.

26 One formula for estimating heat flow through a solid material such as glass, wood, or aluminum involves 5 variables: $R = kA\dfrac{\Delta T}{t}$. The symbols and what they represent are as follows:

R = Rate of heat flow, in BTUs (British thermal units) per hour
k = Thermal conductivity for the specific material
A = Area of the material, in square feet
ΔT = Difference in temperature between outside and inside, in degrees Fahrenheit
t = Thickness of the material, in inches

a. Suppose sheets of glass, wood, and aluminum of the same area and thickness are exposed to the same difference in temperature. How would you expect the rate of heat flow to be different for these sheets of material?

b. Consider a window opening that has an area of 6 square feet. Use the following information about the thermal conductivity of different materials to determine the heat flow rate through the window opening on a day when the outside temperature is 5°F and the inside temperature is 68°F.

   i. The thermal conductivity of glass is 5.8. What is the heat flow rate for a glass window that is 0.5 inches thick?

   ii. Suppose that instead of glass, the same opening is covered with wood having thermal conductivity of 0.78. What is the heat flow for the same temperature conditions if the wood is 0.5 inches thick?

   iii. The thermal conductivity of aluminum is 1,400. What is the heat flow under the same conditions if the opening is covered with aluminum 0.5 inches thick?

   iv. For this window opening, what thickness of glass would be required to achieve the same heat flow rate as the wood in part ii?

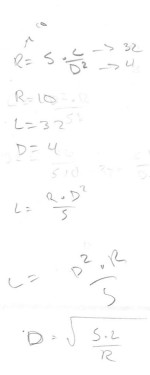

c. Review your response to Part a. Were you correct? If not, modify your response.

d. The symbol k stands for the thermal conductivity of the material. Would you conclude that materials with low k values are good conductors of heat or good insulators? What would you conclude about materials with high k values? Provide evidence to support your answers.

e. What changes in the variables A, ΔT, and t would cause the rate of heat flow to increase or decrease? Why does this make sense based on your thinking about the variables involved?

(27) Many questions about linear functions can be represented by linear inequalities.

a. Refer back to Problems 1 and 2 of Investigation 2 (pages 30–31). The equation $20L + 10C = 2{,}000$ represents the question of how many lawn mowing jobs and how many car washes it will take to achieve Bret's income goal of $2,000. In reality, Bret is probably interested in making *at least* $2,000, not *exactly* $2,000.

   i. Write an inequality to represent the question "How many lawn services and how many car washes will it take to achieve Bret's income goal of *at least* $2,000?"

   ii. How might you represent all possible solutions to the inequality on a graph?

b. Refer back to Applications Task 7 (page 37). Write an inequality that represents the question "How many shirts and hats can the radio station give away for *no more than* $2,400?" Graph all possible solutions of this inequality.

c. Graph the solution sets of the following inequalities.

   i. $6x + 3y \geq 12$

   ii. $2x + 5y \leq 30$

$-12(x-a)$

28 When several electrical appliances are drawing power from the same supply circuit in your home, the current drawn by the total circuit is the sum of the currents drawn by the separate appliances. The voltage remains constant at the standard 120 volts for household use in the United States.

If two appliances with resistances R_1 and R_2 are operating at the same time, the current flow I will be given by the following formula:

$$I = \frac{120}{R_1} + \frac{120}{R_2}$$

a. If a microwave with resistance 25 ohms and a refrigerator with resistance 15 ohms are operating at the same time on a kitchen circuit, what current is used?

b. Suppose a television and a hairdryer are both operating from the same bedroom circuit, drawing a total current of 6.5 amps. If the resistance of the hairdryer is 20 ohms, what is the resistance of the television set?

c. Use a computer algebra system to rewrite the given equation so that it expresses R_1 as a function of I and R_2. Then use that form to explain how R_1 changes as I and R_2 change in various ways (both increasing, both decreasing, one increasing and the other decreasing).

Review

29 Without using a calculator, evaluate the following expressions for the given values of the variables.

a. $3x - 2y$ when $x = 5$ and $y = -1$

b. $\frac{ab}{c}$ when $a = -6$, $b = -2$, and $c = -4$

c. $10 - pq$ when $p = 3$ and $q = -5$

30 Make a copy of isosceles △*ABC*.

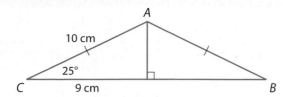

a. Fill in the lengths of as many sides as you can and the measures of as many angles as you can.

b. Find the area of △*ABC*.

31 Recall that the *reciprocal* of any nonzero number *a* is the number $\frac{1}{a}$.

a. Complete the following table showing relationships among a number, its reciprocal, and their product.

Number	2	$\frac{1}{4}$	$\frac{3}{5}$	−6	*n*
Reciprocal	$\frac{1}{2}$				
Product	1				

b. Recall that subtraction can be rewritten as addition: $a - b$ can be rewritten as $a + (-b)$. Division can also be rewritten as multiplication. How can you rewrite $\frac{a}{b}$ as a product?

c. Use your results from Parts a and b to justify each step in simplifying the product below.

$$5 \cdot \frac{3}{5} = 5 \cdot 3 \cdot \frac{1}{5}$$
$$= 5 \cdot \frac{1}{5} \cdot 3$$
$$= 1 \cdot 3$$
$$= 3$$

d. Use similar reasoning to simplify the following products.

i. $2 \cdot \frac{7}{2}$

ii. $3 \cdot \frac{2}{3}$

iii. $b \cdot \frac{a}{b}$ $= b \cdot a \cdot \frac{1}{b} =$
$b \cdot \frac{1}{b} \cdot a = 1 \cdot a = a$

32 Solve each of the following equations for *x*.

a. $17 = 5 - 3x$

b. $\frac{2x}{3} + 5 = 21$ → $\frac{2x}{3} = 16$ → $\frac{x}{3} = 8$ → $x = \frac{8}{3}$

c. $-12(x - 9) = -6(8 - 3x)$ → $-12x + 108 = -48 + 18x$ → $156 = 30x$ $5.2 = x$

33 Find equations for the lines that contain these pairs of points.

a. $(0, -4)$ and $(5, 3)$ $\frac{y_2 - y_1}{x_2 - x_1} = $ slope → plug in one

b. $(-2, 4)$ and $(2, 0)$ of the equations to find y-int.

34 The box plot below shows the price in dollars of 20 models of cordless phones.

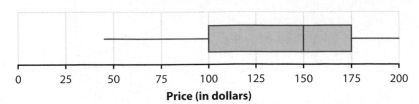

Price (in dollars)

a. What is the approximate median of the data? 150

b. What are the approximate lower and upper quartiles of the data?

c. Write a brief description of the distribution.

35 Solve each of these equations for x.

a. $x^2 + 4 = 13$ $x^2 = 9 \rightarrow x = 3$

b. $3x^2 + 4 = 31$ $3x^2 = 27 \rightarrow x^2 = 9 \rightarrow x = 3$

c. $3x^2 + 12x = 0$ $3x^2 = 12x \rightarrow x^2 = 4x \rightarrow x = 4$

36 Write each of the following radicals in equivalent form with a smaller integer under the radical sign.

a. $\sqrt{20} = 2\sqrt{5}$

b. $\sqrt{48} = 4\sqrt{3}$

c. $\sqrt{72} = 6\sqrt{2}$

d. $\sqrt{18a^2}, a > 0 = 18a$

37 Consider the three equations:

$$3x + 2y = 6 \qquad 9x + 6y = 18 \qquad -6x - 4y = -12$$

a. Show that the ordered pairs $(0, 3)$, $(2, 0)$, $(-2, 6)$, and $(4, -3)$ are solutions to all three equations, but the ordered pair $(-3, 4)$ is not a solution to any of the equations.

b. If another student claimed that any ordered pair (a, b) satisfying one of the three equations is certain to satisfy the other two, would you agree? Why or why not?

c. What result would you expect if you graphed solutions for all three equations? Why?

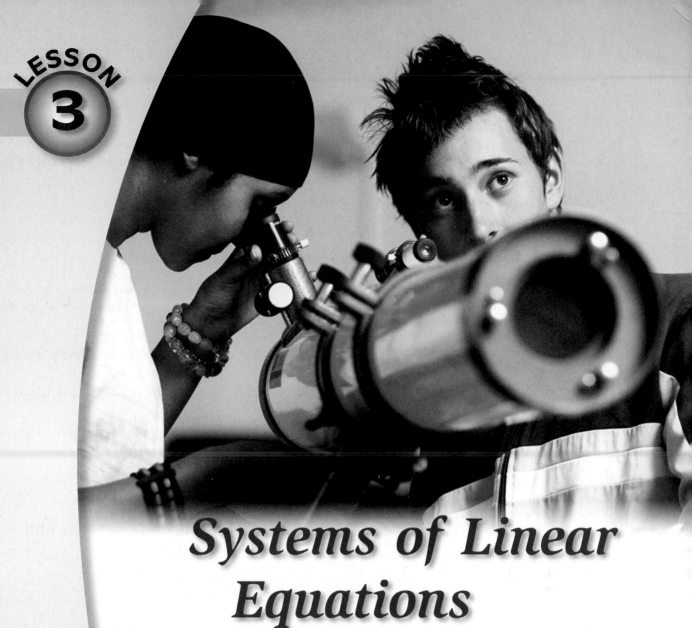

LESSON 3

Systems of Linear Equations

In Lesson 2, you gained experience in writing linear equations with two variables to express a variety of problem conditions. Sometimes, problems involve two linear equations that have to be solved simultaneously. The task is to find one pair (x, y) of values that satisfies both linear equations.

Students in the Hamilton High School science club faced this kind of problem when they tried to raise $240 to buy a special eyepiece for the high-powered telescope at their school. The school PTA offered to pay club members for an after-school work project that would clean up a nearby park and recreation center building.

Because the outdoor work was harder and dirtier, the deal with the PTA would pay $16 for each outdoor worker and $10 for each indoor worker. The club had 18 members eager to work on the project. But, most members would prefer the easier indoor work.

In Lesson 2, you learned that conditions in the science club's situation could be represented by linear equations like these:

$$16x + 10y = 240 \quad \text{and} \quad x + y = 18$$

a What do the variables x and y represent in these equations?

b What problem condition is represented by each equation?

c What are some combinations of numbers of outdoor and indoor workers that will allow the club to earn just enough money to buy the telescope eyepiece? Will any of those combinations also put each willing club member to work?

d What different strategies could you use to find a pair of values for x and y that satisfy both linear equations simultaneously?

Work on the problems of this lesson will develop your skill in writing, interpreting, and solving systems of linear equations.

Investigation 1 — Solving With Graphs and Substitution

There are several different methods for solving systems of linear equations. As you work on the problems of this investigation, look for answers to this question:

> *How can graphs and algebraic substitution be used to solve systems of linear equations?*

As you discussed in the Think About This Situation, a system of linear equations expressing the conditions in the science club's situation is:

$$\begin{cases} 16x + 10y = 240 \\ x + y = 18 \end{cases}$$

The first equation shows that the amount of money they can earn is a linear function of the variables x and y, where x and y represent the number of outdoor and indoor workers, respectively. The second equation shows that the number of club members who will work is also a linear function of those variables.

1 In Lesson 2, you learned that equations in the form $ax + by = c$ are called linear equations because graphs of their solutions are straight lines. The diagram below shows graphs (in the first quadrant) of solutions to the equations $16x + 10y = 240$ and $x + y = 18$.

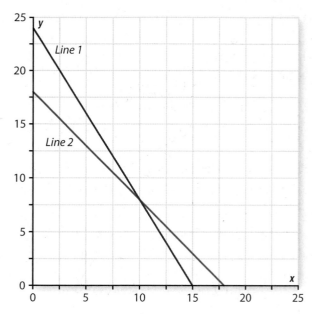

a. Match the graphs to the linear equations they represent. Explain how you know that your answers are correct.

b. Use the graphs to estimate a solution (x, y) for the system of equations—values for x and y that satisfy both equations. Explain what the solution tells about the science club's fund-raising situation.

c. Since graphs give only estimates for solutions of equations, it is important to check the estimates. Show how your graph-based estimate can be checked to see if it is an exact solution to the system.

2 When the date for the work project was set, it turned out that only 13 science club members could participate. The club president talked again with the PTA president and got a new pay deal—$20 per outdoor worker and $15 per indoor worker.

a. Write a system of linear equations in which one equation expresses the new conditions about payment and the other shows the new number of workers.

b. Estimate the solution for this system of equations by using graphs of the two equations. Then check your estimate.

There are many cases when it is easy to solve systems of equations without taking the trouble to produce graphs and estimate the required x and y values. One such symbolic solution strategy, the **substitution method**, combines two equations with two variables into a single equation with only one variable by *substituting* from one equation into the other.

3 Recall the science club's goal of purchasing a $240 telescope eyepiece and the prospect of having 18 workers. Using the substitution method, you could start toward a solution of the system of equations representing this situation by reasoning like this.

> **Step 1:** We have to find values of x and y that satisfy both
> $16x + 10y = 240$ and $x + y = 18$.
>
> **Step 2:** If $x + y = 18$, then $y = 18 - x$.
>
> **Step 3:** That means that $16x + 10(18 - x) = 240$.

This reasoning has led to an equation in Step 3 that contains only one variable, x.

a. Explain each step in the reasoning.

b. Continue the reasoning above to complete a solution of the system.

> **Step 4:** Solve the equation in Step 3 for x. What does this value of x tell you?
>
> **Step 5:** Use the value of x to find a value for y that satisfies the original equation.

c. How could you check your solution from Part b?

d. Explain what your solution for the system of equations tells about the numbers of club members doing outdoor and indoor work necessary to reach their earning goal of $240.

4 Look back at the reasoning used in Problem 3 to solve the system $16x + 10y = 240$ and $x + y = 18$.

a. In Step 2, the equation $x + y = 18$ can also be written as $x = 18 - y$. How can this fact be used to write an alternative to Step 3 that gives a single linear equation involving y alone?

b. Check that this alternative strategy leads to the same solution for the original system.

5 Use reasoning similar to that in Problem 3 or 4 to solve the system of equations that you developed in Problem 2 to model the new conditions—only 13 workers and different pay for each kind of work.

6 Use a graphing method to estimate solutions for three of the following systems of equations. Then use the substitution method to solve the three remaining systems. Check your solutions for all six systems by showing that the values of x and y you find make both equations true statements.

a. $\begin{cases} y = 5x \\ 6x - 2y = -4 \end{cases}$ b. $\begin{cases} 5x - y = -15 \\ x + y = -3 \end{cases}$

c. $\begin{cases} 4x - y = 5 \\ x = 8 - 2y \end{cases}$ d. $\begin{cases} -7x + y = 32 \\ 2x + 3y = 27 \end{cases}$

e. $\begin{cases} 2x + y = 5 \\ 4x - 3y = -10 \end{cases}$ f. $\begin{cases} 4x + 2y = 7 \\ x - 5y = 10 \end{cases}$

Summarize
the Mathematics

Many problem situations can be modeled by systems of linear equations. It is important to know how to solve such systems.

a What does it mean to *solve* a system of linear equations?

b How can you use graphs to estimate the solution for a system of linear equations?

c Explain how to use the substitution method to solve a system of linear equations.

d In what cases would graphing probably be more convenient than substitution? When would substitution be more convenient than graphing?

e Once you have solved a system of equations, how can you check your solution?

Be prepared to share your ideas with others in the class.

✓ Check Your Understanding

As you complete the following tasks, think about the advantages and disadvantages of solving systems of linear equations by graphing and by substitution.

a. Solve the following system of equations in two ways: by graphing the equations and by using substitution.

$$\begin{cases} 2x - y = 2 \\ x + 2y = 18.5 \end{cases}$$

b. The Kesling Middle School booster club is planning a community event to raise money for the school's art department. Based on previous fund-raising events, they estimate that the event will be a sellout—filling all 300 seats in the school auditorium. Plans are to charge adults $8 and children $3 admission. The club wants to earn $2,000 from admission charges.

   i. Write an equation that expresses the relationship among adult attendance, child attendance, and the goal for income from admission charges. Explain what the variables represent.

   ii. Write an equation that expresses the relationship among number of adults, number of children, and total attendance. Explain what the variables represent.

   iii. Describe at least two different ways you could find the numbers of adults and children at the event if the club is to meet their income goal of $2,000 and their attendance estimate of 300 people.

   iv. Solve the system of linear equations and check your solution.

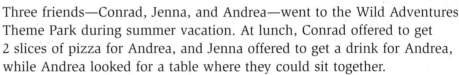

Investigation 2 Solving by Elimination

Graphing and substitution strategies for solving systems of linear equations are convenient in some situations but not in others. The graphing method requires careful plotting of points, and it may give only estimates of solutions. The substitution method is most attractive when it is easy to solve for one variable in terms of the other. As you work on the problems of this investigation, look for an answer to this question:

How can the elimination of a variable be used
to solve a system of linear equations?

Three friends—Conrad, Jenna, and Andrea—went to the Wild Adventures Theme Park during summer vacation. At lunch, Conrad offered to get 2 slices of pizza for Andrea, and Jenna offered to get a drink for Andrea, while Andrea looked for a table where they could sit together.

Prices were not posted at the small pizza stand, but Conrad was charged $10.50 for 4 slices of pizza and 1 large drink, and Jenna was charged $7.50 for 2 slices of pizza and 2 large drinks. Andrea wanted to pay her friends for the food they got her; but at first, all they could figure out was a system of linear equations that expressed the conditions:

$$\begin{cases} 4p + d = 10.50 \\ 2p + 2d = 7.50 \end{cases}$$

1. Analyze and complete the following reasoning used by the friends to solve the system.

   a. What do the variables p and d represent in the given system, and why do the given equations accurately represent the problem conditions?

   b. Conrad claimed that 8 slices of pizza and 2 drinks would cost $21. Is he right? How do you know?

   c. Andrea claimed that 6 slices of pizza and 3 drinks would cost $18. Is she right? How do you know?

   d. Jenna did not find either of these results very useful. She figured that 1 slice of pizza and 1 drink would cost $3.75, so she could solve the problem by solving the system:

   $$\begin{cases} 4p + d = 10.50 \\ p + d = 3.75 \end{cases}$$

   Is she right? How do you know?

   e. By examining the system of equations in Part d, Andrea figured out that 3 slices of pizza would have to cost $6.75, and she wrote the equation $3p = 6.75$. Is she right? How do you know?

   f. Based on this reasoning, what do you think they calculated for the cost of 1 slice of pizza? How about the cost of 1 large drink? How could you check these answers?

Whether they knew it or not, Conrad, Jenna, and Andrea solved the system of linear equations using a strategy called the **elimination method**. That method is based on two key properties of equations:

- If both sides of an equation are multiplied or divided by the same (nonzero) number, then the solutions of the new equation are identical to those of the original.

  For example, the solutions of $2p + 2d = 7.50$ are identical to those of $p + d = 3.75$.

- If you find the sum or difference of two equations in a system, the result often gives useful new information about the unknown values of the variables.

  For example, if $p + d = 3.75$ is subtracted from $4p + d = 10.50$, the result is $3p = 6.75$. From this, we can conclude that $p = 2.25$ and then that $d = 1.50$.

The challenge in using these ideas is finding the multiples, sums, and differences of given equations that lead to a single equation revealing part of the solution.

(2) Consider the system of equations: $\begin{cases} 3x - y = 6 \\ x + 2y = 5.5 \end{cases}$

  a. For each of the steps below, explain what actions have been taken since the previous step. Justify the actions using previous mathematical knowledge and the two properties stated above.

  Start: $\begin{cases} 3x - y = 6 \\ x + 2y = 5.5 \end{cases}$

  Step 1: $\begin{cases} 6x - 2y = 12 \\ x + 2y = 5.5 \end{cases}$

  Step 2: $7x = 17.5$

  Step 3: $x = 2.5$

  Step 4: $2.5 + 2y = 5.5$

  Step 5: $2y = 3$

  Step 6: $y = 1.5$

  Check: $\begin{cases} 3(2.5) - 1.5 = 6 \\ 2.5 + 2(1.5) = 5.5 \end{cases}$

  b. Look back closely at the start of the solution in Part a.

  i. Why was 2 chosen as the constant to multiply both sides of the first equation in the original system to obtain the system in Step 1?

  ii. How could you start the solution process by leaving the first equation as $3x - y = 6$ and multiplying both sides of the *second* equation by a number that makes it easy to eliminate the x variable?

  Show the solution steps that would follow from choosing that multiplier.

③ The steps in solving the system $\begin{cases} 3x - y = 6 \\ x + 2y = 5.5 \end{cases}$ in Part a of Problem 2

produced a total of eight different equations listed below—each telling something about the values of x and y that satisfy the original equations.

$$3x - y = 6 \qquad x + 2y = 5.5 \qquad 6x - 2y = 12 \qquad 7x = 17.5$$

$$x = 2.5 \qquad 2.5 + 2y = 5.5 \qquad 2y = 3 \qquad y = 1.5$$

The diagram below shows graphs of solutions for these eight equations. The scale on each axis is 1.

a. Match each equation with the line that represent its solutions.

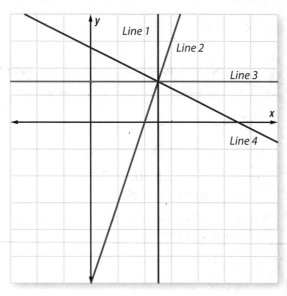

b. Why does it take only four lines on the diagram to represent solutions for all eight equations?

c. What can you conclude from the fact that graphs of all eight equations pass through a single point?

d. Estimate coordinates of the point through which all four lines pass and check to see if those coordinates provide a solution to the original system of equations.

④ Solve these systems of equations using the elimination method. Check each solution.

a. $\begin{cases} 3x + y = 13 \\ x - 5y = 15 \end{cases}$

b. $\begin{cases} 6x - y = 14 \\ 2x + y = 12 \end{cases}$

c. $\begin{cases} 2x + y = -9 \\ 4x + 3y = 1 \end{cases}$

d. $\begin{cases} 3x - 4y = 22 \\ 2x + 5y = 7 \end{cases}$

Summarize
the Mathematics

In this investigation, you examined another method for solving a system of linear equations—the elimination method.

a What would be your first two steps in solving the system $\begin{cases} 2x + y = -9 \\ 4x + 3y = 1 \end{cases}$ using the elimination method? Justify each step.

b You now have worked with three quite different methods for solving systems of linear equations. How do you decide on a method to use in any particular problem?

c Is it possible to use any of the three methods when solving any system of two linear equations?

Be prepared to share your ideas and reasoning with the class.

✔ Check Your Understanding

As you complete these tasks, think about the advantages and disadvantages of using the elimination method for solving a system of linear equations.

a. Use the elimination method to solve each of these systems of linear equations. Check your solutions.

   i. $\begin{cases} 2x + y = 6 \\ 3x + 5y = 16 \end{cases}$

   ii. $\begin{cases} 6x - y = 18 \\ 2x + y = 2 \end{cases}$

   iii. $\begin{cases} 2x + 3y = 5 \\ -3x + 2y = -14 \end{cases}$

b. Given a choice between graphing, substitution, and elimination, which would you choose to solve each of the systems in Part a? Explain the reasoning for your choices.

Investigation 3   Systems with Zero and Infinitely Many Solutions

Systems of linear equations do not always have solutions consisting of a single ordered pair (x, y). As you work through this investigation, look for answers to this question:

> *What are the properties of linear systems that do not*
> *have exactly one ordered pair solution?*

1 Think about the graphing method for estimating solutions to a system of two linear equations.

   a. Suppose (5, 8) is the solution to a system of linear equations. How could you see this by looking at the graphs of the equations?

   b. Sketch graphs of a system of linear equations that has *no* solution.

   c. Is it possible for a system of linear equations to have *infinitely many* solutions? What would the graphs look like in this case?

2 For some people, like athletes and astronauts, selection of a good diet is a carefully planned scientific process. In the case of astronauts, proper nutrition is provided in limited forms. For example, drinks might come in disposable boxes and solid food in energy bars.

   Suppose that, in planning daily diets for a space shuttle team, nutritionists work toward these goals.

- Drinks each provide 30 grams of carbohydrate, energy bars each provide 40 grams of carbohydrate, and the optimal diet should contain 600 grams of carbohydrate per day.

- Drinks each provide 15 grams of protein, energy bars each provide 20 grams of protein, and the optimal diet should contain 200 grams of protein.

The problem is to find a number of drinks and a number of energy bars that will provide just the right nutrition for each astronaut. If we use x to represent the number of drinks and y for the number of energy bars, the goals in diet planning can be expressed as a system of linear equations:

$$\begin{cases} 30x + 40y = 600 \\ 15x + 20y = 200 \end{cases}$$

   a. What does the first equation represent? The second?

Solving that system by the elimination method might follow steps like these.

   Step 1: $\begin{cases} 30x + 40y = 600 \\ 30x + 40y = 400 \end{cases}$

   Step 2: $0 = 200$

b. What properties of equations justify Steps 1 and 2 in this solution method?

c. What does the equation in Step 2 say about the possibility of finding (x, y) values that satisfy the original system?

d. The diagram at the right shows graphs of solutions to the two original linear equations. How does it help explain the difficulty in finding a solution for the system?

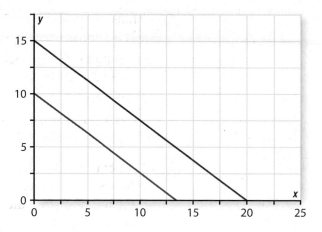

3 Suppose that the condition on protein in Problem 2 was revised to require 300 grams per day.

a. Write the new system of equations expressing the conditions relating number of drink boxes and number of food bars to the required grams of carbohydrate and protein in the diet.

b. Use the elimination method to solve this new system. What happens? What does this tell you?

c. Sketch a graph for each equation in the system in Part a. What do the graphs tell you about solutions for this system?

4 Solve each of the following systems of linear equations in two ways:

- first by estimation using graphs of the two equations,

- then by either substitution or elimination.

In each case, be prepared to explain what the result tells about solutions for the system and how that result is demonstrated in the graph and in the algebraic solution process.

a. $\begin{cases} x - y = 4 \\ 2x - 2y = 8 \end{cases}$

b. $\begin{cases} 3x - 6y = -46.5 \\ -x + 2y = 15.5 \end{cases}$

c. $\begin{cases} y = -5x + 12 \\ y = -5x - 7 \end{cases}$

d. $\begin{cases} x - 2y = 0 \\ 3x - 5y = 2.5 \end{cases}$

5 For which of the systems in Problem 4 could you determine the number of solutions just by examining the equations? How could you tell?

6 In the problems of this lesson, you've studied systems of linear equations that have unique solutions—a single pair of values for the variables that satisfy both equations—and others that have infinitely many solutions or no solutions at all. Find values of a, b, and c that will guarantee the system $\begin{cases} 3x - 2y = 4 \\ ax + by = c \end{cases}$ has:

a. exactly one solution (x, y).

b. infinitely many solutions.

c. no solutions.

Summarize
the Mathematics

In this investigation, you learned how to recognize and solve systems of two linear equations with zero or infinitely many solutions.

a What does it mean to say that a system of linear equations has no solutions? Infinitely many solutions?

b How can you tell from a graph that a system of linear equations has no solutions? Infinitely many solutions?

c If a system of linear equations has infinitely many solutions, does that mean that any ordered pair (x, y) is a solution? Why or why not?

d How can you decide whether a system of linear equations has no solutions, exactly one solution, or infinitely many solutions just by examining the equations?

e You have now had a chance to practice using three methods for solving systems of linear equations. Give specific examples of linear systems that you think are most easily solved by each of these three methods.

    i. graphing   **ii.** elimination   **iii.** substitution

Be prepared to share your ideas, examples, and reasoning with the class.

✔ Check Your Understanding

Use what you have learned about systems of equations to answer these questions.

a. Find values of a and b so that the system $\begin{cases} 3x + 2y = 5 \\ ax + 4y = b \end{cases}$ has infinitely many solutions. Explain how the graphs of the equations will be related.

b. Find values of a and b so that the system $\begin{cases} 3x + 6y = 12 \\ x + ay = b \end{cases}$ has no solutions. Explain how the graphs of the equations will be related.

c. Find values of a and b so that the system $\begin{cases} 3x + 2y = 5 \\ ax + 4y = b \end{cases}$ has exactly one solution. Explain how the graphs of the equations will be related.

Applications

1 To participate in a school trip, Kim had to earn $85 in one week. Kim could earn $8 per hour babysitting and $15 per hour for yard work, but Kim's parents limit work time to 8 hours per week.

  a. Write two equations, one that represents the condition on total number of hours to be worked and the other which relates the number of hours worked at each job toward the fund-raising goal.

  b. Solve the system of equations you wrote in Part a to find out how many hours Kim will have to work at each job to exactly meet the income goal and the time constraint.

2 Solve the following systems of equations, using graphing as the method for one of them.

  a. $\begin{cases} y = 3x - 5 \\ 8x - 4y = 30 \end{cases}$ **b.** $\begin{cases} y = 5x \\ 6x - 2y = 12 \end{cases}$ **c.** $\begin{cases} y = 5x + 15 \\ y = -x - 3 \end{cases}$

3 The relationship between *supply* and *demand* is important in the business world. A supply function indicates the relationship between price of a product and the amount of that product that will be available from suppliers. A demand function indicates the relationship between price of a product and the amount of the product that will be purchased by consumers.

  An electronics store devised the following functions to study supply and demand for its best-selling DVD players.

  Demand: $y = -0.5x + 90$, where y stands for the estimated number sold, and x stands for the price of the DVD player.

  Supply: $y = 1.5x - 30$, where y stands for the number available, and x represents the price of the DVD player.

  a. Why do you suspect the demand function has a negative slope but the supply function's slope is positive?

  b. *Equilibrium* is reached when supply is equal to demand. At what price should the store sell the DVD player in order to reach equilibrium between supply and demand?

(4) Carly is training for an upcoming fitness competition and is trying to find a breakfast combination that meets her nutritional requirements of 950 calories and 25 grams of protein. One serving of her cereal of choice has 200 calories and 2 grams of protein. Her favorite brand of peanut butter contains 180 calories and 8 grams of protein per serving.

   a. Write an equation that relates the number of servings of cereal and the number of servings of peanut butter to the total number of calories she needs for breakfast.

   b. Write another equation that relates the number of servings of cereal and the number of servings of peanut butter to the total amount of protein she needs for breakfast.

   c. What numbers of servings for each type of food would meet both of her nutrition goals?

(5) Laura and Andy are trying to earn money to buy airplane tickets to visit their favorite aunt, Annie. Laura's ticket is going to cost her $280 while Andy found a ticket for $230 on the Internet. To earn their money, they have both decided to mow lawns and babysit. Laura charges $7 per hour for babysitting while Andy charges $5 per hour. To mow a lawn, Laura charges $14 per lawn while Andy charges $16 per lawn.

   a. Write an equation relating income from Laura's work to her ticket cost. Use B to represent number of hours babysitting and L to represent number of lawns mowed. Use the same variables to write another equation relating income from Andy's work to his ticket cost.

   b. Is it possible that Laura and Andy could each reach their ticket price goal by mowing the same number of lawns and babysitting the same number of hours? If so, find those numbers; if not, explain how you know.

(6) Solve the following systems of equations and check your solutions. Show that you know how to use different solution strategies by using each of the three methods once—graphing, substitution, and elimination. Be prepared to explain your choice of solution method in each case.

   a. $\begin{cases} -2x - y = 3 \\ x + 2y = 4 \end{cases}$ b. $\begin{cases} x = 5 - 2y \\ 3x - y = 15 \end{cases}$ c. $\begin{cases} -2x + y = 3 \\ -4x + 2y = 2 \end{cases}$

(7) For each system below, without graphing, determine if the lines represented by the equations are the same, are different and intersect at a point, or are different and parallel. Explain your reasoning in each case. If there is exactly one solution, find and check it.

   a. $\begin{cases} y = -4x + 5 \\ y = -4x - 2 \end{cases}$ b. $\begin{cases} y = 6x - 2 \\ y = 3x - 2 \end{cases}$ c. $\begin{cases} y = 1.5x + 9 \\ y = \frac{3}{2}x + 9 \end{cases}$

8 Without graphing the equations, determine if the lines represented in each system below are the same, are different and intersect at a point, or are different and parallel. Explain your reasoning in each case. If there is exactly one solution, find and check it.

a. $\begin{cases} x + 2y = 8 \\ 2x + y = 4 \end{cases}$

b. $\begin{cases} x - 3y = 6 \\ 3x - 9y = 18 \end{cases}$

c. $\begin{cases} 3x - 2y = 1 \\ 6x - 4y = 10 \end{cases}$

d. $\begin{cases} x + 7 = 10 \\ 2x - 5y = 16 \end{cases}$

9 Find values of a, b, and c that will guarantee that the system

$\begin{cases} x + 2y = 3 \\ ax + by = c \end{cases}$ has:

a. exactly one solution.

b. infinitely many solutions.

c. no solutions.

Connections

10 Imagine yourself as the manager of a local movie theater. Clearly, you are interested in both filling your movie theater and reaching a certain level of revenue.

a. Let x stand for the number of full-price tickets and y for the number of discounted tickets. Then write an equation that relates x and y to the theater capacity of 800 moviegoers.

b. To meet your goal of $6,000 ticket revenue, you have set ticket prices at $9 for full-price tickets and $4 for discounted tickets. Write another equation that relates the number of full-price tickets x and the number of discounted tickets y to the revenue goal of $6,000.

c. How many tickets of each type would you have to sell to fill the theater and meet your financial goal?

11 Suppose a system of two linear equations has the indicated number of solutions. In each case, what can you say about the slopes and y-intercepts of their graphs?

a. infinitely many solutions

b. no solutions

c. exactly one solution

12 The work below shows a solution of the system of equations
$\begin{cases} 6x - 3y = 12 \\ x + y = 5 \end{cases}$ by the method of elimination. The diagram shows
graphs of lines representing solutions of the various separate linear equations that occur in steps of the solution. The scales on both axes are 1.

Start: $\begin{cases} 6x - 3y = 12 \\ x + y = 5 \end{cases}$

Step 1: $\begin{cases} 6x - 3y = 12 \\ 3x + 3y = 15 \end{cases}$

Step 2: $9x = 27$

Step 3: $x = 3$

Step 4: $6(3) - 3y = 12$

Step 5: $y = 2$

Step 6: Solution is $(3, 2)$.

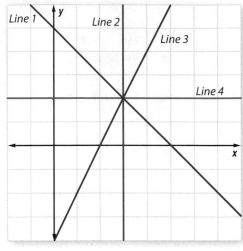

a. Match each equation with the line representing its solutions.

$6x - 3y = 12$ $\qquad$ $x + y = 5$ $\qquad$ $3x + 3y = 15$

$9x = 27$ $\qquad$ $x = 3$ $\qquad$ $y = 2$

b. Explain why the four different lines have exactly one point in common.

13 Solve the following linear equations. Explain how the results are similar to or different from the possibilities for solutions of systems with two linear equations and two variables.

a. $3(5 + 2x) = 6x - 10$

b. $3(5 + 2x) = 6x + 15$

c. $2(5 + 2x) = 6x - 10$

14 Solve the following quadratic equations. Explain how the results are similar to or different from the possibilities for solutions of systems with two linear equations and two variables.

a. $2x^2 + 1 = 9$ $\qquad$ b. $2x^2 + 1 = 1$ $\qquad$ c. $2x^2 + 10 = 6$

15 The quadratic formula says that any equation of the form $ax^2 + bx + c = 0$ has solutions given by

$$x = \frac{-b}{2a} + \frac{\sqrt{b^2 - 4ac}}{2a} \quad \text{and} \quad x = \frac{-b}{2a} - \frac{\sqrt{b^2 - 4ac}}{2a}.$$

Solve the quadratic equations below. Explain how the results are similar to or different from the possibilities for solutions of systems with two linear equations and two variables.

a. $x^2 + 5x + 6 = 0$

b. $x^2 + 10x + 25 = 0$

c. $x^2 + 5x + 7 = 0$

Reflections

16 Changes in the cost of the telescope eyepiece and the number of club members willing to work required solving the system $16x + 10y = 245$ and $x + y = 20$, where x and y represent the number of workers on outdoor and indoor work, respectively.

Robin produced this graph to use in estimating the solution. She estimated that $x = 7.5$ and $y = 12.5$ was the solution.

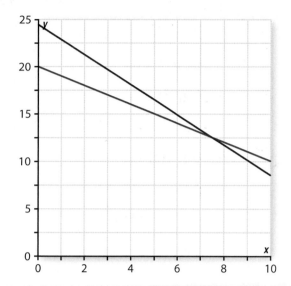

a. Is that an accurate estimate?

b. Does the solution make sense in the problem situation? Why or why not?

17 Faced with the following system of equations, two students, Lincoln and Claire, both decided to use the substitution method to find the solution(s).

$$\begin{cases} 5x - y = -15 \\ x + y = -3 \end{cases}$$

Lincoln's method:

$x = -3 - y$

So, $5(-3 - y) - y = -15$.
$-15 - y - y = -15$
$-15 - 2y = -15$
$-2y = 0$
$y = 0$

So, $x + (0) = -3$.
$x = -3$

The solution is $(-3, 0)$.

Claire's method:

$y = -3 - x$

So, $5x - (-3 - x) = -15$.
$5x + 3 - x = -15$
$4x + 3 = -15$
$4x = -18$
$x = -4.5$

So, $(-4.5) + y = -3$.
$y = 1.5$

The solution is $(-4.5, 1.5)$.

There are errors in the work of both Lincoln and Claire, but one of them was "lucky" and got the correct solution.

a. What are the errors in each case?

b. Which student got the correct solution? How do you know?

18 After examining the methods used by Lincoln and Claire in Task 17, Danil decided to try using the elimination method to make sure that he got the correct solution. In which steps, if any, did Danil make an error in reasoning?

Danil's method:

Start: $\begin{cases} 5x - y = -15 \\ x + y = -3 \end{cases}$

Step 1: $6x = -18$

Step 2: $x = -3$

Step 3: $5(-3) - y = -15$

Step 4: $-15 - y = -15$

Step 5: $y = -30$

Step 6: Solution is $(-3, -30)$.

19 What kinds of errors have you made in using substitution or elimination methods to solve systems of linear equations. What do you think you can do to reduce those errors?

20 Consider the following system of linear equations.

$$\begin{cases} 2x - 3y = -80 \\ 6x - y = 160 \end{cases}$$

a. How can you tell by examining the two equations that the system has one solution?

b. Which method would you use to solve the system? Why?

21 What are the strengths and weaknesses of each method for solving systems of linear equations? Which method do you prefer to use?

Extensions

22 Even after learning three methods for solving systems of linear equations, Jamie likes to use a guess-and-test method first. To help with the testing part of that strategy, Jamie wrote a simple spreadsheet program that could be modified for any given system. Here is what the formulas in Jamie's spreadsheet looked like as he worked on one system.

◇	A	B	C	∧
Systems of Linear Equations.xls			▢ ⧉ ☒	
1	$x =$			
2	$y =$			≣
3	Condition 1:	=2*B1−3*B2	6	
4	Condition 2:	=B1+2*B2	10	∨
5				

a. What system of linear equations does it seem Jamie was trying to solve?

b. What numbers will appear in cells B3 and B4 if Jamie enters 5 in B1 and −3 in B2?

c. How will Jamie know when the spreadsheet shows a solution?

23 In the Think About This Situation problem (page 50) that involved raising money for the science club to buy a new eyepiece for the Hamilton High School telescope, it is not very realistic to aim at earning exactly $240. More than that amount would certainly be acceptable.

a. How do you think you could find all combinations of indoor and outdoor work by 18 workers that would give at least $240 of income from the PTA?

b. What would the graph of this situation look like, in comparison with the graph of the two linear equations?

24 Given the linear equations $ax + by = c$ and $dx + ey = f$ and nonzero numbers m and n, the equation $m(ax + by) + n(dx + ey) = mc + nf$ is called a **linear combination** of the two equations.

a. If (h, k) is a solution of the system $ax + by = c$ and $dx + ey = f$, is (h, k) also a solution to the linear combination of the two equations? Explain your reasoning.

b. How must a, d, m, and n be related in order to eliminate the x term from the linear combination?

c. How must b, e, m, and n be related in order to eliminate the y term from the linear combination?

25 Modify the substitution or elimination methods for solving systems of linear equations with two equations and two variables to solve these systems involving *three* equations and *three* variables.

a. $\begin{cases} 2x - y + z = 15 \\ x + y - 3z = -6 \\ z = 3 \end{cases}$

b. $\begin{cases} x + y + z = 6 \\ 2x + y - z = 7 \\ x - y + z = 2 \end{cases}$

26 During Investigation 1, you had the opportunity to find out ways that the Hamilton High School science club could meet two requirements, total number of workers and total money earned, in saving money to buy a new telescope eyepiece. Both of their requirements (sometimes called *constraints*) involved equality. In other words, they had to use *exactly* 18 workers, and they had to earn *exactly* $240.

a. Suppose that the club could provide *at most* 18 workers. Write an inequality that expresses this constraint.

b. The club needs to earn *at least* $240 for the new eyepiece. Write an inequality that expresses this constraint.

c. Find some ordered pairs that meet the new constraints.

d. Solve the new system of inequalities by finding the region on a graph where coordinates of points meet both inequality constraints.

Review

27 Recall that in the formula $I = Prt$, I represents the amount of simple interest earned, P represents the initial amount invested (sometimes called the principal), r represents the interest rate, and t represents the number of years of investment. Use the formula to complete the following tasks.

a. Write three new formulas, expressing each of the variables (P, r, and t) as a function of the other three.

In Parts b–e, select and use the formula that makes required calculations easiest.

b. How much simple interest is gained on $1,500 invested for 12 years at a rate of 6%?

c. What interest rate is needed to earn $80 in simple interest on an initial investment of $600 over 3 years?

d. How much would you need to invest to earn $125 in simple interest in 5 years at a 4% interest rate?

e. How long would it take to earn $100 interest on an investment of $500 earning simple interest at a rate of 12% per year?

28 In the diagram at the right, $\overline{CD} \cong \overline{CB}$ and $\angle ACD \cong \angle ACB$.

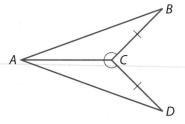

a. Explain why $\triangle ACD \cong \triangle ACB$.

b. Is $\overline{AD} \cong \overline{AB}$? Why?

29 Without using a calculator, sketch graphs you would expect from the following functions.

a. $y = \dfrac{3}{x}$ **b.** $y = 3.5x - 3.5$

c. $y = 3.5^x$ **d.** $y = 3.5x^2 + 3.5$

30 Which of the following vertex-edge graphs contain an Euler circuit? Explain your answer.

Graph I   **Graph II**   **Graph III**

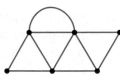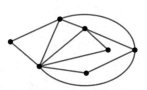

31 For each of the following equations, solve for x and check your solution(s).

a. $x^2 = 0.16$ **b.** $x^2 - 10 = 111$

c. $x^2 - 4x = 21$ **d.** $3x^2 + 11.5x = 2$

Looking Back

The lessons in this unit involved many different situations in which two or more variables were related to each other. Comparing the functions and equations that were used to express relationships and conditions in those settings, you discovered a few common patterns occurring again and again.

As a result of your work on problems in the lessons, you should be better able to recognize situations in which variables are related by direct and inverse variation and by systems of linear equations. You should also be better able to use symbolic expressions and equations to describe and reason about those kinds of relationships.

The tasks in this final lesson will help you review and organize your thinking about functions, equations, and systems.

1 **Sound Wave Properties** The sounds that you hear—from musical instruments and human voices to whispering wind and crackling lightning—are all carried by vibrating waves of air. The pitch of a sound is determined by the frequency of vibration. A higher frequency gives a higher pitch. Sound frequency depends on properties of the sound source.

For example, when the drummer in a band hits one of the cymbals, the frequency F of its sound is related to the density d of the material and the square of the cymbal's radius r.

The pattern of that relationship is expressed well by the formula $F = 0.25\dfrac{d}{r^2}$.

   a. What changes in the variables d and r would lead to higher-pitched sound? What changes would lead to lower-pitched sound?

   b. How are the patterns of change you described in Part a supported by your experience playing or listening to cymbals, drums, or other similar musical instruments?

   c. What formula expresses the relationship among frequency, density, and radius in equivalent form showing d as a function of F and r?

   d. What formula expresses the relationship among frequency, density, and radius in equivalent form showing r as a function of F and d?

2 **Karate Physics** Karate is an impressive form of the martial arts. Highly-trained men, women, boys, and girls break bricks and boards with chops from their hands, feet, or even heads. Some of you may have even attempted a karate chop and discovered that, without proper training and technique, it can hurt.

Karate chops break bricks and boards by applying carefully-aimed bursts of energy. Different targets require different amounts of energy. Think about the four target boards pictured here:

1   2   3   4

a. Which board do you think would require the greatest energy to break?

b. The target boards differ in length and thickness. How would you expect those variables to affect the required breaking energy?

c. Breaking energy E depends on board length L and thickness T. Which of the following do you think expresses the relationship between those variables?

$$E = T \cdot L \qquad E = T + L \qquad E = T - L \qquad E = \frac{T}{L}$$

3 **Production-Line Economics** Since Henry Ford introduced assembly line ideas for production of the Model T Ford car in 1908, manufacturing companies of all kinds have found ways to use machines and human workers in schemes for mass production of many different products.

Suppose that the Kalamazoo bottling company has two machines for filling spring water bottles—one that fills 3,000 per hour and another that fills 4,000 per hour. The factory has orders for 64,000 bottles per day, so the plant manager wants to set a production schedule that just fills the demand. The plant operates two 8-hour shifts per day, but neither machine can run longer than 14 hours per day, allowing time for start-up, cleanup, and repairs.

a. Find four combinations of operating hours for the two machines that will just meet the production quota within the two working shifts.

b. Sketch a graph showing all possibilities for x hours of operation for machine 1 and y hours of operation for machine 2 that meet the daily production quota of 64,000 bottles.

c. If machine 1 is operated for only 8 hours, how long must machine 2 operate?

d. If machine 2 is operated for only 12 hours, how long must machine 1 operate?

e. Use a system of linear equations to find a combination of operating hours for machines 1 and 2 that meet the production quota and use a total of 18 hours of operator labor.

④ Counting Coins Collecting coins is a popular hobby. Some people try to locate rare coins; others simply try to collect lots of coins as a saving strategy.

Grocery chains have now placed coin-counting machines in their stores. The machines provide customers with an easy way to turn their loose change into cash. The machines are programmed to make a profit for their owners by giving the customer credit for only about 90% of the value they deposit.

Kris had been saving quarters and dimes for the past few months and had decided that it was time to cash in her savings. After gathering all of her coins, she took them down to her favorite grocery store to place them in the coin-counting machine.

The receipt stated that she had placed a total of $54.35 worth of dimes and quarters into the machine. She knew that she had collected 275 coins, but she did not know how many of each coin she had deposited.

a. See if you can figure out the missing information.

  i. Write a system of linear equations in which one equation expresses the condition about the number of coins that Kris deposited and the other relates the numbers of dimes and quarters to the total value of the money deposited.

  ii. Find the solution of the system of equations you wrote in part i using any of the methods you have learned in this unit. How many of each coin did she deposit?

  iii. Solve the system you wrote in part i using a different method than the one you chose in part ii.

b. Kris decided it would be much more efficient to save only quarters. She and her friend Tim, another quarter collector, decided to combine their quarters and deposit them into the same coin-counting machine Kris had used before. They combined their quarters before they each counted their individual contribution but later found there were a total of 65 coins.

  i. How much total money did they deposit?

  ii. Write a system of linear equations in which one equation expresses the condition about Tim's and Kris's total quarters and the other relates each of their contribution to the amount of money deposited.

  iii. Find and check the solution of the system of equations you wrote in part ii using any of the methods you have learned in this unit. How many quarters did each of the students deposit?

  iv. How would knowing that Kris contributed 4 times as many coins as Tim help you find how much each contributed?

Summarize
the Mathematics

In this unit, you investigated a variety of situations in which two or more variables were related to each other by functions or equations. In some cases, the relationship could be expressed by a single function or equation; in other cases, the key questions required work with a system of two equations.

a Consider relations expressed in forms like $z = \dfrac{kx}{y}$, where x, y, and $k > 0$.

    i. How does z change as x increases or decreases?

    ii. How does z change as y increases or decreases?

    iii. What equivalent forms express x as a function of z and y? y as a function of x and z?

    iv. How, if at all, will your answers change if $k < 0$?

b Consider equations expressed in the form $ax + by = c$.

    i. How many pairs of (x, y) values can be found to satisfy such equations?

    ii. What pattern can be expected in graphs of the solution pairs?

c Consider systems of linear equations like $\begin{cases} ax + by = c \\ dx + ey = f \end{cases}$.

    i. What are the possible numbers of solutions for such systems? How are those possibilities illustrated in graphs of the solutions for the separate equations?

    ii. How can you tell the number of solutions by examining the symbolic forms of the equations?

    iii. What methods can be used to find the solution(s)?

    iv. What are the key steps in application of each method?

    v. How do you decide which method to use in a particular case?

Be prepared to share your responses and reasoning with the class.

✔ Check Your Understanding

Write, in outline form, a summary of the important mathematical concepts and methods developed in this unit. Organize your summary so that it can be used as a quick reference in future units and courses.

UNIT 2

MATRIX METHODS

Often, you need to organize information before you can use it to solve related problems. A *matrix* is a rectangular array consisting of rows and columns that can help you organize, display, and analyze information, and solve many types of problems. For example, *matrices* can be used to track consumer demand, create computer graphics, rank tournaments, classify ancient pottery, study friendship and trust, and solve systems of linear equations.

In this unit of *Core-Plus Mathematics*, you will study matrices and their properties and use matrices to solve problems in a wide variety of settings, both real-world and purely mathematical. The necessary concepts and skills are developed in three lessons.

Lessons

1 *Constructing, Interpreting, and Operating on Matrices*

Construct matrices to represent and analyze information, interpret given matrices, and operate on matrices using row sums, matrix addition, and multiplication of a matrix by a number.

2 *Multiplying Matrices*

Use matrix multiplication, including powers of matrices, to solve a variety of problems.

3 *Matrices and Systems of Linear Equations*

Compare properties of matrices and matrix operations to corresponding properties of real numbers and their operations. Use matrices and their properties to solve systems of linear equations.

Constructing, Interpreting, and Operating On Matrices

In prior *Core-Plus Mathematics* units, you studied several useful mathematical objects, including linear, exponential, and quadratic functions, polygons and polyhedra, data and probability distributions, and vertex-edge graphs. In this unit, you will investigate another important mathematical object, a *matrix*. Matrices and matrix methods provide powerful tools for solving a wide variety of problems. Matrices also provide useful ways for representing ideas in geometry, algebra, statistics, and discrete mathematics. For example, matrices can be used to represent geometric transformations and vertex-edge graphs, to solve systems of linear equations, and to organize and analyze data.

Matrices even relate to the shoes you wear. Many people wear athletic shoes these days, whether on the job, in school, or on the playing field. Marketing and sales of athletic shoes is big business. There are huge megastores in many cities, such as the one in Chicago pictured here. In 2004, Americans bought 361,929,000 pairs of athletic shoes. (Source: ShoeStats 2005, American Apparel & Footware Association)

Managing an athletic shoe store is a complicated job. Sales need to be tracked, inventory must be controlled carefully, and changes in the market must be anticipated. In particular, the store manager needs to know which shoes will sell. Think about the shoe store where you bought your last pair of athletic shoes.

a What information might the manager of the store want to know about the kinds of shoes the customers prefer? Make a list.

b It is not enough just to have information. The manager needs to organize and manage the information in order to make good decisions. What are some ways the manager might organize the information?

In this lesson, you will learn how to construct matrices to help organize and analyze data; how to interpret matrices in diverse settings; and how to combine and operate on matrices in several ways, including using row sums, matrix addition, and multiplication of a matrix by a number.

Investigation 1 There's No Business Like Shoe Business

There are many different brands of athletic shoes, and each brand of shoe has many different styles and sizes. Shoe-store managers need to know which shoes their customers prefer so they can have the right shoes in stock. As you explore shoe data in this investigation, look for answers to this question:

> *How can you construct and use a rectangular array of numbers (a matrix) to organize, display, and analyze information?*

1 Work together with the whole class to find out about the brands of athletic shoes preferred by students in your class.

   a. Make a list of all the different brands of athletic shoes preferred by students in your class.

   b. How many males prefer each brand? How many females prefer each brand?

(2) One way to organize and display these data is to use a kind of table. You can do this by listing the brands down one side, writing "Men" and "Women" across the top, and then entering the appropriate numbers.

   a. Complete a table like the one below for your class data. Add or remove rows as needed. A rectangular array of numbers like this is called a **matrix**.

Athletic-Shoe Brands

	Men	Women
Converse	____	____
Nike	____	____
Reebok	____	____

   b. The matrix above has 3 **rows** and 2 **columns**. How many rows were needed in the matrix you constructed to display your class data? How many columns?

   c. Could you organize your class data using a matrix with 2 rows? If so, how many columns would it have?

   d. The **size of a matrix** is written as $m \times n$, where m is the number of *rows* and n is the number of *columns*. Thus, the sample matrix in Part a is a 3×2 matrix (which is read as "3 by 2"). What is the size of the matrix you constructed to display your class data?

(3) Knowing the brands of shoes that customers prefer certainly will help a store manager decide which shoes to stock. Other information that will be useful to a store manager can also be organized in *matrices* (plural of matrix).

   a. What other information might the manager of the store want to know about the kinds of shoes the customers prefer? Look back at the list you generated for the Think About This Situation and add to it if necessary.

   b. Construct a matrix that could be used to organize some of the information from your list in Part a. (You probably won't be able to organize all of the information on your list in one matrix.) Don't worry about actually collecting the information; just set up the matrix, label the rows and columns, and give the matrix a title according to the information that it will show.

   c. Compare your matrix with those made by others.

   i. Do all of the matrices make sense? If not, explain why not.

   ii. Are the row and column labels and titles appropriate? If not, how would you modify them?

   iii. How many different variables can be represented in one matrix? Explain your thinking.

4 Suppose you are a manager of a local FleetFeet shoe store. Data on monthly sales of Converse, Nike, and Reebok shoes are shown in the matrix below. Each entry represents the number of pairs of shoes sold.

Monthly Sales

	J	F	M	A	M	J	J	A	S	O	N	D
Converse	40	35	50	55	70	60	40	70	40	35	30	80
Nike	55	55	75	70	70	65	60	75	60	55	50	75
Reebok	50	30	60	80	70	50	10	75	40	35	40	70

a. Describe any patterns you see in the data.

b. Describe any general trends over time that you observe. Which months have the highest sales? What could be a reason for the high sales?

c. Are there any outliers in the data? If so, explain why you think they could have occurred.

d. How many pairs of Nikes were sold over the year?

e. How many pairs of all three brands together were sold in February?

Summarize
the Mathematics

In this investigation, you explored how matrices can be used to organize and display data.

a The Shoe Outlet sells women's shoes, sizes 5 to 11, and men's shoes, sizes 6 to 13. The manager would like to have an organized display of the number of pairs sold this year for each shoe size. Describe a matrix that could be used to organize these data. What is the size of your matrix?

b What are some advantages of using matrices to organize and display data? What are some disadvantages?

c Explain how the same information can be displayed in a matrix in different ways.

Be prepared to share your explanations and thinking with the entire class.

✓Check Your Understanding

Suppose that the FleetFeet shoe store chain has stores in Chicago, Atlanta, and San Diego; the top-selling brands are Nike and Reebok; and in 2007, the average sales figures per month were as follows: 250 pairs of Nike and 195 pairs of Reebok in Chicago, 175 pairs of Nike and 175 pairs of Reebok in Atlanta, and 185 pairs of Nike and 275 pairs of Reebok in San Diego.

a. Organize these data using one matrix. Label the rows and columns and give the matrix a title.

b. What is the size of the matrix?

c. How many pairs of Reebok shoes are sold in all three cities combined?

d. In which city were the most shoes sold?

Investigation 2 · Analyzing Matrices

Matrices can be used to organize all sorts of data, not just sales data. In this investigation, you will analyze some situations in archeology, sociology, and sports. As you explore these different situations, look for answers to this question:

> *How can you interpret and operate on a matrix to help understand and analyze data?*

Archeology Archeologists study ancient civilizations and their cultures. One way they study these cultures is by exploring sites where the people once lived and analyzing objects that they made. Archeologists use matrices to classify and then compare the objects they find at various archeological sites. For example, suppose that pieces of pottery are found at five different sites. The pottery pieces have certain characteristics: they are either glazed or not glazed, ornamented or not, colored or natural, thin or thick.

1 Information about the characteristics of the pottery pieces at all five sites is organized in the matrix below. A "1" means the pottery piece has the characteristic and a "0" means it does not have the characteristic.

Pottery Characteristics

	Glaze	Orn	Color	Thin
Site A	0	1	0	0
Site B	1	0	0	0
Site C	1	0	1	0
Site D	1	1	1	1
Site E	0	1	1	1

a. What does it mean for pottery to be "glazed"? "Ornamented"?

b. What does the "1" in the third row and the first column mean?

c. Is the pottery at site E thick or thin?

d. Which site has pottery pieces that are glazed and thick but are not ornamented or colored?

e. How many of the sites have glazed pottery? Explain how you used the rows or columns of the matrix to answer the question.

2 You can use the matrix in Problem 1 to determine how much the pottery pieces differ between sites. For example, the pieces found at sites A and B differ on exactly two characteristics—glaze and ornamentation. So, you can say that the **degree of difference** between the pottery pieces at sites A and B is 2.

a. Explain why the degree of difference between pottery pieces at sites A and C is 3.

b. Find the degree of difference between the pottery pieces at sites D and E.

3 You can construct a new matrix that summarizes the degree of difference information.

Degree of Difference

	A	B	C	D	E
A	_	2	3	_	_
B	2	_	_	_	_
C	3	_	_	_	_
D	_	_	_	_	_
E	_	_	_	_	_

a. What number would best describe the difference between site A and site A?

b. What number should be placed in the third row, fourth column? What does this number tell you about the pottery at these two sites?

c. Complete a copy of the degree of difference matrix shown above.

d. Describe one or two patterns you see in the degree of difference matrix.

4 Archeologists want to learn about the civilizations that existed at the sites. For instance, they would like to know whether different sites represent different civilizations and whether one civilization was more advanced than another. A lot of evidence is needed to make such decisions. However, you can make some conjectures based just on the pottery data in the matrices from Problems 1 and 3.

a. Find two sites that you think might be from the same civilization. Explain how the pottery evidence supports your choice.

b. Find two sites that you think might be from different civilizations. Give an argument defending your choice.

c. Give an argument supporting the claim that the civilization at site D was more advanced than the others. What assumptions are you making about what it means for a civilization to be "advanced"?

Sociology Sociologists study human social behavior. They may use matrices in their analysis of social relations.

5 Suppose a sociologist is studying friendship and trust among five classmates at a certain high school. The classmates are asked to indicate with whom they would like to go to a movie and to whom they would loan $10. Their responses are summarized in the following two matrices. Each matrix is read from row to column. "1" means "yes" and "0" means "no". For example, the "1" in the first row and fourth column of the movie matrix means that student A would like to go to a movie with student D.

Movie Matrix

		with				
		A	B	C	D	E
Would Like to Go to a Movie	A	0	1	1	1	1
	B	0	0	1	1	1
	C	1	0	0	1	0
	D	0	1	1	0	1
	E	1	0	0	1	0

Loan Matrix

		to				
		A	B	C	D	E
Would Loan Money	A	0	0	0	1	1
	B	1	0	0	1	0
	C	0	0	0	1	0
	D	1	1	0	0	1
	E	0	1	1	0	0

a. Would student A like to go to a movie with student B? Would student B like to go to a movie with student A?

b. With whom would student D like to go to a movie?

c. What does the "0" in the fourth row, third column of the loan matrix mean?

d. To whom would student A loan $10?

e. A **square matrix** has the same number of rows and columns. The **main diagonal** of a square matrix is the diagonal line of entries running from the top-left corner to the bottom-right corner.

   i. Why do you think there are zeroes for each entry in the main diagonals of the movie matrix and the loan matrix?

   ii. Could you use different entries, or no entries, in the diagonals? Explain.

6 Now consider additional information conveyed by the matrices on page 80.

    a. Explain why the movie matrix could be used to describe *friendship*, while the loan matrix could describe *trust*.

    b. Write two interesting statements about friendship and trust among these five students, based on the information in the matrices.

7 Discuss with your classmates how you can use the rows or columns of the movie and loan matrices to answer the following questions.

    a. How many students does student C consider as friends?

    b. How many students consider student C as a friend?

    c. Who seems to be the most trustworthy student?

    d. Who seems to be the most popular student?

Summarize
the Mathematics

In the previous investigations, you analyzed matrices to get useful information about the situations being studied.

a Give three examples from your analysis of shoe sales, pottery, or friendship and trust that show how you performed computations on or compared the entries of the given matrix to get useful information. For each example, describe the situation, the computation or comparison, and the information obtained.

b One common and useful matrix operation is to sum all of the numbers in a row or in a column. The resulting number is called a *row sum* or *column sum*. Describe one example in which you computed a row sum. Give another example where you found a column sum.

Be prepared to share your examples and thinking with the class.

✓ Check Your Understanding

In 2006, the University of Maryland won the NCAA women's basketball championship. They defeated Duke 78-75 in overtime, completing a 34-4 season.

   Crystal Langhorne was Maryland's high scorer for the season with 654 points. Teammate Shay Doron was the next highest scorer with 511 points, followed by Marissa Coleman with 510 points, and Laura Harper with 413 points.

The matrix below shows some of the nonshooting performance statistics for the top four Maryland scorers for the entire 2005–06 season.

Nonshooting Performance Statistics

	Assists	Steals	Rebounds	Blocked Shots	Turnovers	Fouls
Langhorne	77	27	325	14	98	70
Doron	149	67	143	11	118	80
Coleman	115	48	299	52	115	83
Harper	26	31	258	70	80	108

Source: umterps.cstv.com/sports/w-baskbl/stats/2005-2006/teamcume.html

a. A "turnover" is when an action (other than a foul, steal, or scoring a basket) gives the other team control of the ball. How many turnovers did Coleman have?

b. How many rebounds were made by all of these players combined during the season?

c. Which of the performance factors do you think are positive, that is, they contribute to winning a game? Which performance factors do you think are negative?

d. Describe how you could give a "nonshooting performance score" to each player. The score should include both positive and negative factors. Compute this score for each player. Which player do you think contributed most to the team over the season in the area of nonshooting performance? Explain your choice.

e. Below are two possible methods to compute the "nonshooting performance scores." For each method, did you use that method when you computed the scores in Part d? If not, use the method now. Compare the method and the score you get to your work in Part d.

  i. Modify the matrix above so that the entries reflect positive and negative performance factors. Then use row sums to find the "nonshooting performance scores."

  ii. Construct two matrices–one for the positive factors and one for the negative factors. Then combine these two matrices to find the "nonshooting performance scores."

f. There were 38 games in the season. Langhorne and Doron played all 38 games. Coleman played in 37 games, and Harper played in 36 games. Adjust the numbers in the given matrix to give estimates if all 4 players had played in all 38 games. Then compute a "nonshooting performance score" for each player. Compare to what you found in Part e.

Investigation 3 — **Combining Matrices**

You have seen that a matrix can be used to store and organize data. You also have seen that you can operate on the numbers in the rows or columns of a matrix to get additional information and draw conclusions about the data. In this investigation, you will consider situations in which it is also useful to combine two matrices. As you work through the problems, keep track of answers to the following question:

What are some other useful methods for operating on a matrix or combining two matrices?

1 The movie and loan matrices from the previous investigation are shown below. You can analyze these matrices together to see how friendship and trust are related in this group of five students.

Movie Matrix

with

Would Like to Go to a Movie

$$\begin{array}{c} \\ A \\ B \\ C \\ D \\ E \end{array} \begin{array}{ccccc} A & B & C & D & E \\ \left[\begin{array}{ccccc} 0 & 1 & 1 & 1 & 1 \\ 0 & 0 & 1 & 1 & 1 \\ 1 & 0 & 0 & 1 & 0 \\ 0 & 1 & 1 & 0 & 1 \\ 1 & 0 & 0 & 1 & 0 \end{array}\right] \end{array}$$

Loan Matrix

to

Would Loan Money

$$\begin{array}{c} \\ A \\ B \\ C \\ D \\ E \end{array} \begin{array}{ccccc} A & B & C & D & E \\ \left[\begin{array}{ccccc} 0 & 0 & 0 & 1 & 1 \\ 1 & 0 & 0 & 1 & 0 \\ 0 & 0 & 0 & 1 & 0 \\ 1 & 1 & 0 & 0 & 1 \\ 0 & 1 & 1 & 0 & 0 \end{array}\right] \end{array}$$

    a. Who does student A consider a friend and yet does not trust enough to loan $10?

    b. Do you think it is reasonable that a student could have a friend who he or she does not trust enough to loan $10?

    c. Who does student B trust and yet does not consider that person to be friends?

    d. Who does student D trust and also consider to be a friend?

2 A friend you trust is a *trustworthy friend*.

    a. Combine the movie and loan matrices to construct a new matrix that shows who each of the five students considers to be a trustworthy friend.

    b. Write down a systematic procedure explaining how to construct the trustworthy-friend matrix.

    c. Compare your procedure with that of others.

    d. Write two interesting observations about the information in this new matrix.

In the next situation, you will explore other methods of combining matrices and learn some of the ways in which matrix operations are used in business and in industry.

Motor vehicles are produced in many regions around the world. Production levels significantly affect the economies of many countries, and thus they are tracked carefully.

The two matrices below show the production of passenger vehicles (PV), light commercial vehicles (LCV), and heavy trucks (HT) in three regions for the years 2004 and 2005. The numbers shown are in thousands.

2004 Vehicle Production

	PV	LCV	HT
North America	6,359	9,406	470
South America	1,992	419	122
European Union	14,664	1,637	522
Japan	8,720	1,009	770

2005 Vehicle Production

	PV	LCV	HT
North America	6,667	9,087	549
South America	2,290	522	138
European Union	14,178	1,680	550
Japan	9,016	1,047	724

Source: www.oica.net

3 Analyze the two matrices above.

　a. How many vehicles are represented by the entry in the first row and first column of the matrix for 2004?

　b. According to these data, how many heavy trucks were produced in South America in 2005?

　c. Explain the meaning of the entry in the third row and second column of the 2004 matrix.

　d. Describe any patterns you see in these data.

4 Additional information can be derived by combining the two matrices.

　a. According to these data, by how much did passenger vehicle production increase in South America from 2004 to 2005?

　b. Construct a new matrix with the same row and column labels that shows how much vehicle production changed from 2004 to 2005 for each region and each type of vehicle.

　　i. How did you obtain entries in the new matrix from the two given matrices?

　　ii. Explain any trends or unusual entries in the new matrix.

5 Construct a matrix with the same row and column labels as the given matrices that shows the total number of motor vehicles produced over the two-year period 2004 through 2005. How did you obtain the new matrix?

6 Suppose that the auto industry projected a 10% increase in vehicle production from 2005 to 2006 over all regions and all types of vehicles. Construct a matrix that shows the projected 2006 production figures for each region and each type of vehicle. How did you obtain the new matrix?

Summarize
the Mathematics

In this investigation, you explored ways to operate on matrices or combine two matrices in order to derive new information.

a Which of the problems about vehicle production involved *adding matrices*?

b If $A = \begin{bmatrix} 3 & 7 \\ 2 & 10 \end{bmatrix}$ and $B = \begin{bmatrix} 5 & -5 \\ 11 & 0 \end{bmatrix}$, how would you compute $A + B$?

   i. Write a general description (definition) of matrix addition, that is, how to add two matrices M and N to get the sum matrix, $M + N$.

   ii. What must be true about the sizes of matrices M and N in order to compute $M + N$?

c Which of the problems about vehicle production involved *subtracting matrices*?

d Which of the problems about vehicle production involved multiplying each entry of a matrix by the same number?

e If $A = \begin{bmatrix} -3 & 7 & \frac{1}{2} \\ 1 & 8 & 15 \end{bmatrix}$, how would you compute $3A$?

   i. Write a general description (definition) for how to multiply a matrix M by a number k to get a new matrix, kA. (This is also called **scalar multiplication**.)

   ii. How are the sizes of matrices M and kM related?

f Consider all the situations you have analyzed so far. What other operations have you performed on matrices?

Be prepared to share your examples and descriptions with the class.

✓ Check Your Understanding

Below are two matrices showing the 2004 and 2005 regular season passing statistics for three top NFL quarterbacks. "Att" is an abbreviation for passes attempted; "Cmp" refers to passes completed; "TD" refers to passes thrown for a touchdown; and "Int" refers to passes that were intercepted.

2004 Passing Statistics

	Att	Cmp	TD	Int
Manning	497	336	49	10
Roethlisberger	295	196	17	11
Hasselbeck	474	279	22	15

2005 Passing Statistics

	Att	Cmp	TD	Int
Manning	453	305	28	10
Roethlisberger	268	168	17	9
Hasselbeck	449	294	24	9

Source: www.nfl.com/stats/2004/regular;
www.nfl.com/stats/2005/regular

Let A represent the 2005 matrix and B represent the 2004 matrix.

a. Compute $A + B$. What does $A + B$ tell you about the passing performance of the three quarterbacks?

b. Compute $A - B$. What does $A - B$ tell you about the passing performance of the three quarterbacks?

c. Compute $B - A$.

   i. How do the numbers in $B - A$ differ from the numbers in $A - B$?

   ii. What does a negative number in the "Cmp" column of $A - B$ tell you about the trend in completions from 2004 to 2005?

   iii. What does a negative number in the "Cmp" column of $B - A$ tell you about the trend in completions from 2004 to 2005?

d. Compute $\frac{1}{2}A$. What could $\frac{1}{2}A$ mean for this situation?

e. The matrices you have been operating on in this task are related to a specific context. For practice, perform the indicated operations on the following matrices which have no context.

   i. $\begin{bmatrix} 2 & -1 & 5 \\ 6 & 3 & 4 \end{bmatrix} + \begin{bmatrix} 5 & -7 & 6 \\ 4 & -2 & -4 \end{bmatrix}$

   ii. $3 \begin{bmatrix} 4 & \frac{2}{3} \\ -2 & 0 \end{bmatrix}$

   iii. $5 \begin{bmatrix} 1 & 2 & 3 \\ 0 & 4 & 5 \\ 2 & 1 & -8 \end{bmatrix} + \begin{bmatrix} 2 & -1 & 2 \\ 3 & 3 & 3 \\ 4 & 1 & 4 \end{bmatrix}$

   iv. $\begin{bmatrix} 23 & 15 \\ -5 & 42 \\ 0 & 33 \end{bmatrix} - \begin{bmatrix} 18 & 20 \\ 8 & -3 \\ 0 & 10 \end{bmatrix}$

Applications

1 The movie matrix from Investigation 2 is reproduced below. Recall that the movie matrix can be thought of as describing friendship, and it is read from row to column. Thus, for example, student A considers student B as a friend since there is a "1" in the first row and second column.

Movie Matrix

Would Like to Go to a Movie

	A	B	C	D	E
A	0	1	1	1	1
B	0	0	1	1	1
C	1	0	0	1	0
D	0	1	1	0	1
E	1	0	0	1	0

with (column labels A B C D E)

a. *Mutual friends* are two people who consider each other as friends.

   i. Are students A and D mutual friends?

   ii. Find at least two pairs of mutual friends.

   iii. How do mutual friends appear in the matrix?

b. Construct a new matrix that shows mutual friends. To do this, list all five people across the top and also down the side. Write a "1" or a "0" for each entry, depending on whether or not the two people corresponding to that entry are mutual friends.

c. Who has the most mutual friends?

d. Compare the first row of the mutual-friends matrix to the first column. Compare each of the other rows to its corresponding column. What relationship do you see? Explain why the mutual-friends matrix has this relationship between its rows and columns. (See Connections Task 8 for more information about matrices with this property.)

2 Spreadsheets are one of the most widely-used software applications. A spreadsheet displays organized information in the same way a matrix does. One common use of spreadsheets is to itemize loans. For example, suppose that you are going to buy your first car. The one you decide to buy needs some work, but you can get it for $500. You borrow the $500 at 9% annual interest and agree to pay it back in 12 monthly payments. The following spreadsheet summarizes all the information about this loan.

Car Loan.xls   ⊟ ⧉ ☒

◇	A	B	C	D	E
1	Loan Amt. =	$500	Interest Rate/Mo.	9%/12 =	0.0075
2					
3	Month (end)	Payment	To Interest	To Principal	Balance
4	1	$43.73	$3.75	$39.98	$460.02
5	2	$43.73	$3.45	$40.28	$419.74
6	3	$43.73	$3.15	$40.58	$379.16
7	4	$43.73	$2.84	$40.89	$338.27
8	5	$43.73	$2.54	$41.19	$297.08
9	6	$43.73	$2.23	$41.50	$255.58
10	7	$43.73	$1.92	$41.81	$213.76
11	8	$43.73	$1.60	$42.13	$171.64
12	9	$43.73	$1.29	$42.44	$129.19
13	10	$43.73	$0.97	$42.76	$86.43
14	11	$43.73	$0.65	$43.08	$43.35
15	12	$43.68	$0.33	$43.35	$0.00
16	Totals	$524.71	$24.71	$500.00	
17					

a. How much principal will you still owe after the sixth payment?

b. How much interest will you pay in the fourth month?

c. In any given row, how do the entries in the "To Interest" and "To Principal" columns compare to the entry in the "Payment" column? Why are the entries related in this way?

d. Why do the entries in the "To Principal" column get bigger throughout the loan period?

e. How can you use nearby entries to compute the entries in the Month 10 row?

f. How much money will you save if you pay for the car in cash instead of borrowing the $500 and paying off the loan over a year?

③ Music is an enjoyable part of many people's lives. The format in which you get music has changed over the years and continues to change. The following matrices show shipments of music albums in the United States in different formats during the transition period between the 1990s and 2000s. The formats shown are: CD–compact disc; CASS–cassette; VNL–vinyl; DL–download. The 0* in the DL column means that it was not possible to download music at that time, or data were not yet being gathered for this format, or the number is so small that it rounds to 0. The numbers shown are millions of units.

Album Shipments Late 1990s

(in millions)

	CD	CASS	VNL	DL
1994	662.1	345.4	1.9	0*
1995	722.9	272.6	2.2	0*
1996	778.9	225.3	2.9	0*
1997	753.1	172.6	2.7	0*
1998	847.0	158.5	3.4	0*
1999	938.9	123.6	2.9	0*

Album Shipments Early 2000s

(in millions)

	CD	CASS	VNL	DL
2000	942.5	76.0	2.2	0*
2001	881.9	45.0	2.3	0*
2002	803.3	31.1	1.7	0*
2003	746.0	17.2	1.5	0*
2004	767.0	5.2	1.36	4.6
2005	705.4	2.5	1.02	13.6

Source: Recording Industry Association of America; www.riaa.com/news/marketingdata/facts.asp

a. Describe any patterns you see in these data.

b. Analyze these matrices. Use the rows or columns of the matrices to help answer the following questions.

  i. How many millions of albums were shipped in all 4 formats in 2005?

  ii. How many CDs were shipped in the early 2000s?

  iii. How many more cassettes were shipped in the late 1990s than in the early 2000s?

c. Do these two matrices have appropriate sizes so they could be added? Would it make sense to add these matrices? Explain.

d. Construct a 2 × 4 matrix that shows the total number of album shipments in each of the 4 formats for the late 1990s and the early 2000s. Use the matrix to help answer the following questions.

  i. Which format shows the greatest number of increased shipments from the late 1990s to the early 2000s? Which format shows the most dramatic increase in shipments from the late 1990s to the early 2000s?

  ii. Consider the total albums shipped in all 4 formats. Which period, the late 1990s or the early 2000s, had more albums shipped? How many more?

  iii. Describe the general trends shown in this matrix. Do you think these trends will continue? Explain.

(4) An automotive manufacturer produces several styles of sport wheels. One of the styles is available in two finishes (chrome-plated and silver-painted) and three wheel sizes (15-inch, 16-inch, and 17-inch).

  In October, a retailer in the Midwest purchased sixteen 15-inch chrome wheels, twenty-four 16-inch chrome wheels, eight 17-inch chrome wheels, eight 15-inch silver wheels, twelve 16-inch silver wheels, and four 17-inch silver wheels. In November, the retailer ordered twelve 15-inch chrome wheels, thirty-two 16-inch chrome wheels, sixteen 17-inch chrome wheels, twelve 15-inch silver wheels, and twenty 16-inch silver wheels.

a. Construct two matrices that show the wheel orders—one for October and one for November. Label the matrices and the rows and columns.

b. How many of each type of wheel were ordered by the retailer during these two months combined? Represent this information in a matrix. Label the matrix and the rows and columns.

c. Suppose that over the entire fourth quarter (October, November, and December) the retailer has agreed to order the number of wheels shown in the following matrix.

Fourth-Quarter Order

	15-in.	16-in.	17-in.
Chrome	40	52	36
Silver	28	32	16

   i. Construct a matrix that shows how many of each type of wheel must be ordered in December to meet this agreement.

   ii. Explain any unusual entries in the matrix.

d. In October of the next year, the retailer orders twice the number of each type of wheel ordered the previous October. November's order is three times the number of each type of wheel ordered the previous November. Construct a matrix that shows the number of each type of wheel ordered in the two months combined.

⑤ The first matrix below presents combined monthly sales for three types of men's and women's jeans at JustJeans stores in three cities. The second matrix below gives the monthly sales for women's jeans.

Combined Sales

	Levi	Lee	Wrangler
Chicago	250	195	105
Atlanta	175	175	90
San Diego	185	210	275

Women's Jeans Sales

	Levi	Lee	Wrangler
Chicago	100	90	70
Atlanta	80	85	50
San Diego	105	50	150

a. Construct a matrix that shows the monthly sales for men's jeans for each brand and each city. Which matrix operation did you use to construct this matrix?

b. Organizing the data in different ways can highlight different information. Copy and complete the following matrices to show sales of men's and women's jeans in each city, for each of the three brands. Label the rows.

Chicago

M   W

$$\begin{bmatrix} & \\ & \\ & \end{bmatrix}$$

Atlanta

M   W

$$\begin{bmatrix} & \\ & \\ & \end{bmatrix}$$

San Diego

M   W

$$\begin{bmatrix} & \\ & \\ & \end{bmatrix}$$

c. Refer to the matrices in Part b. Construct one matrix that shows the total sales of men's and women's jeans for each of the three brands, that is, sales in all three cities combined. Label the rows of the matrix with the brands and the columns with "M" and "W". Which matrix operation did you use to construct this matrix?

d. For the first quarter, the managers of the Chicago, Atlanta, and San Diego stores have placed jeans orders with the main warehouse as indicated in the matrices below. Let C represent the matrix for the store in Chicago, A for the store in Atlanta, and S for the store in San Diego.

Chicago

	M	W
Levi	300	330
Lee	345	300
Wrangler	120	240

Atlanta

M	W
300	255
300	270
135	165

San Diego

M	W
252	315
513	162
405	450

For the second quarter, the managers' orders for each brand are tripled in Chicago, stay the same in Atlanta, and are $\frac{2}{3}$ as big in San Diego.

  i. Think about how to calculate the total-orders matrix T of men's and women's jeans in all three cities combined, for each of the three brands for the second quarter. Write a rule for calculating T in terms of C, A, and S.

  ii. Compute the second row, second column entry of T. What does this entry tell you about jeans orders placed with the warehouse?

6 Consider the following matrices.

$$A = \begin{bmatrix} 2 & -4 & 6 \\ 0 & 1.5 & 3 \\ 7 & -3.5 & 8 \\ 1 & -1 & 6 \end{bmatrix} \qquad B = \begin{bmatrix} 2 & 3 & -6 \\ 0 & 1 & 6.5 \\ 11 & -3 & 6 \end{bmatrix}$$

$$C = \begin{bmatrix} 1 & 0 \\ 0 & 1 \\ 1 & 1 \end{bmatrix} \qquad D = \begin{bmatrix} -1 & 1.25 & 0 \\ 8 & -12 & 5 \\ 0 & 0 & 18 \end{bmatrix}$$

a. Compute $B + D$.

b. Compute $6C$.

c. Compute $-A$.

d. Compute $B + B$.

e. Compute $2B - 3D$.

f. Compute $D - B$.

g. Construct a new matrix E that could be added to A. Then compute $A + E$.

Connections

7 In Problem 4 of Investigation 1, you examined some monthly shoe sales data, reproduced below. Each entry represents the number of pairs of shoes sold.

Monthly Sales

	J	F	M	A	M	J	J	A	S	O	N	D
Converse	40	35	50	55	70	60	40	70	40	35	30	80
Nike	55	55	75	70	70	65	60	75	60	55	50	75
Reebok	50	30	60	80	70	50	10	75	40	35	40	70

a. What is the mean number of pairs of Reeboks sold per month?

b. Which brand has more variability in its monthly sales? Explain how you determined variability.

c. Identify at least two types of data plots that could be used to represent the monthly sales data.

d. Create a plot that you think would be most informative.

8 Symmetry is an important concept in mathematics. In prior units of *Core-Plus Mathematics,* you examined geometric shapes and graphs of functions in terms of their symmetries. Symmetry also applies to matrices, but only to square matrices. A square matrix is said to be **symmetric** if it has reflection (or mirror) symmetry about its main diagonal. (Recall that the main diagonal of a square matrix is the diagonal line of entries running from the top-left to the bottom-right corner.) So, a square matrix is symmetric if the numbers in the mirror-image positions, reflected in the main diagonal, are the same. For example, consider the three matrices below. Matrices A and B are symmetric, but matrix C is not symmetric.

$$A = \begin{bmatrix} 0 & 1 & 0 & 1 \\ 1 & 0 & 1 & 1 \\ 0 & 1 & 0 & 0 \\ 1 & 1 & 0 & 0 \end{bmatrix} \quad B = \begin{bmatrix} 25 & 3 & 4 & 5 \\ 3 & 36 & 6 & 7 \\ 4 & 6 & 9 & 8 \\ 5 & 7 & 8 & 10 \end{bmatrix} \quad C = \begin{bmatrix} 0 & 0 & 1 & 1 \\ 1 & 0 & 1 & 0 \\ 0 & 1 & 0 & 0 \\ 1 & 1 & 1 & 0 \end{bmatrix}$$

a. Identify two square matrices from this lesson.

b. Which of the following matrices are symmetric? For those that are not symmetric, explain why not.

   i. The pottery matrix (page 79)

   ii. The degree-of-difference matrix (page 79)

   iii. The movie matrix (page 80)

   iv. The loan matrix (page 80)

   v. The nonshooting-performance-statistics matrix (page 82)

   vi. The mutual-friends matrix (Applications Task 1 Part b, page 87)

c. For those matrices in Part b that are symmetric, what is it about the situations represented by the matrix that causes the matrix to be symmetric?

d. Create your own symmetric matrix with four rows and four columns.

   i. Compare the first row to the first column. Compare the second row to the second column. Do the same for the remaining two rows and columns.

   ii. Make a conjecture about the corresponding rows and columns of a symmetric matrix.

e. Test your conjecture from Part d on the symmetric matrices you identified in Part b.

9 You may recall from Course 1 that a vertex-edge graph is a collection of vertices with edges joining some of those vertices. A *directed graph* or **digraph** is a vertex-edge graph in which the edges have a direction, shown by arrows. An **adjacency matrix** for a digraph is a matrix where each entry tells how many direct connections (directed edges) there are from the vertex corresponding to the row to the vertex corresponding to the column. A digraph with four vertices and its adjacency matrix are shown below.

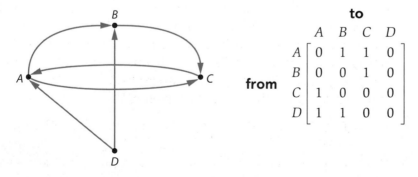

Notice that the *B-C* entry is "1" because there is a directed edge from *B* to *C* in the digraph. The *C-B* entry is "0" since there is no directed edge from *C* to *B*.

The movie (friendship) matrix from Investigation 2 can be thought of as an adjacency matrix for some digraph.

Movie Matrix

Would Like to Go to a Movie

$$
\begin{array}{c}
\text{with} \\
\begin{array}{c c}
 & \begin{array}{ccccc} A & B & C & D & E \end{array} \\
\begin{array}{c} A \\ B \\ C \\ D \\ E \end{array} &
\left[\begin{array}{ccccc}
0 & 1 & 1 & 1 & 1 \\
0 & 0 & 1 & 1 & 1 \\
1 & 0 & 0 & 1 & 0 \\
0 & 1 & 1 & 0 & 1 \\
1 & 0 & 0 & 1 & 0
\end{array}\right]
\end{array}
\end{array}
$$

a. What should the vertices and directed edges of the digraph represent?

b. Draw a digraph for the movie matrix.

c. Mutual friends were defined in Applications Task 1 as two people who consider each other as friends. How can you use the digraph for the friendship matrix to find pairs of mutual friends?

d. Does each of the five students have at least one mutual friend?

e. Write down one interesting statement about friendship among these five people that is illustrated by the friendship digraph.

10 You may have seen distance charts like the one below on maps or in brochures or atlases.

Miles / Kilometers	Barstow	Bishop	Eureka	Fresno	Los Angeles	Needles	Palm Springs	Redding	Sacramento	San Diego	San Francisco
Barstow		215	685	245	115	145	125	575	415	175	420
Bishop	350		560	225	265	360	300	432	270	355	295
Eureka	1,100	900		450	645	825	750	147	290	765	275
Fresno	390	360	725		220	385	325	330	170	340	190
Los Angeles	185	430	1,040	350		260	110	540	385	125	385
Needles	230	575	1,330	620	415		210	715	455	320	560
Palm Springs	200	485	1,210	520	175	335		650	490	140	485
Redding	925	695	235	535	880	1,150	1,045		165	665	215
Sacramento	665	440	645	275	620	895	760	260		505	85
San Diego	285	570	1,235	545	200	515	225	1,075	815		505
San Francisco	675	475	440	300	615	900	785	350	135	815	

Source: visitorinfo.bookcalifornia.com/mileage.html

a. Use this chart to find the following distances in both miles and kilometers.

i. distance between Needles and Palm Springs

ii. distance between Needles and Los Angeles

b. Explain why the given chart is not symmetric.

c. Suppose you only want to know distances in kilometers. Does the bottom half of this chart (below the main diagonal) provide enough information to find the distance between any two cities shown? Explain.

d. Think about constructing an 11 × 11 matrix in which each entry shows the distance in kilometers between the respective cities. Explain why this matrix must be symmetric about the main diagonal.

(11) In Applications Task 2, you investigated a computer-generated spreadsheet summarizing payments on a $500 loan borrowed at 9% annual interest. Payments were made every month for one year. All the information about this loan was summarized in the spreadsheet below. Consider how the spreadsheet entries were computed.

Car Loan.xls

◇	A	B	C	D	E
1	Loan Amt. =	$500	Interest Rate/Mo.	9%/12 =	0.0075
2					
3	Month (end)	Payment	To Interest	To Principal	Balance
4	1	$43.73	$3.75	$39.98	$460.02
5	2	$43.73	$3.45	$40.28	$419.74
6	3	$43.73	$3.15	$40.58	$379.16
7	4	$43.73	$2.84	$40.89	$338.27
8	5	$43.73	$2.54	$41.19	$297.08
9	6	$43.73	$2.23	$41.50	$255.58
10	7	$43.73	$1.92	$41.81	$213.76
11	8	$43.73	$1.60	$42.13	$171.64
12	9	$43.73	$1.20	$42.44	$129.19
13	10	$43.73	$0.97	$42.76	$86.43
14	11	$43.73	$0.65	$43.08	$43.35
15	12	$43.68	$0.33	$43.35	$0.00
16	Totals	$524.71	$24.71	$500.00	
17					

a. Let P represent the entries in the "Payment" column, TI represent the entries in the "To Interest" column, and TP represent the entries in the "To Principal" column. Write an equation that shows the relationship among P, TI, and TP.

b. If *NOW* is the current balance and *NEXT* is the balance next month, which of the rules below show how to compute the balance next month if you know the balance this month? If a rule does not represent the *NEXT* balance, explain why it doesn't work.

   i. $NEXT = NOW - (43.73 - 0.0075NOW)$

   ii. $NEXT = NOW + \frac{0.09}{12}NOW - 43.73$

   iii. $NEXT = 1.0075NOW - 43.73$

Reflections

12 Mariah claims that only two variables can be represented in a matrix. Scott claims that three variables are represented in a matrix. For each of these claims, give a supporting argument.

13 A *matrix* is defined in this lesson as "a rectangular array of numbers" (see page 76). Many of the arrays that you have worked with in this lesson have additional features, such as labels on the rows and columns or entries that are numbers with units. An array with additional features is technically not a matrix, instead we would call it a *table*. The difference between a matrix and a table is not vitally important in this unit, so we have worked with the underlying array of numbers and not worried about this technical detail. In more advanced courses in mathematics, like Linear Algebra, it will be important to be very precise about matrices. For now, describe and give examples of some of the ways a table can be different from a matrix. In your description and examples, be sure to include differences related to at least these three characteristics: entries, labels, and operations.

14 In Connections Task 11, you found that the pattern of change in the monthly balance of a 12-month $500 loan at 9% annual interest could be described by three rules:

Rule I: $NEXT = NOW - (43.73 - 0.0075NOW)$

Rule II: $NEXT = NOW + \frac{0.09}{12}NOW - 43.73$

Rule III: $NEXT = 1.0075NOW - 43.73$

   a. Use algebraic reasoning to show that Rule I and Rule II are equivalent expressions.

   b. Use algebraic reasoning to show that Rule I and Rule III are equivalent expressions.

   c. Based on your work in Parts a and b, what can you conclude about Rule II and Rule III? Why?

15 In Connections Task 9, you modeled friendship with a matrix and a digraph. What do you think are the advantages of each representation?

16 For any matrix A, can you always compute $A + A$? Why or why not?

17 Tables of information are similar to matrices and are often used in newspapers and at Internet sites for reporting data. Find an example of a table in a newspaper or at an Internet site. Then complete the following tasks.

   a. Briefly describe the information displayed in the table.

   b. Describe some other way that the information could have been displayed. Why do you think the newspaper writer or Web site designer decided to display the information using a table?

c. How is the table similar to a matrix? How is it different? (See Reflections Task 13 for more about the difference between a table and a matrix.)

d. Think of the table as a matrix. Describe an operation on the rows, columns, or entries of the matrix that will yield additional information about the situation being represented. Perform the operation and report the information gained.

Extensions

18 One characteristic of spreadsheets that makes them so useful is that you can define the entries in the spreadsheet so that when you change one entry, related entries automatically change accordingly. Using spreadsheet software, create a spreadsheet that generates loan information like that in Applications Task 2. Build the spreadsheet so you can enter any loan amount, payment amount, and interest rate. Experiment with some different loan scenarios, as follows.

a. Change the annual interest rate to 19% (which could correspond to a credit card interest rate). Will a $43.73 monthly payment be sufficient to pay off a $500 loan in one year? If not, try different loan payments in the spreadsheet to find one that works.

b. Change the borrowed amount to $1,000. Assuming a 9% annual interest rate, will an $85 monthly payment be sufficient to pay off the loan in one year? If not, try different loan payments in the spreadsheet to find one that works.

c. Change the length of the loan to 24 months, but keep the loan amount at $500 and the rate at 9%. What is the smallest monthly payment that would pay off the loan in 24 months?

19 One of the purposes of the penal system is to rehabilitate people in prison. Unfortunately, many people released from prison are reconvicted and return to prison within a few years after their release. Professionals working to solve this problem use data like those summarized in the following matrix. The entries of the matrix show the status of prisoners and released prisoners *next* year given their status *this* year. For example, look at the fourth row, fifth column. The "93%" entry means that 93% of those released from prison who are in their third year of freedom this year will remain free and be in their fourth year of freedom next year.

Freedom Status

| | | Next Year | | | | | |
		in prison	1st yr. of freedom	2nd yr. of freedom	3rd yr. of freedom	4th yr. of freedom	> 4 yrs. of freedom
This Year	in prison	76%	24%	0%	0%	0%	0%
	1st year of freedom	19%	0%	81%	0%	0%	0%
	2nd year of freedom	12%	0%	0%	88%	0%	0%
	3rd year of freedom	7%	0%	0%	0%	93%	0%
	4th year of freedom	3%	0%	0%	0%	0%	97%
	> 4 years of freedom	0%	0%	0%	0%	0%	100%

Source: Indiana State Reformatory Data from *Cost Benefit Evaluation of Welfare Demonstration Projects*. Bethesda, MD: Resources Management Corporation, 1968.

a. Most of the 0% entries refer to impossible events. For example, look at the third row, second column. Why is it impossible for a person released from prison who is in his or her second year of freedom this year to be in his or her first year of freedom next year?

b. What percentage of people released from prison who are in their second year of freedom this year will remain free and enter a third year of freedom next year?

c. Explain what the "7%" entry in the fourth row, first column means.

d. What is the sum of each row of the matrix? Why does this make sense?

e. Based on your analysis of the matrix, describe at least one trend related to released prisoners returning to prison. What might explain the trend?

20 Did you ever stop to think about your genes? (Not the jeans you wear, but the genes that determine your physical characteristics.) Geneticists are always trying to find ways to analyze genes more precisely. A commonly used method of analyzing genetic structure is to study *mutations*, or alterations, of a gene. One famous experiment examined the virus called *phage T4*. The genetic

structure of the virus was studied by looking at mutations of the gene which result when one segment of the gene is missing. As part of this experiment, it was possible to gather data showing how the segments of the gene overlap each other. The results were expressed in the form of a matrix called the *overlap matrix*. One part of that matrix showing the overlaps for nineteen segments, is shown on page 99. The segments are labeled by the codes displayed across the top and down the side. A "1" means that there is an overlap between the two segments associated with the row and column.

Gene Segments

	184	215	221	250	347	455	459	506	749	761	782	852	882	A103	B139	C4	C33	C51	H23
184	0	1	0	1	0	1	0	0	0	0	1	0	0	0	0	0	1	1	1
215	1	0	0	0	0	0	0	0	0	0	0	0	0	0	0	0	0	0	1
221	0	0	0	0	1	0	1	1	1	1	1	1	1	1	1	1	1	0	1
250	1	0	0	0	0	0	0	0	0	0	0	0	0	0	0	0	1	1	1
347	0	0	1	0	0	0	0	0	0	0	1	0	0	0	0	0	0	0	1
455	1	0	0	0	0	0	0	0	0	0	0	0	0	0	0	0	0	0	1
459	0	0	1	0	0	0	0	0	1	1	1	1	0	0	0	1	0	0	1
506	0	0	1	0	0	0	0	0	0	0	1	0	0	0	0	0	0	0	1
749	0	0	1	0	0	0	1	0	0	1	1	1	0	0	0	1	0	0	1
761	0	0	1	0	0	0	1	0	1	0	1	1	0	0	0	1	0	0	1
782	1	0	1	0	1	0	1	1	1	1	0	1	1	1	1	1	1	0	1
852	0	0	1	0	0	0	1	0	1	1	1	0	0	0	0	1	0	0	1
882	0	0	1	0	0	0	0	0	0	0	1	0	0	0	0	1	0	0	1
A103	0	0	1	0	0	0	0	0	0	0	1	0	0	0	1	0	0	0	1
B139	0	0	1	0	0	0	0	0	0	0	1	0	0	1	0	0	0	0	1
C4	0	0	1	0	0	0	1	0	1	1	1	1	1	0	0	0	0	0	1
C33	1	0	1	1	0	0	0	0	0	0	1	0	0	0	0	0	0	0	1
C51	1	0	0	1	0	0	0	0	0	0	0	0	0	0	0	0	0	0	1
H23	1	1	1	1	1	1	1	1	1	1	1	1	1	1	1	1	1	1	0

Source: On the topology of the genetic fine structure, *Proc Nat Sci Acad USA* 45 (1959).

a. Does segment 882 overlap segment 221? What does the entry in the sixth row and tenth column mean? How many segments overlap segment 749?

b. In Connections Task 8, a symmetric matrix was defined as a matrix that is symmetric about the main diagonal. Is the overlap matrix a symmetric matrix? If not, explain why not. If so, what is it about the situation being modeled that causes the matrix to be symmetric?

c. Which segments have the smallest number of overlaps?

d. Which segment do you think is the longest? Why?

21 Consider the freedom-status matrix from Extensions Task 19.

a. Often in mathematical modeling, some assumptions are made so that the situation is easier to model. In this case, it is assumed that if someone has remained out of prison for more than four years, then that person will continue to stay out of prison.

   i. Which entry or entries of the matrix correspond to this assumption?

   ii. Why do you think this assumption was made? Does it seem reasonable?

b. Is a person who has been out of prison for two years more or less likely to return to prison than someone who has been out for four years? Explain.

c. What percentage of all those released from prison remain free for more than one year after release? For more than two years after release? For more than three years?

d. What percentage of people released from prison get reconvicted and sent back to prison within three years of their release? Compare your answer to your results in Part c.

e. If a prison has 500 inmates, how many can be expected to be released in a given year? Of these, how many can be expected to remain out of prison for more than four years?

f. Construct a digraph (see Connections Task 9) that represents the matrix from Extensions Task 19.

Review

22 Evaluate each of the following expressions without using your calculator.

a. $-4(5) + 2(-6)$

b. $-10 + 15 - 20 \cdot 2$

c. $12 \cdot 4 + 2(-3)$

d. $(-7)(-4) - 4(-2)$

e. $2(-4)^2$

23 For each of the following tables of (x, y) values, determine if the pattern of change is linear, exponential, or neither. For those that are linear or exponential, write a function rule that would match the table and a *NOW-NEXT* rule that describes the pattern of change.

a.
x	2	3	4	5	6
y	4	16	64	256	1,024

b.
x	2	3	4	5	6
y	6	10	14	18	22

c.
x	2	3	4	5	6
y	12	27	48	75	108

d.
x	2	3	4	5	6
y	75	60	45	30	15

24 The matrix below is an adjacency matrix for a vertex-edge graph.

$$\begin{array}{c} \\ A \\ B \\ C \\ D \\ E \end{array} \begin{array}{c} \begin{array}{ccccc} A & B & C & D & E \end{array} \\ \left[\begin{array}{ccccc} 0 & 2 & 0 & 0 & 0 \\ 2 & 0 & 1 & 1 & 1 \\ 0 & 1 & 0 & 1 & 0 \\ 0 & 1 & 1 & 0 & 2 \\ 0 & 1 & 0 & 2 & 0 \end{array}\right] \end{array}$$

a. Create a graph that matches this adjacency matrix.

b. Does your graph have an Euler circuit? Explain your reasoning.

25 Solve each of the following equations. Check your solutions.

a. $7x^2 = 252$

b. $3^x = 243$

c. $\frac{3}{4}x = 60$

d. $7x^2 + 20 = 252$

e. $2(3^x) = 486$

f. $\frac{3 + x}{4} = 60$

26 Trace each figure onto your paper. Then draw all lines of symmetry and identify all rotational symmetries of the figure.

a.

b.

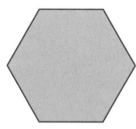

c.

d.

27 If possible, draw a triangle that meets each set of constraints. Then determine if any other triangle meeting the constraints will be congruent to the one you drew.

a. The angles have measure 30°, 70°, and 80°.

b. The sides have length 5 cm, 3 cm, and 6 cm.

c. Two sides have length 2.5 inches, and the angle between those sides has measure 140°.

d. The sides have length 5 cm, 7 cm, and 14 cm.

e. Two sides have lengths 3.5 cm and 6 cm, and one angle is 90°.

Jeremy is creating a spinner for a game he made up. He wants the spinner to meet the following conditions.

- The spinner has red, blue, green, and yellow sections on it.
- The probability of spinning red is 0.25.
- The probability of spinning blue is twice the probability of spinning red.
- The probability of spinning green is $\frac{1}{8}$.

  a. Draw a spinner that meets these conditions.

  b. What is the probability of spinning yellow?

 Examine the net of a solid shown at the right.

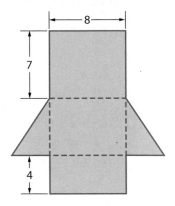

a. What is the name of such a solid?

b. Sketch the solid and find its volume.

c. What is the surface area of the solid?

d. Sketch two other possible nets for this solid.

 Write each of the following expressions in a simpler equivalent form.

a. $(5x^2y)(2x^3y^5)$

b. $\dfrac{6x^3y^6}{2x^5y^2}$

c. $(3x^3)^4$

d. $4x^2(2x^5)^2$

LESSON

2

Multiplying Matrices

I n Lesson 1, you constructed and interpreted matrices and
operated on them in several different ways. You used matrices
and their operations to analyze many different situations
and solve problems in a variety of contexts. You have
seen that matrices are particularly useful to businesses
and manufacturers in tracking production levels and
inventories of products.

Matrices can also be used to help consumer-oriented companies detect trends and make forecasts based on trends. Think about brand-switching trends in buying athletic shoes.

a Have you ever switched shoe brands? Maybe you bought Reebok one year and Adidas the next year. If you have switched athletic shoe brands, what were your reasons for switching?

b Why do you think shoe stores and shoe companies would want to know about trends in brand switching? How do you think they could gather information and analyze trends in brand switching?

c How might matrices and matrix operations be used in making inventory forecasts based on brand-switching trends?

In this lesson, you will learn about an important new matrix operation—matrix multiplication. You will use matrix multiplication, including powers of matrices, to solve a variety of problems. You will also discover an important connection between paths in digraphs and powers of the corresponding adjacency matrices.

  Investigation 1   Brand Switching

Manufacturers and big shoe stores carry out market research to gather information about brand switching. Suppose that the results of such market research at one large shoe store are shown in the following matrix. You can use this matrix to estimate how many customers will buy each shoe brand in the future. As you work on the problems in this investigation, look for answers to this question:

How can you multiply matrices to help make predictions based on trend data?

Brand-Switching Matrix

		Next Brand		
		Nike	Reebok	Fila
Current Brand	Nike	40%	40%	20%
	Reebok	20%	50%	30%
	Fila	10%	20%	70%

1 Each entry in the brand-switching matrix is the percentage of customers who will buy a certain brand of shoe, given the brand they currently own. We can use this information to make estimates about future purchases. For example, the entry in the second row and third column means that we can estimate that 30% of the people who now own Reebok will buy Fila as their next pair of shoes.

a. What percentage of people who now own Nike will buy Reebok next?

b. What percentage of people who now own Reebok will stay with Reebok on their next shoe purchase?

c. Based on these data, to which shoe brand do you think the customers are most loyal? Why?

2 Assume that buyers purchase a new pair of shoes every year; suppose that this year 700 people bought Nike, 500 people bought Reebok, and 400 people bought Fila. Answer the following questions by using those assumptions and the brand-switching matrix on page 104.

a. How many people will buy Nike next year? Explain your method.

b. How many people will buy Reebok next year?

3 One way to answer the questions in Problem 2 is to use a new matrix operation. A one-row matrix for this year's numbers is written to the left of the brand-switching matrix. A one-row matrix is often more simply called a **row matrix**.

Buyers This Year			Brand-Switching Matrix			
				N	R	F
N	R	F	N	40%	40%	20%
[700	500	400]	R	20%	50%	30%
			F	10%	20%	70%

a. Complete the computation below for the number of people who will buy Fila next year.

$$\begin{pmatrix} Number\ of\ people \\ who\ will\ buy \\ Fila\ next\ year \end{pmatrix} = 700 \times (\underline{\quad}) + 500 \times (\underline{\quad}) + 400 \times (\underline{\quad}) = \underline{\quad}$$

b. To which column of the brand-switching matrix do the numbers in the blanks correspond?

c. Which column of the brand-switching matrix can you combine with the row matrix to get the number of people who will buy Reebok next year? Explain how the row and the column are combined.

4 You have just performed a new matrix operation. This method of combining the row matrix with the columns of the brand-switching matrix is called **matrix multiplication**. When the row matrix is combined with each column of the brand-switching matrix, the result can be written as another row matrix.

a. Using the computations you have already done, list the entries of the row matrix on the right below.

Buyers This Year				Brand-Switching Matrix				Buyers Next Year		
					N	R	F			
N	R	F		N	40%	40%	20%	N	R	F
[700	500	400]	×	R	20%	50%	30%	= [__	__	__]
				F	10%	20%	70%			

We say that the row matrix for this year is *multiplied* by the brand-switching matrix to get the row matrix for next year. The term "matrix multiplication" always refers to this type of multiplication and not to any of the other multiplications that you have done.

b. Based on the matrix multiplication in Part a, describe the trend for shoe sales next year. If you were the store manager, would you adjust your shoe orders for next year from what you ordered this year? Explain.

5 Now that you have learned how to multiply matrices, practice it a few times before you continue analyzing the brand-switching situation. Perform the indicated matrix multiplications.

a. $[2 \quad 3 \quad 5] \begin{bmatrix} 6 & 0 & 2 \\ 5 & 3 & 1 \\ 3 & 6 & 2 \end{bmatrix}$
　　　　　　　　b. $[8 \quad -2] \begin{bmatrix} 2 & 4 \\ 3 & -4 \end{bmatrix}$

6 The brand-switching matrix can be used to estimate how many people will buy each brand of shoe farther into the future. In Problem 4, you found that:

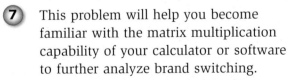

$$\begin{bmatrix} \text{the row matrix} \\ \text{showing how many} \\ \text{people bought each} \\ \text{brand this year} \end{bmatrix} \times \begin{bmatrix} \text{the brand} \\ \text{switching} \\ \text{matrix} \end{bmatrix} = \begin{bmatrix} \text{the row matrix} \\ \text{showing how many} \\ \text{people will buy each} \\ \text{brand next year} \end{bmatrix}$$

a. How many people will buy each brand two years from now? Show which matrices you could multiply to estimate this answer.

b. Use matrix multiplication to estimate the number of people who will buy each brand three years from now.

The way you have been multiplying matrices in this investigation is so useful that all calculators and software with matrix capability are designed with this kind of multiplication built in. All you have to do is enter the matrices and then multiply using the usual multiplication key or command.

7 This problem will help you become familiar with the matrix multiplication capability of your calculator or software to further analyze brand switching.

a. Enter the brand-switching matrix and the matrix for the number of buyers this year. When entering a matrix, the first thing you do is enter its size; that is, you enter (*number of rows*) × (*number of columns*). When entering the brand-switching matrix, enter all percentages as decimals.

b. Use your calculator or computer software to check your computations from Problems 4 and 6 for the number of people who are expected to buy each brand one, two, and three years from now.

c. Let *NOW* represent the matrix showing how many people buy each brand this year and *NEXT* represent the matrix showing how many people will buy each brand next year. Write a rule that shows how *NOW* and *NEXT* are related.

d. Use this *NOW-NEXT* rule and the last-answer function on your calculator or computer software to estimate how many people will buy each brand four, five, ten, and twenty years from now.

   i. Describe the trend of sales over time.

   ii. If you were the shoe store manager, what would be your long-term strategy for ordering shoes? Explain.

Summarize the Mathematics

In this investigation, you learned how to multiply a row matrix by another matrix and how to interpret the product.

a Describe how to multiply a row matrix by another matrix. What must be true about the size of the other matrix?

b Describe any limitations you see for using the brand-switching matrix to estimate long-term shoe sales.

Be prepared to share your descriptions and thinking with the class.

✓Check Your Understanding

Use your understanding of matrix multiplication to complete these tasks.

a. To prepare for a dance, a school needs to rent 40 chairs, 3 large tables, and 6 punch bowls. There are two rental shops nearby that rent all of these items, but they have different prices as shown in the matrix below. Prices shown are per item.

Rental Prices

	U-Rent	Rent-All
Chairs	$2	$2.50
Tables	$20	$15
Bowls	$6	$4

   i. What is the size of the rental-prices matrix?

   ii. Put the information about how many chairs, tables, and bowls the school needs into a row matrix.

   iii. Use matrix multiplication to find a matrix that shows the total cost of renting all of the equipment from each of the two shops.

   iv. From which shop should the school rent?

b. Perform the following matrix multiplications without using a calculator or computer. Then check your answers using technology.

i. $[-3 \quad 1] \begin{bmatrix} 4 & 5 & 2 \\ 0 & 6 & -2 \end{bmatrix}$

ii. $[3 \quad 6 \quad -5 \quad 9] \begin{bmatrix} 3 & 5 \\ 7 & 8 \\ 3 & -9 \\ 6 & 7 \end{bmatrix}$

iii. $[20 \quad 5 \quad 3] \begin{bmatrix} 1 & 2 & 4 \\ 1 & 0 & 4 \\ 1 & 3 & 5 \end{bmatrix}$

 **More Matrix Multiplication**

Matrix multiplication can be used to help analyze many different situations. As you work on the following problems, look for answers to these questions:

How do you multiply two matrices, each of which has several rows and columns?

Under what conditions is it possible and sensible to multiply two matrices?

Sports Uniforms Suppose that three Little League baseball teams are considering two suppliers for their team uniforms, Uniforms Plus and Sporting Supplies, Inc. Since they consider the quality and delivery from each supplier to be the same, their only objective is to spend the least amount of money.

Each team will order three different sizes of uniforms, small, medium, and large. Each supplier charges different prices for these three sizes, as shown in the matrix below.

Cost per Uniform

	S	M	L
Uniforms Plus	$28	$36	$41
Sporting Supplies	$34	$35	$36

All three of the Little League teams—the Kalamazoo Zephyrs, the Fairfield Fliers, and the Prescott Pioneers—have the same number of players. However, they need different quantities of small-, medium-, and large-size uniforms, as shown in the matrix below.

Quantity of Uniforms

	Zephyrs	Fliers	Pioneers
S	6	6	9
M	11	4	6
L	3	10	5

1 Each team wants to choose a supplier so that they spend the least amount of money possible on their uniforms. Matrix multiplication can help them make the right decision.

   a. Recall how you multiplied the row matrix by the brand-switching matrix in Investigation 1. Use a similar method to multiply the cost matrix by the quantity matrix, as laid out below, without using the matrix feature of your calculator or computer software. Be prepared to explain your method.

$$
\begin{array}{c}
\\
\text{Uniforms Plus} \\
\text{Sporting Supplies}
\end{array}
\begin{array}{c}
\text{S} \quad \text{M} \quad \text{L} \\
\left[\begin{array}{ccc}
\$28 & \$36 & \$41 \\
\$34 & \$35 & \$36
\end{array}\right]
\end{array}
\times
\begin{array}{c}
\quad \text{Zephyrs} \;\; \text{Fliers} \;\; \text{Pioneers} \\
\begin{array}{c} \text{S} \\ \text{M} \\ \text{L} \end{array}
\left[\begin{array}{ccc}
6 & 6 & 9 \\
11 & 4 & 6 \\
3 & 10 & 5
\end{array}\right]
\end{array}
=
\left[\begin{array}{ccc}
\underline{\quad} & \underline{\quad} & \underline{\quad} \\
\underline{\quad} & \underline{\quad} & \underline{\quad}
\end{array}\right]
$$

   b. The entries in the product matrix provide very useful information. Explain what the number in the first row and second column of your product matrix means. What does the number in the second row and third column mean?

   c. Labels on matrices can help you interpret them more easily. Label the rows and columns of your product matrix. Describe how the row and column labels of the cost matrix, the quantity matrix, and the answer matrix are related.

   d. Give your product matrix a title.

   e. Now use your product matrix to solve the problem: Which supplier should each of the teams use in order to spend the least amount of money on uniforms?

2 Compare your answers and explanations in Problem 1 with those of other students in your class. Discuss and resolve any differences.

3 Call the cost matrix *C* and the quantity matrix *Q*. So far, you have multiplied *C* × *Q*. Try multiplying *Q* × *C*. Can you do it? Explain why or why not.

Roofing Crews A roofing contractor has three crews, X, Y, and Z, working in a large housing development of similar homes. The contractor wants to keep track of the crews' work. Matrices can help. Ultimately, the contractor wants to know the total labor cost for all three crews. To help find the total cost, the contractor organizes information on houses roofed, time, and cost in the following matrices.

The first matrix below shows the number of houses roofed by each of the three crews during the second and third quarters of the year.

Number of Houses Roofed

	X	Y	Z
Apr–June	10	12	9
July–Sept	11	14	10

$= Q$

The following matrix shows the time required (in days) for each crew to roof one house and clean up.

Time Required per House (in days)

	Roof	Cleanup
X	3.5	0.5
Y	3.0	0.5
Z	4.0	0.5

$= D$

The matrix below shows the total crew labor cost per day to apply the roof and clean up.

Labor Cost per Day

	Cost
Roof	$520
Cleanup	$160

$= C$

4 The total labor cost for the three crews depends on the total time spent by the crews to apply roofs and clean up during the second and third quarters.

a. What is the total time spent for all three crews combined to apply roofs from April through June?

b. Enter your answer from Part a into a total-time matrix like that below. Then complete the rest of the entries.

Total Time (in days)

	Roof	Cleanup
Apr–June	___	___
July–Sept	___	___

$= T$

c. What two matrices can be multiplied to give the total-time matrix?

5 In Problem 4, you created a matrix showing total labor *time* for the crews. Now consider labor *cost*.

a. Multiply the total-time matrix (T) by the labor-cost-per-day matrix (C).

b. Label the rows and columns of the product matrix. What information is given by the entries in the product matrix?

c. What do you notice about the labels of all three matrices? Compare your answers and labeling with those of your classmates and resolve any differences.

6 What is the grand total labor cost of all three crews, for roofing all the houses and cleaning up, from April through September?

7 As you may have noticed, it does not always make sense to multiply two matrices. Consider the three roofing matrices on the previous page: Q, D, and C.

a. Can you multiply $C \times D$? Explain.

b. Consider $Q \times D$.

i. Can you multiply $Q \times D$?

ii. If so, does the information in the product matrix make sense? If possible, label the rows and columns and describe the information contained in the product matrix.

c. Try multiplying in the reverse order.

i. Can you multiply $D \times Q$?

ii. If so, what does the number in the first row and first column mean? If possible, label the rows and columns and describe the information contained in the product matrix.

Summarize
the Mathematics

Matrix multiplication can be useful but only in certain situations.

a Describe how to multiply two matrices. Give an example.

b Explain why it may not be possible to multiply two matrices. Give an example.

c Explain why it may be possible to multiply two matrices, but the multiplication does not make sense. Give an example.

d Does the order of matrix multiplication matter? Explain.

e If two matrices can be multiplied, how are the labels on the product matrix related to the labels of the matrices being multiplied? Give an example.

Be prepared to share your descriptions and thinking with the entire class.

✓Check Your Understanding

Apply your understanding of matrix multiplication to help solve the following problems.

a. A toy company in Seattle makes stuffed toys, including crabs, ducks, and cows. The owner designs the toys, and then they are cut out, sewn, and stuffed by independent contractors. For the months of September and October, each contractor agrees to make the number of stuffed toys shown in the following matrix.

Number of Toys Made

	Sept	Oct
Crabs	10	20
Ducks	25	30
Cows	10	30

Two of the contractors, Elise and Harvey, know from experience how many minutes it takes them to make each type of toy, as shown in this matrix:

Time per Toy (in minutes)

	Crab	Duck	Cow
Elise	55	60	90
Harvey	80	50	100

   i. Use matrix multiplication to find a matrix that shows the total number of minutes each of the two contractors will need in order to fulfill their contracts for each of the two months.

   ii. Convert the minute totals to hours. What matrix operation could you use to do this conversion? Does your calculator or computer software have the capability to perform this type of matrix operation?

b. Perform the following matrix multiplications without using a calculator or computer. Then check your answers using technology.

i. $\begin{bmatrix} 2 & 3 \\ 4 & 5 \end{bmatrix}\begin{bmatrix} 6 \\ 7 \end{bmatrix}$

ii. $\begin{bmatrix} 1 & 3 \\ 6 & 5 \end{bmatrix}\begin{bmatrix} 0 & 2 \\ 3 & 3 \end{bmatrix}$

iii. $\begin{bmatrix} 1 & 0 & 2 \\ 2 & 3 & 0 \\ -4 & 1 & 1 \end{bmatrix}\begin{bmatrix} 2 & 3 \\ 4 & 0 \\ 1 & 2 \end{bmatrix}$

iv. $\begin{bmatrix} 2 & 5 \\ 1 & 3 \end{bmatrix}\begin{bmatrix} x \\ y \end{bmatrix}$

v. $\begin{bmatrix} 0 & 1 & 1 \\ 1 & 0 & 2 \\ 0 & 0 & 1 \end{bmatrix}\begin{bmatrix} 0 & 1 & 1 \\ 1 & 0 & 2 \\ 0 & 0 & 1 \end{bmatrix}$

Investigation 3   The Power of a Matrix

In this investigation, you will learn a new way to represent and analyze information using matrices. As you explore problems about ecosystems and tennis tournaments, look for answers to these questions:

How can you represent a vertex-edge graph with a matrix?

If you multiply such a matrix by itself, what information do you get about the vertex-edge graph and the situation represented by the graph?

Pollution In an Ecosystem An ecosystem is the system formed by a community of organisms and their interaction with their environment. The diagram below shows the predator-prey relationships of some organisms in a willow forest ecosystem.

Willow Forest Ecosystem

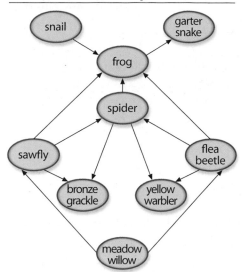

Such a diagram is called a **food web**. An arrow goes from one organism to another if one is food for the other. So, for example, the arrow from spider to yellow warbler means that spiders are food for yellow warblers.

Pollution can cause all or part of the food web to become contaminated. In the following problems, you will explore how matrix multiplication can be used to analyze how contamination of some organisms spreads through the rest of the food web.

1. Using the willow forest ecosystem food web, discuss answers to the following questions.

   a. How are predator-prey relationships represented in the food web diagram? What do the arrows mean?

   b. Think about the effect on the ecosystem if a pollutant is introduced at some point in the forest. Assume the pollution does not kill any organisms, but the contamination is spread by eating.

      i. What might happen when a toxic chemical washes into a stream in which the frogs live?

      ii. What might happen if a pesticide contaminates the sawflies?

   c. The food web can be viewed as a vertex-edge graph, where the vertices are the organisms and the edges are the arrows. Since the edges have a direction, this type of vertex-edge graph is sometimes called a **directed graph**, or **digraph**.

   How can paths through the digraph help to analyze the spread of contamination? Illustrate your answer in the case where the contamination first effects the flea beetle.

2 Matrices can be used to help find paths through digraphs. The first step in finding paths is to construct an *adjacency matrix* for the food web digraph. You may recall from *Core-Plus Mathematics* Course 1 that an adjacency matrix is constructed by using the vertices of the digraph as labels for the rows and columns of a matrix. Each entry of the matrix is a "1" or a "0" depending on whether or not there is an arrow in the digraph (directed edge) *from* the row vertex *to* the column vertex.

a. Below is a partially completed adjacency matrix for the food web digraph. Complete the adjacency matrix by filling in all the blank entries. For consistency in this investigation, we will always list the organisms alphabetically across the columns and down the rows.

Adjacency Matrix

	Bg	Fb	Fr	Gs	Mw	Sa	Sn	Sp	Yw
Bronze grackle	0	0	0	0	0	0	0	0	0
Flea beetle	—	—	—	—	—	—	—	—	—
Frog	0	0	0	1	0	0	0	0	0
Garter snake	0	0	0	0	0	0	0	0	0
Meadow willow	0	1	0	0	0	1	0	0	0
Sawfly	1	0	1	0	0	0	0	1	0
Snail	0	0	1	0	0	0	0	0	0
Spider	1	0	—	—	0	—	0	0	1
Yellow warbler	—	—	—	—	—	—	—	—	—

$= A$

b. Compare your matrix with the matrices constructed by other students. Discuss and resolve any differences in your matrices so that everyone agrees upon the same matrix. Label this adjacency matrix *A*.

3 The adjacency matrix tells you if there is an edge from one vertex to another. An edge from one vertex to another is like a path of length one. Now think about paths of length two. A **path of length two** from one vertex to another means that you can get from one vertex to the other by moving along two consecutive directed edges.

a. The partially completed matrix below shows the number of paths of length two in the food web digraph. Complete the matrix.

Number of Paths of Length Two

	Bg	Fb	Fr	Gs	Mw	Sa	Sn	Sp	Yw
Bronze grackle	0	0	0	0	0	0	0	0	0
Flea beetle	—	—	—	—	—	—	—	—	—
Frog	0	0	0	0	0	0	0	0	0
Garter snake	0	0	0	0	0	0	0	0	0
Meadow willow	1	0	2	0	—	—	—	—	—
Sawfly	1	0	1	1	0	0	0	0	1
Snail	0	0	0	1	0	0	0	0	0
Spider	0	0	0	1	0	0	0	0	0
Yellow warbler	0	0	0	0	0	0	0	0	0

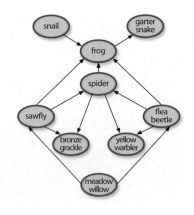

b. Compare your matrix to those constructed by others. Discuss and resolve any differences so that everyone has the same matrix.

4 What matrix operation(s) could be used to get the paths-of-length-two matrix from the paths-of-length-one matrix *A*? Make and test some conjectures. (You may find it helpful to use vertex-edge graph software or other technology for this problem.)

5 Suppose that the meadow willows are contaminated by polluted ground water. In turn, they contaminate other organisms that feed directly or indirectly on them. However, at each step of the food chain, the concentration of contamination decreases.

a. Suppose that organisms more than two steps from the meadow willow in the food web are no longer endangered by the contamination. Using the digraph, find one organism that is safe.

b. How can the matrices be used to help find all the safe organisms? Explain your reasoning.

c. Compare your methods for finding all the safe organisms with others.

Tournament Rankings You have seen that powers of an adjacency matrix give you information about paths of certain lengths in the corresponding vertex-edge graph. This connection between graphs and matrices is useful for solving a variety of problems. For example, it is often very difficult to rank players or teams in a tournament accurately and systematically. A vertex-edge graph can give you a good picture of the status of the tournament. The corresponding adjacency matrix can help determine the ranking of the players or teams. Consider the following tournament situation.

The second round of a city tennis tournament involved six girls, each of whom was to play every other girl. However, the tournament was rained out after each girl had played only four matches. The results of play were the following:

- Erina beat Keadra.
- Akiko beat Julia.
- Keadra beat Akiko and Julia.
- Julia beat Erina and Maria.
- Maria beat Erina, Cora, and Akiko.
- Cora beat Erina, Keadra, and Akiko.

6 Using the information above, can you decide anything about how the girls should be ranked at this stage of the tournament? Explain your reasoning.

A digraph and an adjacency matrix can be used to help rank the girls at this stage of the tournament with no ties.

a. Represent the status of the tournament by completing a copy of the digraph and adjacency matrix below.

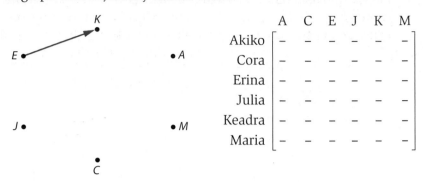

	A	C	E	J	K	M
Akiko	–	–	–	–	–	–
Cora	–	–	–	–	–	–
Erina	–	–	–	–	–	–
Julia	–	–	–	–	–	–
Keadra	–	–	–	–	–	–
Maria	–	–	–	–	–	–

b. Rank the girls as clearly as you can. Use the information shown in the digraph and adjacency matrix to explain your ranking.

c. If you did not use row sums of the adjacency matrix in Part b, what additional information do these sums provide?

d. Compute the square of the adjacency matrix and discuss what the entries tell you about the tournament. How could you use this information to help rank the girls?

e. Use further operations on the adjacency matrix to rank the players with no ties. Explain the ranking system you used. Compare your method and results with others.

Summarize the Mathematics

In this investigation, you explored how powers of an adjacency matrix for a digraph and sums of the powers could be used to analyze the digraph and the situation it models.

a Consider paths in a digraph.

   i. How do paths in a food web help you track the spread of contamination through the ecosystem?

   ii. What do paths in a tournament digraph tell you about the tournament?

b What do powers of the adjacency matrix tell you about paths in the digraph?

c Explain how you can use powers and sums of matrices to track pollution through an ecosystem and to rank the players in a tournament.

Be prepared to share your thinking and tournament-ranking plan with the class.

✓ Check Your Understanding

In any group of people, some are leaders and some are followers. This relationship of leaders and followers is called *social dominance*. The following digraph shows social dominance within a group of five people in an advertising agency. An arrow from one vertex to another means that the first person is socially dominant (is the "leader") over the other.

Social Dominance Graph

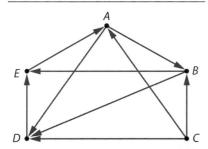

a. Describe and explain at least one interesting or unusual feature that you see in the dominance digraph.

b. Construct an adjacency matrix *M* for this digraph.

c. Using your adjacency matrix, identify an overall leader of this group. Can you rank, with no ties, all five people in terms of social dominance? Explain.

d. Compute M^3.

   i. Explain what the "1" for the *A-E* entry means in terms of social dominance.

   ii. Trace the paths in the digraph that correspond to the "2" for the *C-A* entry.

e. Use powers of your adjacency matrix and sums of rows to break any ties and rank the five people in terms of social dominance.

f. What do the arrows between *A*, *B*, and *E* indicate about these three people? Explain how this could be possible in a social group.

On Your Own

Applications

1 The Fairfield Hobbies and Games store sells two types of ping-pong sets. A Standard Set contains two paddles and one ball, and a Tournament Set contains four paddles and six balls. This information is summarized in the matrix below.

Ping-Pong Sets

	Balls	Paddles
Tourn	6	4
Std	1	2

a. This month, the store orders 35 Tournament Sets and 50 Standard Sets. Organize this information in a row matrix. Label the columns of the matrix.

b. Use matrix multiplication to find another matrix that shows the total number of balls and paddles in all of the ping-pong sets ordered this month. Label the rows and columns.

2 Matrix multiplication can be used to help rank choices. For example, suppose that in your Social Studies class during an election year you are supposed to rank four U.S. presidential candidates based on three issues—environment, gun control, and minimum wage. Based on this ranking, you will choose who you would vote for.

| John Kerry | George W. Bush | Ralph Nader | Michael Badnarik |

You, like most voters, probably care more about some of these issues than others. Here's one way to do the ranking. First, rate each candidate on a scale of 1 to 5 on each issue where 5 is the best and the rating indicates your opinion of the candidate's strength on that issue. The ratings for one student, Tonya, are shown in the following matrix. (The rows correspond to the candidates, A, B, C, D, and the columns correspond to the issues, environment (E), gun control (G), and minimum wage (M).)

$$
\begin{array}{c}
\quad\quad E \quad G \quad M \\
\begin{array}{c} A \\ B \\ C \\ D \end{array}
\left[\begin{array}{ccc}
4 & 5 & 2 \\
1 & 5 & 3 \\
3 & 3 & 4 \\
4 & 2 & 4
\end{array}\right] = S
\end{array}
$$

a. Based on the information shown so far, which candidate is top rated by Tonya? Explain your reasoning.

b. Tonya cares more about some issues than others. For her, the environment is twice as important as gun control, and minimum wage is three times as important as gun control. Thus, she would really rate a candidate based on this formula: R = 2E + G + 3M. Use this formula to compute Tonya's rating for candidate A.

c. Consider the following one-column matrix that shows the "weight" of Tonya's concern for each issue. (A one-column matrix is often simply called a **column matrix**.)

$$
W = \left[\begin{array}{c} 2 \\ 1 \\ 3 \end{array}\right]
$$

Multiply $S \times W$. What information is contained in the product matrix? Which candidate does Tonya rate the highest?

d. Adjust W to show your relative weight of concern for each issue. Then multiply $S \times W$ to examine the resulting candidate ratings.

3 The owners of a local gas station want to evaluate their business. They decide to examine sales, prices, and gross profits for the first two weeks in each of the last two years. This information is summarized in the following matrices.

Number of Gallons Sold in 1st Two Weeks of Year 1

	Regular	Super	Ultimate
Week 1	3,410	850	870
Week 2	3,230	810	780

= Q1

Revenue and Profit per Gallon in 1st Two Weeks of Year 1

	Rev/gal	Profit/gal
Regular	$2.80	$0.17
Super	$2.95	$0.19
Ultimate	$3.03	$0.20

= P1

Number of Gallons Sold in 1st Two Weeks of Year 2

	Regular	Super	Ultimate
Week 1	3,350	870	850
Week 2	3,240	780	790

= Q2

Revenue and Profit per Gallon in 1st Two Weeks of Year 2

	Rev/gal	Profit/gal
Regular	$2.86	$0.17
Super	$3.01	$0.18
Ultimate	$3.12	$0.21

= P2

a. What information would be provided by the product $Q1 \times P1$?

b. Multiply $Q1 \times P1$. Label the rows and columns of the product matrix.

c. Use matrix multiplication to find the total revenue and profit for all three types of gasoline combined for each of the two weeks in Year 2.

d. Were the first two weeks of Year 2 better than the first two weeks of Year 1? Explain.

e. Consider *P2* × *Q2*. Is it possible to carry out this matrix multiplication? If so, what do the entries in the product matrix tell you, if anything, about sales, prices, and profits at the gas station?

4 Perform the following matrix multiplications without using a calculator or computer. Then check your answers using technology.

a. $\begin{bmatrix} 2 & 3 \end{bmatrix} \begin{bmatrix} 4 & 1 & 0 \\ -5 & 3 & 1 \end{bmatrix}$

b. $\begin{bmatrix} 18 & -4 \\ 6 & -1 \\ \frac{1}{2} & 2 \end{bmatrix} \begin{bmatrix} 2 \\ 3 \end{bmatrix}$

c. $\begin{bmatrix} 6 & 5 \\ 4 & 3 \end{bmatrix} \begin{bmatrix} 2 \\ 7 \end{bmatrix}$

d. $\begin{bmatrix} 2 & 0 \\ 1 & 1 \end{bmatrix} \begin{bmatrix} 3 & -3 \\ 4 & 5 \end{bmatrix}$

e. $\begin{bmatrix} 2 & 1 & 2 \\ 3 & 5 & 4 \\ 1 & 0 & 3 \end{bmatrix} \begin{bmatrix} -1 & 0 \\ 2 & 4 \\ 3 & 3 \end{bmatrix}$

f. $A = \begin{bmatrix} 0 & 1 & 2 \\ 0 & 0 & 1 \\ 1 & 1 & 0 \end{bmatrix}$. Compute A^2.

5 Five students played in a round-robin ping-pong tournament. That is, every student played everyone else. The results were the following:

- Anna beat Darien.
- Bo beat Anna, Chan, and Darien.
- Chan beat Anna, Emilio, and Darien.
- Darien beat Emilio.
- Emilio beat Anna and Bo.

a. Represent the tournament results with a digraph by letting the vertices represent the players and drawing an arrow from one player to another if the first beats the second.

b. Construct an adjacency matrix for the digraph. Remember that you write "1" for an entry if there is an arrow from the player on the row to the player on the column.

c. Use sums and powers of the adjacency matrix to rank the five students in the tournament. Explain your method and report the rankings.

6 In Lesson 1, you investigated matrices that described friendship among a group of people. The friendship matrix below is for a different group of five people. Recall that an entry of "1" means that the person on the row considers the person on the column as a friend.

Friendship Matrix

	A	B	C	D	E
A	0	1	1	1	1
B	0	0	1	1	1
C	1	0	0	1	0
D	1	1	1	0	1
E	1	0	0	1	0

Two people are *mutual friends* if they consider each other friends. Thus, person A and person B are not mutual friends, but C and D are.

In this task, you will investigate *cliques*. A *clique* is a group of people who are all mutual friends of each other. (For this problem, consider only three-person cliques.)

a. Find one other pair of mutual friends and one clique.

b. To use powers of a matrix to find cliques, first build a *mutual-friends matrix M* by listing the five people across the top and down the side of a new matrix. Then write a "1" for each entry where the two people represented by that entry are mutual friends. If the people are not mutual friends, enter a "0".

c. Think of M as an adjacency matrix for a digraph and construct the digraph.

d. Compute M^3. What do the entries of M^3 tell you about mutual friends?

e. What do the entries in the main diagonal of M^3 tell you about cliques? Explain.

f. Consider the three-person cliques of each person.

   i. List all of the cliques of C. List all of the cliques of A.

   ii. How many cliques is B in? How many cliques is D in?

7 For any given case that comes before the Michigan Supreme Court, one judge is designated to write an opinion on the case (although any judge can choose to write an opinion on any case). All of the judges then sit together, discuss the case, and each written opinion is passed around and signed by all who approve of it. A case is decided when a majority of judges sign one opinion. The information in the following matrix, taken from historical court records, shows how often judges on the court from 1958–60 agreed with (and signed) one another's written opinions. As always, the matrix is read from row to column. For example, Judge Carr agreed with 61% of Judge Black's opinions.

Judge Agreements

	Ka	V	D	S	C	E	B	Ke
Kavanagh	—	76%	80%	85%	81%	88%	83%	77%
Voelker	81%	—	60%	90%	59%	86%	99%	63%
Dethmers	66%	65%	—	75%	99%	77%	72%	95%
Smith	78%	79%	63%	—	57%	81%	84%	64%
Carr	63%	58%	100%	66%	—	70%	61%	100%
Edwards	61%	68%	66%	76%	65%	—	70%	65%
Black	75%	84%	48%	77%	44%	68%	—	55%
Kelly	60%	53%	86%	63%	91%	61%	62%	—

Source: Leadership in the Michigan Supreme Court. *Judicial Decision Making.* New York: Free Press of Glencoe, 1963.

There are several ways you could analyze these data. Complete the analysis that follows.

a. Examine the matrix and write down two interesting statements about this particular Michigan Supreme Court.

b. Now, convert all the entries into 0s and 1s according to this rule:

Whenever a judge agrees with 75% or more of another judge's opinions, say that the one judge "agrees with" the other judge, and that entry should be a "1". Otherwise, enter a "0".

c. Interpret the new matrix by answering the following questions.

   i. Does Kavanagh agree with Dethmers?

   ii. Does Dethmers agree with Kavanagh?

   iii. Which judge agrees with the most other judges?

   iv. Which judge is agreed with by the most other judges?

d. Two judges who agree with each other are called *allies*.

   i. Find two allies among the judges.

   ii. Examine the *ally matrix* below. How are the two allies you found in part i indicated in the matrix? Identify another pair of allies.

Ally Matrix

	Ka	V	D	S	C	E	B	Ke
Kavanagh	0	1	0	1	0	0	1	0
Voelker	1	0	0	1	0	0	1	0
Dethmers	0	0	0	0	1	0	0	1
Smith	1	1	0	0	0	1	1	0
Carr	0	0	1	0	0	0	0	1
Edwards	0	0	0	1	0	0	0	0
Black	1	1	0	1	0	0	0	0
Kelly	0	0	1	0	1	0	0	0

   iii. Think of the ally matrix as an adjacency matrix for a digraph and then construct the digraph.

e. A group of three judges who are all allies with each other is called a *coalition*.

   i. Find one coalition.

   ii. Call the ally matrix A and compute A^3. What do the entries of A^3 tell you about allies?

   iii. What do the entries in the main diagonal of A^3 tell you about coalitions? Explain.

   iv. Describe some similarities and differences between coalitions and cliques. (See Applications Task 6.)

f. Three of these judges were Republicans and five were Democrats. Can you pick out the Republicans and Democrats from these data? Explain your reasoning.

Connections

8 Recall that the size of a matrix is number of rows m by number of columns n, written $m \times n$. Suppose you want to multiply two matrices, $A \times B$, and suppose the size of A is $m \times n$.

   a. State the size of B as completely as you can.

   b. Given the size of B from Part a, state the size of the product matrix, $A \times B$.

9 There are at least two different types of multiplication that involve matrices:

   • multiplying two matrices using the standard row-by-column method that you learned in this lesson (matrix multiplication), and

   • multiplying each entry in a matrix by the same number (scalar multiplication).

   Each method is useful in certain contexts. For each method, find one situation from this lesson where that multiplication method can be used to better understand the situation.

10 Two matrix operations that you have used quite often are matrix multiplication and finding row sums. There is a connection between these two operations. Consider the following square matrix.

$$A = \begin{bmatrix} 0 & 1 & 0 & 0 \\ 0 & 0 & 1 & 0 \\ 1 & 0 & 0 & 1 \\ 1 & 1 & 0 & 0 \end{bmatrix}$$

   a. Multiply A on the right by a column matrix filled with 1s. That is, multiply:

$$\begin{bmatrix} 0 & 1 & 0 & 0 \\ 0 & 0 & 1 & 0 \\ 1 & 0 & 0 & 1 \\ 1 & 1 & 0 & 0 \end{bmatrix} \begin{bmatrix} 1 \\ 1 \\ 1 \\ 1 \end{bmatrix}$$

   Compare the product matrix to the row sums of A. Explain why this makes sense.

   b. What matrix multiplication would have the same effect as summing the rows of A^2? Summing the rows of A^3?

   c. Let E be the 4×1 matrix filled with 1s, and suppose that A represents the results of a tournament with four players. Explain the meaning of the following expression in terms of ranking the tournament:

$$AE + \frac{1}{2}A^2E + \frac{1}{3}A^3E$$

11 In music, a change of key sounds more natural if only a few notes are changed. If two keys differ by too many notes, then a change from one key to the other is "remote" and sounds "unnatural" to people in our culture. Each key has five closely related keys, that is, keys that do not differ by very many notes. A vertex-edge graph can be used to model this situation, as follows. (The symbol ♭ is read "flat." For example, B♭ is read "B flat.")

Related Key Graph

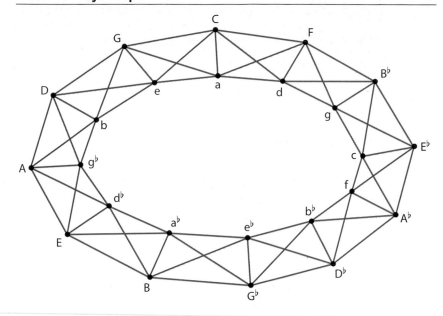

The twelve vertices in the outer circle represent the major keys: C, F, B flat, E flat, A flat, D flat, G flat, B, E, A, D, G. The vertices in the inner circle represent the twelve minor keys (written in lower-case letters): a, d, g, c, f, b flat, e flat, a flat, d flat, g flat, b, e. Each vertex is joined to the five vertices that represent the five closely related keys.

a. Suppose that key changes between keys that are one or two edges apart on the graph are thought to sound "natural," but key changes between keys that are farther apart sound "unnatural."

   i. Does a key change from C to g♭ sound natural? How about from G to A?

   ii. How many natural key changes are there from B?

b. What would be the size of an adjacency matrix for this graph?

c. How could you use operations on the adjacency matrix to answer Part a? Explain your thinking.

Reflections

12 In Investigation 2, you learned how to multiply matrices using a specific "row-column procedure." Think about this multiplication procedure.

  a. Show with hand movements how this procedure works.

  b. Reflect on your experience learning this procedure and then write a few sentences explaining your answer to the following questions: Did you find this procedure easy or difficult to learn? Was it surprising or strange in any way? Did you at first think about multiplying matrices in a different way?

13 What must be true about the size of matrix *A* to allow you to compute $A \times A$?

14 Think about the matrix analysis of the ecosystem in Investigation 3.

  a. What happens if you keep computing powers of the adjacency matrix for the food web digraph? What does this mean in terms of paths through the food web?

  b. What do the entries in the last matrix, before you reach all 0s, tell you about path lengths? About the possible spread of contamination?

Willow Forest Ecosystem

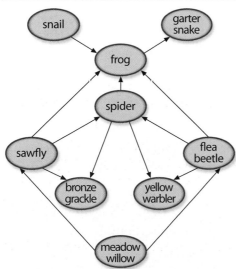

  c. Explain how to compute a single matrix that will show, for each organism, all the organisms that are farther up the food chain. Compute the matrix and check it by examining the graph.

  d. Contamination of which organisms has the potential to impact the ecosystem most? What matrix computation addresses this question?

  e. Do you think that the adjacency matrix for any digraph will eventually have a power for which all of its entries are 0s? Explain why or why not.

15 Describe a situation, different from those in this lesson, where matrix multiplication would be useful. Write and answer two questions about the situation that involve using matrices.

16 Look back at your work for Applications Task 7. Compare the judge-agreements matrix with the matrix you constructed in Part b of that task. What useful information is lost in the matrix you prepared?

Extensions

17 Sometimes it is useful to interchange the rows and columns of a matrix. For a matrix M, the matrix obtained by interchanging the rows and columns of M is called the **transpose of M**, denoted M^T.

For example, if $M = \begin{bmatrix} 3 & 4 & 5 \\ 6 & 7 & 8 \end{bmatrix}$, then $M^T = \begin{bmatrix} 3 & 6 \\ 4 & 7 \\ 5 & 8 \end{bmatrix}$. Consider some

of the matrices you have studied—see the cost and quantity matrices from Investigation 2, reproduced below.

Cost per Uniform

$$C = \begin{array}{c} \\ \text{Uniforms Plus} \\ \text{Sporting Supplies} \end{array} \begin{array}{ccc} S & M & L \\ \begin{bmatrix} \$28 & \$36 & \$41 \\ \$34 & \$35 & \$36 \end{bmatrix} \end{array}$$

Quantity of Uniforms

$$Q = \begin{array}{c} \\ S \\ M \\ L \end{array} \begin{array}{ccc} \text{Zephyrs} & \text{Fliers} & \text{Pioneers} \\ \begin{bmatrix} 6 & 6 & 9 \\ 11 & 4 & 6 \\ 3 & 10 & 5 \end{bmatrix} \end{array}$$

a. Consider multiplication of matrices C and Q.

   i. Which multiplication is possible: $C \times Q$ or $Q \times C$? Why?

   ii. Carry out the possible multiplication.

b. Now consider the transpose matrices.

   i. Compute C^T and Q^T. Label the rows and columns of the two transpose matrices. Are the matrix titles still appropriate?

   ii. Think about which multiplications are possible: $C^T \times Q^T$ and $Q^T \times C^T$? Check your thinking by carrying out the possible multiplication(s).

c. Compare the product matrices from Parts a and b. Do the two product matrices contain the same information? Explain.

d. Consider the brand-switching matrix equation on page 105, reproduced below.

| Buyers This Year | Brand-Switching Matrix | Buyers Next Year |

		N	R	F			
		N	R	F		N	R	F
N R F		N [40%	40%	20%]		[__	__	__]
[700 500 400] ×	R	20%	50%	30%	=			
	F	10%	20%	70%				

Sometimes, equations like this are written with column matrices (see Applications Task 2 Part c) instead of row matrices. Using *transpose matrices*, rewrite this equation as an equivalent equation with column matrices.

18 The most general definition of an adjacency matrix for a digraph is that it is a matrix with entries that indicate *how many* edges there are from the vertex on the row to the vertex on the column. In this lesson, an adjacency matrix was defined as a matrix with entries that indicate *if* there is an edge from the vertex on the row to the vertex on the column. Thus, the adjacency matrices in this lesson always had entries that were 1s or 0s. Using the more general definition of an adjacency matrix, an adjacency matrix can have entries that are larger than 1.

a. Draw a digraph that has an adjacency matrix in which some entries are larger than 1.

b. Describe the kinds of digraphs that have adjacency matrices with only 1s and 0s as entries.

19 Consider the brand-switching matrix from Investigation 1, reproduced below.

Brand-Switching Matrix

		Next Brand		
		Nike	Reebok	Fila
	Nike	[40%	40%	20%]
Current Brand	Reebok	20%	50%	30%
	Fila	10%	20%	70%]

This matrix models a type of process called a *Markov process*, named after the Russian mathematician A. A. Markov. There are two key components of a Markov process: *states* and a *transition matrix*. In the brand-switching example, the states are the row matrices that show how many people buy each shoe brand in a given year. The transition matrix is the brand-switching matrix, which shows how the states change from year to year. Powers of this matrix give you information about the long-term behavior of the Markov process.

a. Call the brand-switching matrix B. Enter B into your calculator or computer software, entering the percents as decimals, and use the last answer function to compute all the powers of B up to B^{20}. Describe what happened. Explain the meaning of the entries of B^{20}.

b. In Investigation 1, you assumed that the numbers of people buying each brand of shoe this year were as follows: 700 people bought Nike, 500 people bought Reebok, and 400 people bought Fila. Do the following matrix multiplications using powers of B:

$[700 \quad 500 \quad 400] \times B^4$

$[700 \quad 500 \quad 400] \times B^{10}$

$[700 \quad 500 \quad 400] \times B^{20}$

c. Explain the meaning of $[700 \quad 500 \quad 400] \times B^n$ for a positive integer n.

20 Look back at the information on the extent of agreement among Michigan Supreme Court judges in Applications Task 7. The matrix summarizing that information is reproduced below. In Task 7, you looked for allies and coalitions among the judges. In this task, you will rank the judges according to how much influence they exert upon one another.

Judge Agreements

	Ka	V	D	S	C	E	B	Ke
Kavanagh	—	76%	80%	85%	81%	88%	83%	77%
Voelker	81%	—	60%	90%	59%	86%	99%	63%
Dethmers	66%	65%	—	75%	99%	77%	72%	95%
Smith	78%	79%	63%	—	57%	81%	84%	64%
Carr	63%	58%	100%	66%	—	70%	61%	100%
Edwards	61%	68%	66%	76%	65%	—	70%	65%
Black	75%	84%	48%	77%	44%	68%	—	55%
Kelly	60%	53%	86%	63%	91%	61%	62%	—

a. Look at the data for Kavanagh and Edwards. If you were to choose one of these judges as being dominant over the other, who would you pick as the dominant judge? Why?

b. Judge X is said to *dominate* Judge Y if Y agrees with X more than X agrees with Y. The goal now is to rank the judges according to dominance. To begin, think of a way to construct a *dominance matrix* using 0s and 1s. Construct the dominance matrix.

Call the dominance matrix D. Direct dominance, as shown in D, is more powerful than the indirect dominance of second-stage, third-stage, or further-removed dominance, as shown in powers of D. The powers of D can be *weighted* with an appropriate multiplier to reflect the degrees of dominance.

c. Use weighted powers of the dominance matrix, along with row sums and matrix sums, to rank the eight judges according to dominance. Give the entries in D full weight (multiply by 1) and the entries in D^2 half weight (multiply by $\frac{1}{2}$). Continue in this manner. Multiply the entries in D^3 by $\frac{1}{3}$ and so on up through D^7, which would be multiplied by $\frac{1}{7}$.

   (You may not need powers of D up through D^7 to get a clear ranking. A general rule, however, is to include powers up through D^{n-1}, where n is the number of vertices. The reason for stopping at $n-1$ is that the longest possible path from a vertex without returning to that vertex has length $n-1$.)

d. Use this weighted ranking method to rank the players in the tennis tournament in Investigation 3 (pages 115–116). How does this ranking compare with your previous ranking?

21 You have seen that matrices have some properties that are similar to properties of real numbers. Consider the idea of square root. Every non-negative real number has a square root. For example, to find a square root of 7, you must find a number x such that $x^2 = 7$. Consider a similar matrix situation. Find a matrix A that satisfies the matrix equation below.

$$A^2 = \begin{bmatrix} 1 & 1 & 0 \\ 0 & 0 & 1 \\ 1 & 0 & 1 \end{bmatrix}$$

Hints: You might find A by thinking how to multiply a matrix by itself, just as you can find the square root of 9 by thinking about how to multiply a number by itself to get 9. Another way to find A is to think of A as an adjacency matrix for a digraph. Given the information in matrix A^2, what does the diagraph for matrix A look like?

22 In Investigation 3, you found the number of paths of length two in the food web digraph by squaring the adjacency matrix for the digraph.

a. Explain as precisely as you can why multiplying the adjacency matrix by itself gives you information about the number of paths of length two.

b. In general, how could you use an adjacency matrix to find paths of length n ($n \geq 1$) in a digraph?

Review

23 Sketch graphs of the following linear equations.

   a. $3x + 4y = 12$

   b. $x - 3y = 5$

   c. $5x + y = 7$

24 Rewrite each of the following expressions in simplest equivalent form.

   a. $3(4x - 5) + 10x - 1$

   b. $4 + (x + 9)x + 5x$

   c. $7n(2n + 4) - 6(3n - 12)$

   d. $15 - 4(2p + 1) - (6 - p)$

25 Without using a protractor, try to draw an angle with each given measure. Then use a protractor to check your angle estimates.

   a. $45°$

   b. $120°$

   c. $30°$

   d. $170°$

   e. $135°$

26 Sketch a rectangular prism that has edges of length 3 cm, 5 cm, and 8 cm.

   a. Find the volume of the rectangular prism you sketched.

   b. Find the surface area of the rectangular prism you sketched.

   c. Find the volume and surface area of a prism that is similar to the one you sketched but has been enlarged by a scale factor of 3.

27 Solve each system of equations.

   a. $\begin{cases} y = 4x + 1 \\ 2x + 5 = y \end{cases}$

   b. $\begin{cases} 4x + y = 10 \\ 14x - y = -1 \end{cases}$

   c. $\begin{cases} 10x + 2y = 46 \\ 3x - 3y = 21 \end{cases}$

28 Place the following in order from smallest to largest.

   a. $\frac{1}{3}$, 0.33, 30%, 0.033, $\frac{303}{1,000}$

   b. $\frac{3}{4}$, 50%, 0.7, $\frac{6}{9}$, 0.25

29 The shaded triangle below has been used to create a pattern. For each of the positions 1–4, describe the transformation that will map the shaded triangle onto that position.

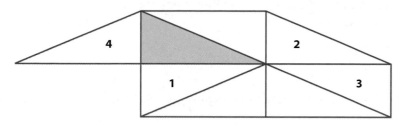

30 Fill in each blank with the appropriate number.

a. 1.5 ft = _____ in. = _____ yd

b. 3,750 mm = _____ cm = _____ m

c. 360 minutes = _____ seconds = _____ hours = _____ days

31 Mr. Hernandez gave each of his students a 20-item test. The table below gives some statistics associated with the number of items answered correctly by the students in his class.

Mean	15.5
Median	17
Standard Deviation	2.4

a. After returning the tests, Mr. Hernandez found an error on his answer key and so decided to add one point to each person's score. What would the new mean, median, and standard deviation be?

b. The scores on this test can be turned into percentages by multiplying each score by 5. What are the mean and median percentages of the original test results? What would the standard deviation of those percentages be?

32 Real numbers and operations on real numbers satisfy certain properties.

a. What real number x satisfies each equation below? (The variable a represents any real number.) Illustrate your answers for some sample values of a.

  i. $a + x = a$

  ii. $a \cdot x = a$

b. What real number x satisfies each equation? Illustrate your answers for some sample real number values of a.

  i. $a + x = 0$

  ii. $a \cdot x = 1$

The whiteboard shows:

$$x + 2y = 8$$
$$2x + y = 4 \Rightarrow$$

$$x = 8 - 2y$$
$$2x + y = 4$$
$$\Downarrow$$
$$2(8 - 2y) + y = 4$$
$$16 - 4y + y =$$

Matrices and Systems of Linear Equations

In previous lessons, you learned about matrices and matrix operations, and you used matrix methods to solve a variety of problems. One of the most common uses of matrices is to solve systems of linear equations. You already know several methods for solving a system of two linear equations. Using matrices will be another powerful method to add to your toolkit.

Think about systems of linear equations, their solutions, and how they might be represented with matrices.

a Consider a system of two linear equations like the one below.

$$x - 2y = 8$$
$$3x + 6y = 18$$

  i. What, if anything, can you say about the solution of this system just by examining the equations?

  ii. What are some methods you know for solving such a system?

  iii. Which method would you use? Why?

b Can you think of a way you might represent the system of equations above with matrices and matrix multiplication?

The first step in learning how to use matrices to solve linear systems is to examine more closely some properties of matrices. In Investigation 1, you will learn about some important properties of matrices and compare those properties to properties of real numbers. In Investigation 2, you will learn how to represent a system of two linear equations as a matrix equation and then how to solve the matrix equation. Finally, in Investigation 3, you will compare this matrix method for solving linear systems to methods you have previously learned.

Investigation 1   Properties of Matrices

Matrices and matrix operations obey certain properties. The matrix methods you use to solve problems often depend upon these properties. As you investigate properties of matrices, look for answers to these questions:

What are some important properties of operations with matrices?

How are these properties similar to, and different from, properties of operations with real numbers?

In arithmetic, you studied numbers and operations on numbers. In algebra, you have studied expressions and operations on expressions. In both settings, you found that certain properties were obeyed. For example, one such property is the *Associative Property of Addition:* $a + (b + c) = (a + b) + c$.

  In this unit, you have been studying matrices and matrix operations, including matrix addition and matrix multiplication. You will now investigate properties of matrices and their operations. This situation occurs frequently in mathematics—certain mathematical objects (like numbers or matrices) along with operations on those objects are studied and then their properties are examined.

1 **Matrix Addition** To begin this exploration of properties of matrices, first consider matrix addition.

a. Suppose $A = \begin{bmatrix} 3 & 4 & 2 \\ 1 & 0 & 9 \end{bmatrix}$. To which of the following matrices can A be added?

$$B = \begin{bmatrix} 1 & 9 & 8 \\ 6 & 5 & 4 \end{bmatrix} \quad C = \begin{bmatrix} 2 & 3 & 4 \\ 1 & 5 & 6 \\ 8 & 5 & 6 \end{bmatrix} \quad D = \begin{bmatrix} 4 & 2 \\ 3 & 5 \\ 8 & 7 \end{bmatrix} \quad E = \begin{bmatrix} -7 & 798 & 87.9 \\ 0 & 0 & \frac{2}{3} \end{bmatrix}$$

b. Under what conditions can two matrices be added? State the conditions as precisely as you can and explain your reasoning.

2 **Commutative Property of Addition** You know from your previous studies that the order in which you add two numbers does not matter. That is, for all real numbers a and b, $a + b = b + a$.

a. Give three examples of the Commutative Property of Addition for real numbers.

b. Suppose $A = \begin{bmatrix} 3 & 4 & 2 \\ 1 & 0 & 9 \end{bmatrix}$ and $B = \begin{bmatrix} 1 & 9 & 8 \\ 6 & 5 & 4 \end{bmatrix}$. Is it true that

$A + B = B + A$?

c. Do you think $A + B = B + A$ for all matrices A and B (assuming that A has the same number of rows and columns as B)? Defend your answer.

3 **Additive Identity** The number 0 has a unique property with respect to addition: adding 0 to any real number leaves the number unchanged. That is, $a + 0 = a$, for all real numbers. The number 0 is called the *additive identity*. Consider a similar situation for matrices.

a. Suppose $A = \begin{bmatrix} 4 & 2 \\ -3 & 5 \\ 8 & 7 \end{bmatrix}$. Find a matrix C so that $A + C = A$.

b. Suppose matrix B has 4 rows and 3 columns. Find a matrix E such that $B + E = B$.

c. Look at the matrices you found in Parts a and b. Write a description (definition) of the **additive identity matrix** for $m \times n$ matrices. Such a matrix is also called a **zero matrix**.

4 **Additive Inverse** Every real number has an *additive inverse*. A number and its additive inverse sum to zero.

a. What is the additive inverse of 17? Of $\frac{3}{4}$? Of -356.76?

b. A matrix and its *additive inverse matrix* sum to the zero matrix.

Let $C = \begin{bmatrix} 2 & 4 & -3 \\ 3 & -5 & -7 \end{bmatrix}$. Find the additive inverse matrix for C by filling in the blanks for the matrix below:

$$C + \begin{bmatrix} — & — & — \\ — & — & — \end{bmatrix} = \begin{bmatrix} 0 & 0 & 0 \\ 0 & 0 & 0 \end{bmatrix}$$

c. For any matrix A, describe (define) the **additive inverse matrix** for A.

5 **Matrix Multiplication** Another important matrix operation is matrix multiplication. As you discovered in Lesson 2, only matrices of "compatible" sizes can be multiplied.

a. Suppose A is a matrix that has 4 rows and 2 columns, and matrix B has 3 rows and 4 columns. Is it possible to multiply $A \times B$? How about $B \times A$? Explain.

b. Suppose C is a matrix that can be multiplied on the right by a 3×2 matrix, D. That is, you can multiply $C \times D$. What could be the size of C? What would be the size of the product matrix?

c. Suppose matrix A has size $m \times n$. What must be the size of B so that it is possible to multiply $A \times B$? What must be the size of the product matrix?

d. Sometimes it is possible to multiply two matrices in either order. What are the conditions on the sizes of two matrices A and B so that it is possible to multiply $A \times B$ and also $B \times A$?

6 **Commutative Property of Multiplication** In the case of real numbers, you know that the order of multiplication of numbers does not matter. That is, $ab = ba$, for all real numbers a and b.

a. Give three examples illustrating the Commutative Property of Multiplication for real numbers.

b. Check to see if the commutative property is true for multiplication of 2×2 matrices. Explain your reasoning. Compare with other students and resolve any differences.

c. In Part d of Problem 5, you found a condition on the sizes of two matrices A and B so that it is possible to multiply both $A \times B$ and also $B \times A$. But just because it is possible to multiply in both orders, do you necessarily get the same answer? You just explored this question with 2×2 matrices in Part b. Check it out for some other size matrices of your choice. For example, construct a 2×3 matrix and a 3×2 matrix, then multiply in both orders and see if you get the same answer.

d. Based on your work above, is the commutative property true for matrix multiplication? Explain.

Recall that a matrix with the same number of rows and columns is called a *square matrix*. Square matrices have several important properties with respect to matrix multiplication.

7 **Multiplicative Identity** The number 1 has the unique property that multiplying any real number by 1 does not change the number. That is, $a \times 1 = 1 \times a = a$, for all real numbers a. The number 1 is called the *multiplicative identity*. A square matrix that acts like the number 1 in this regard is called an *identity matrix* (or *multiplicative identity matrix*). Multiplying a matrix by the identity matrix does not change the matrix. That is, an **identity matrix** I has the property that $A \times I = I \times A = A$. Identity matrices are always square.

a. Find the identity matrix for 2×2 square matrices by filling in the blanks for the matrix below.

$$\begin{bmatrix} 5 & 4 \\ 2 & 6 \end{bmatrix} \begin{bmatrix} — & — \\ — & — \end{bmatrix} = \begin{bmatrix} 5 & 4 \\ 2 & 6 \end{bmatrix}$$

Compare your answer with those of other students. Resolve any differences.

b. Multiply $\begin{bmatrix} 5 & 4 \\ 2 & 6 \end{bmatrix}$ on the left by the identity matrix you found in Part a. Check that you get $\begin{bmatrix} 5 & 4 \\ 2 & 6 \end{bmatrix}$ as the answer.

c. Suppose matrix A has 3 rows and 3 columns. Find the identity matrix I such that $A \times I = A$.

d. Write a description of an identity matrix.

8 **Multiplicative Inverse** The product of a number and its *multiplicative inverse* is 1. Every nonzero number has a multiplicative inverse. For example, the multiplicative inverse of 5 is $\frac{1}{5}$ since $5 \times \frac{1}{5} = 1$.

a. What is the multiplicative inverse of 3? Of $\frac{1}{2}$? Of $\frac{5}{3}$?

b. Just as the product of a number and its multiplicative inverse is 1, the product of a matrix and its multiplicative inverse matrix is I. That is, the **multiplicative inverse matrix** for the square matrix D is the matrix written D^{-1}, such that $D \times D^{-1} = I$, where I is the identity matrix.

Suppose $D = \begin{bmatrix} 5 & 3 \\ 3 & 2 \end{bmatrix}$. Make and test a conjecture about the entries of D^{-1}.

c. There are several systematic methods for finding the entries of D^{-1}. One way is to use the fact that if D^{-1} is a matrix $\begin{bmatrix} a & b \\ c & d \end{bmatrix}$ such that $D \times D^{-1} = I$, then

$$\begin{bmatrix} 5 & 3 \\ 3 & 2 \end{bmatrix} \begin{bmatrix} a & b \\ c & d \end{bmatrix} = \begin{bmatrix} 1 & 0 \\ 0 & 1 \end{bmatrix}.$$

Test the following strategy for finding numbers a, b, c, and d that make this matrix equation true.

Step 1: Perform the indicated matrix multiplication to create a system of four linear equations.

$$\begin{bmatrix} 5 & 3 \\ 3 & 2 \end{bmatrix} \begin{bmatrix} a & b \\ c & d \end{bmatrix} = \begin{bmatrix} 1 & 0 \\ 0 & 1 \end{bmatrix}$$

One equation is $5a + 3c = 1$. Write down the other three equations.

Step 2: Solve the system of two equations involving a and c for a and c. Then solve the other system of two equations for b and d.

Step 3: Use the results from Step 2 to write D^{-1}. Check that $D \times D^{-1} = I$.

9 A multiplicative inverse matrix is often simply an **inverse matrix**. Other methods for finding an inverse matrix include using technology, using a formula, and using special matrix manipulations. A particular method using technology is provided below. (Other methods are examined in the On Your Own tasks.)

a. Consider again the matrix $D = \begin{bmatrix} 5 & 3 \\ 3 & 2 \end{bmatrix}$. Compute the inverse

matrix for D using your calculator or computer software. On most calculators, this can be done by entering matrix D into your calculator and then pressing the [x^{-1}] key. Compare this matrix with what you found in Problem 8. Resolve any differences.

b. An inverse matrix should work whether multiplied from the right or the left. That is,

$$D \times D^{-1} = D^{-1} \times D = I.$$

i. For matrix D from Part a, check that $D^{-1} \times D = I$.

ii. Also check that $D \times D^{-1} = I$.

c. Use your calculator or computer software to find A^{-1}, where

$$A = \begin{bmatrix} -8 & -10 \\ 2 & 3 \end{bmatrix}.$$

Check that $A^{-1} \times A = A \times A^{-1} = I$.

10 Every real number (except 0) has a multiplicative inverse. Check to see if square matrices have this property.

a. Consider $A = \begin{bmatrix} 0 & 9 \\ 0 & 4 \end{bmatrix}$. Without using your calculator or computer

software, try to find entries a, b, c, and d that will make the matrix equation true.

$$\begin{bmatrix} a & b \\ c & d \end{bmatrix} \begin{bmatrix} 0 & 9 \\ 0 & 4 \end{bmatrix} = \begin{bmatrix} 1 & 0 \\ 0 & 1 \end{bmatrix}$$

Does A have an inverse? That is, does the matrix A^{-1} exist?

b. Find a square matrix with all nonzero entries that does not have an inverse.

Summarize
the Mathematics

In this investigation, you examined properties of matrices and their operations and compared them with corresponding properties of real numbers.

a What are the conditions on the sizes of two matrices that allow them to be added? Describe the size of the sum matrix.

b What are the conditions on the sizes of two matrices that allow them to be multiplied? Describe the size of the product matrix.

c Describe and give an example of each of the following:

   i. a matrix and its additive inverse
   ii. a (multiplicative) identity matrix
   iii. a square matrix and its (multiplicative) inverse
   iv. a square matrix that does not have a (multiplicative) inverse
   v. two matrices A and B for which $A \times B$ and $B \times A$ are both defined, but $A \times B \neq B \times A$

d List some properties of real numbers and their operations that are shared by matrices and their operations.

e List some properties of real numbers that are not shared by matrices.

Be prepared to share your descriptions, examples, and thinking with the class.

✓ Check Your Understanding

Investigate other similarities and differences between operations on real numbers and the corresponding operations on matrices.

a. An important property of multiplication of numbers concerns products that equal zero. If x and y are real numbers and if $xy = 0$, what can you conclude about x or y? Is it possible that $xy = 0$, and yet $x \neq 0$ and $y \neq 0$?

b. Do you think the property in Part a is true for matrix multiplication? Make a conjecture, and then consider Part c below.

c. Suppose

$$A = \begin{bmatrix} 2 & 3 \\ 4 & 6 \end{bmatrix} \text{ and } B = \begin{bmatrix} 6 & 9 \\ -4 & -6 \end{bmatrix}.$$

Compute $A \times B$. Is it true for matrices that if $A \times B = 0$, then either $A = 0$ or $B = 0$?

d. Think of another property of addition or multiplication of real numbers, and investigate whether matrices also have this property. Prepare a brief summary of your findings.

Investigation 2 **Smart Promotions, Smart Solutions**

An expansion baseball team is planning a special promotion at its first game. Fans who arrive early will get a team athletic bag or a cap, as long as supplies last. Suppose the promotion manager for the team can buy athletic bags for $9 each and caps for $5 each. The total budget for buying bags and caps is $25,500. The team plans to give a bag or a cap, but not both, to the first 3,500 fans. The promotion manager wants to know: *How many caps and bags should be given away?* As you solve this problem, look for answers to this question:

How can a system of linear equations be represented and solved using matrices?

1 First, solve the problem any way you can. That is, find the number of bags and caps that can be given away to 3,500 fans so that the entire budget of $25,500 is spent.

2 If you have not done so already, set up and solve a system of linear equations to solve this problem.

3 A system of linear equations like you used in Problem 1 or 2 can be represented with matrices. The matrix representation leads to another useful method for solving this problem.

a. Write the two equations, one above the other, in a form like that below.

$$\underline{\quad} b + \underline{\quad} c = 3{,}500$$
$$\underline{\quad} b + \underline{\quad} c = 25{,}500$$

b. This system of equations can be represented by a single matrix equation. Determine the entries of the matrix below so that when you do the matrix multiplication, you get the two equations in Part a.

$$\begin{bmatrix} \underline{\quad} & \underline{\quad} \\ \underline{\quad} & \underline{\quad} \end{bmatrix} \begin{bmatrix} b \\ c \end{bmatrix} = \begin{bmatrix} 3{,}500 \\ 25{,}500 \end{bmatrix}$$

c. Your matrix equation is of the form $A \times X = D$, or simply $AX = D$.

   i. Which matrix corresponds to A? This matrix is called the *matrix of coefficients*. Explain why that is a sensible name for the matrix.

   ii. Which matrix corresponds to X?

   iii. Which matrix corresponds to D?

d. Compare the matrix equation $AX = D$ to the linear equation $3x = 6$. How are these equations similar? How are they different?

(4) Thinking about how to solve the linear equation $3x = 6$ can help you figure out how to solve the matrix equation $AX = D$.

a. Solve the equation $3x = 6$. Describe your method and explain why it works.

b. Many students solve this equation by dividing both sides by 3, which is a good method. Juan solved it slightly differently. Here is his explanation:

> I solved $3x = 6$ by multiplying both sides by $\frac{1}{3}$.
> This is essentially the same as dividing by 3.

Explain what Juan means and why his method works.

c. When solving $3x = 6$, you can divide both sides by 3 or multiply both sides by $\frac{1}{3}$. To solve similar matrix equations like $AX = D$, we use the multiplication method, since we know how to multiply matrices. Explain each step of the following comparison. Supply any missing details.

Solving a Linear Equation	Solving a Matrix Equation
$3x = 6$	$AX = D$
(*inverse of 3*) $\times 3x =$ (*inverse of 3*) $\times 6$	(*inverse of A*) $\times AX =$ (*inverse of A*) $\times D$
$x = $ (*inverse of 3*) $\times 6$	$X = $ (*inverse of A*) $\times D$
$x = \frac{1}{3} \times 6$	$X = A^{-1} \times D$

(5) Now you are ready to solve the matrix equation you completed in Problem 3 Part b:

$$\begin{bmatrix} 1 & 1 \\ 9 & 5 \end{bmatrix} \begin{bmatrix} b \\ c \end{bmatrix} = \begin{bmatrix} 3{,}500 \\ 25{,}500 \end{bmatrix}$$

a. What matrices should you multiply to solve the matrix equation?

b. Solve the matrix equation. Record the matrix solution. What values for b and c do you get? How many bags and how many caps should the team give away?

c. Compare your values for b and c with your solutions in Problems 1 and 2. Resolve any differences.

d. To solve the matrix equation as above, you needed to find an inverse matrix. What method did you use to find the inverse matrix?

(6) As usual, it is a good idea to check your solution.

a. Use your equations in Part a of Problem 3 to check your solution.

b. Use the matrix equation in Problem 5 to check your solution.

c. When checking your solution, you should always check it against the original problem. Reread the original problem in the introduction to this investigation. Verify that your solution solves this problem.

(7) The two equations below could represent the relationship between quantities of other promotional items.

$$x + y = 5{,}000$$
$$8x + 12y = 42{,}000$$

a. Using the context of promotional products for a team, describe a situation that could be modeled by this system of equations.

b. Represent the two equations with a matrix equation. Then solve the matrix equation. Check that your values for x and y satisfy the original system of equations and the matrix equation.

c. Interpret your solution in terms of the situation you described in Part a.

Summarize
the Mathematics

In this investigation, you learned how to solve a system of two linear equations using matrices. For a given system of two linear equations:

a Describe how to represent the system as a matrix equation.

b Describe how to use inverse matrices to solve the corresponding matrix equation $AX = C$. Explain how this inverse-matrix method is similar to how you solve equations like $5x = 20$.

c Describe at least two ways to check the solution.

Be prepared to share your descriptions and thinking with the entire class.

✓ Check Your Understanding

Cultivating the good will of fans is important for professional sports teams. Suppose that the promotions manager of the baseball team in Problem 1 decides to enhance the promotion by giving better prizes to more fans. The team owner agrees to increase the promotion budget to $37,500 and give a cap or jacket to the first 4,500 fans. The caps still cost $5 each; the jackets now cost $10 each.

a. Write a system of linear equations that represents this situation.

b. Write and solve the matrix equation representing this situation.

c. How many caps and how many jackets should be given away?

d. Show at least one way to check your answer.

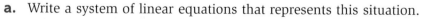

e. The matrix equation you solved in Part b relates to a particular context. For practice, solve the following matrix equations which do not relate to a particular context.

i. $\begin{bmatrix} 6 & 3 \\ 1 & 2 \end{bmatrix} \begin{bmatrix} x \\ y \end{bmatrix} = \begin{bmatrix} 4 \\ 5 \end{bmatrix}$

ii. $\begin{bmatrix} -2 & 4 \\ 0 & 3 \end{bmatrix} \begin{bmatrix} x \\ y \end{bmatrix} = \begin{bmatrix} 7 \\ -1 \end{bmatrix}$

Investigation 3 — Analyzing and Comparing Methods

Setting up and solving a matrix equation is a useful method for solving a system of linear equations. However, there are some limitations to this method, and there are other methods that might be better depending on the situation. As you work through this investigation, look for answers to these questions:

For solving systems of linear equations:

What are some limitations of the inverse-matrix method?

What are some advantages and disadvantages of each of the other methods you know?

1 **Getting Equations in the Right Form** In order to represent a system of linear equations as a matrix equation, as you did in the last investigation, the linear equations must be in the form $ax + by = c$. Consider the following linear system:

$$3x = 4 - 2y$$
$$x + 7 = 6y$$

a. Rewrite each of these equations in the form $ax + by = c$.

b. Solve this system of linear equations by setting up and solving a matrix equation.

2 **Limitations of the Inverse-Matrix Method** Setting up a matrix equation and then multiplying by the inverse of a matrix is a very useful method for solving a system of linear equations. But there are some limitations!

a. Try the method on each of the systems below. What happens when you try to calculate the inverse matrix?

$$2x + 5y = 10 \qquad\qquad 6x + 4y = 12$$
$$2x + 5y = 20 \qquad\qquad 3x + 2y = 6$$

b. Think about connections among the graph of a linear system, the number of solutions, and the inverse-matrix method.

   i. The inverse-matrix method fails for both of the linear systems in Part a. What does the graph of each linear system look like? How many solutions are there for each system?

   ii. What would the graph look like for a system in which calculating the inverse matrix is possible? How many solutions would such a system have?

c. Describe at least two patterns in the equations of a system of linear equations that indicate you will not be able to use an inverse matrix to solve the system. Check your conjecture by writing and solving systems of linear equations that exhibit these patterns.

Comparing Methods You now have several methods for solving a system of two linear equations: matrices, graphs, tables, elimination (combining the equations in a way that eliminates one of the variables), and substitution (rewriting the system as a single equation). In addition, you can sometimes solve the system just by examining the symbolic forms of the equations (like in Problem 2).

3 Consider the systems of equations below. You could solve each system in several ways. Sometimes one solution strategy is better than another. It is important to develop skill in selecting an appropriate method.

System I: $y = x - 4$
$2x - y = -2$

System II: $s + p = 5$
$2s - p = 4$

System III: $x + 3y = 12$
$4x + 12y = 48$

System IV: $y = 0.85 + 0.10x$
$y = 0.50 + 0.15x$

System V: $6x + 8y = 22$
$40x + 30y = 100$

a. For each system above, identify one of these five methods for solving it: matrices, graphs, elimination, substitution, and examining the equations in the system. *Choose a different method for each system.* Explain why you think the method you chose is a good method for solving that particular system.

b. Solve each system using the method you identified in Part a.

Summarize
the Mathematics

In this investigation, you explored limitations of the inverse-matrix method for solving systems of linear equations. You also considered the question of choosing among the methods you know for solving such a system.

a Describe at least one limitation of the inverse-matrix method for solving a system of linear equations. Give an example illustrating this limitation.

b List all the methods you know for solving a system of linear equations. For each method:

   i. Describe the key steps in using the method.

   ii. Describe some advantages and disadvantages of the method.

c You have learned that for a system of two linear equations in two variables, there are 0 solutions, 1 solution, or infinitely many solutions. For each of these three possibilities for the number of solutions:

   i. Describe the graph of the linear system.

   ii. Describe what happens when you solve the system using elimination or substitution.

   iii. Describe what happens when you try to solve the system using the inverse-matrix method.

Be prepared to share your descriptions, examples, methods, and analysis with the class.

✔ Check Your Understanding

Set up and solve a matrix equation to solve each of the following systems of linear equations. Then check your solution by solving each system using one other method.

a. $y = 2x - 1$
      $3x - y = 1$

b. $x + y = -2$
      $6x + y = 0.5$

Applications

1 In Investigation 1, you reviewed some properties of real numbers and their operations and investigated corresponding properties of matrices and their operations. The Distributive Property of Multiplication over Addition links multiplication and addition of numbers. That is, the distributive property guarantees that for all real numbers k, a, and b:

$$k(a + b) = ka + kb$$

a. Give two examples of the distributive property with numbers.

b. Suppose k is any number and A and B are any two matrices with the same size. Is it true that $k(A + B) = kA + kB$? Explain your reasoning.

2 In Investigation 1 Problem 7, you learned that the identity matrix for 2×2 matrices satisfies this equation:

$$\begin{bmatrix} 5 & 4 \\ 2 & 6 \end{bmatrix} \begin{bmatrix} a & b \\ c & d \end{bmatrix} = \begin{bmatrix} 5 & 4 \\ 2 & 6 \end{bmatrix}$$

a. Perform the matrix multiplication to create four linear equations.

b. Separate these four equations into two systems of two equations. Solve each system of two equations.

c. The results from Part b give you the entries for the identity matrix. Check that you get the same answer as when you found the identity matrix in Problem 7 on page 136.

3 At a school basketball game, the box office sold 400 tickets for a total revenue of $1,750. Tickets cost $6 for adults and $4 for students. In the rush of selling tickets, the box office did not keep track of how many adult and student tickets were sold. The school would like this information for future planning.

a. Let a represent the number of adult tickets sold and s represent the number of student tickets sold.

    i. Write an equation showing the relationship among a, s, and the number of tickets sold.

    ii. Write an equation showing the relationship among a, s, and the total revenue from ticket sales.

b. Write a matrix equation that represents the system of linear equations from Part a.

c. Solve the matrix equation. How many adult and student tickets were sold?

d. Check your solution using graphs.

e. Describe another way that you could check your solution.

4   A designer plans to inlay the brick design below into a concrete patio.

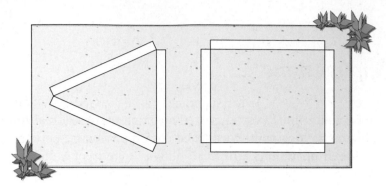

The design must meet the following specifications.

- The figures are outlined by rows of bricks, one brick wide (as represented in the drawing by the white strips).
- The rectangular design requires 50 bricks.
- The isosceles triangle design requires 40 bricks.
- The length of the longer sides of the rectangle is the same length as the longer sides of the triangle, and the length of the shorter sides of the rectangle is the same length as the shorter side of the triangle.

Find the number of bricks needed for each side of each figure by setting up and solving a system of linear equations.

5   A school principal and the local business community have devised an innovative plan to motivate better school attendance and achievement. They plan to give gift certificates to students who score high in each category. Students with high attendance will be awarded $25 gift certificates, and those with good grades will receive $20 gift certificates. The total budget for this plan is $1,500, and the planning committee would like to award 65 gift certificates. The next step is to determine the number of each type of gift certificate to be printed.

a. If x is the number of gift certificates for attendance and y is the number of gift certificates for good grades, write equations that model this situation.

b. Set up and solve a matrix equation that represents the system of linear equations from Part a. How many of each type of gift certificate can be awarded?

c. Verify your solution using a different method for solving a system of linear equations.

6 Solve the following matrix equations.

a. $\begin{bmatrix} 2 & 1 \\ 3 & 5 \end{bmatrix} \begin{bmatrix} x \\ y \end{bmatrix} = \begin{bmatrix} 8 \\ 7 \end{bmatrix}$

b. $\begin{bmatrix} -2 & 1 \\ 6 & -3.5 \end{bmatrix} \begin{bmatrix} x \\ y \end{bmatrix} = \begin{bmatrix} 12.75 \\ -6.5 \end{bmatrix}$

c. Choose one of the matrix equations above and rewrite it as a system of linear equations. Then solve the system using any nonmatrix method. Check that you get the same solution(s) as you found before.

Connections

7 In Investigation 1, you used two methods to find an inverse matrix: solving two systems of linear equations and using a calculator or computer software. Another method that works for 2 × 2 matrices is to use the following formula.

If $A = \begin{bmatrix} a & b \\ c & d \end{bmatrix}$, the inverse of A (when it exists) is given by:

$$A^{-1} = \frac{1}{ad - bc} \begin{bmatrix} d & -b \\ -c & a \end{bmatrix}$$

a. Use this formula to find the inverse of the following matrix. Then use matrix multiplication to check that the product of the inverse and the original matrix is the identity matrix.

$$\begin{bmatrix} 6 & 8 \\ 2 & 3 \end{bmatrix}$$

b. Use the formula to find the inverses of the following matrices. Check your answers.

i. $\begin{bmatrix} 5 & 3 \\ 3 & 2 \end{bmatrix}$

ii. $\begin{bmatrix} -8 & -10 \\ 2 & 3 \end{bmatrix}$

c. You discovered in Investigation 1 that not all matrices have an inverse.

i. Examine the formula for A^{-1} given above. What do you think will go wrong when you try to use the formula to compute the inverse of a 2 × 2 matrix that has no inverse?

ii. In Problem 10 (page 137), you discovered that $\begin{bmatrix} 0 & 9 \\ 0 & 4 \end{bmatrix}$ does not have an inverse. What happens when you try to use the formula above?

iii. Use what you've discovered about limitations of the formula to construct two matrices with all nonzero entries but no inverse.

8 The brand-switching matrix from Lesson 2 on page 105 is reproduced below with decimal entries instead of percents. This matrix provides information for projecting how many people will buy certain brands of athletic shoes on their next purchase given the brand they currently own. Matrix multiplication along with its properties can help you analyze this situation.

Brand-Switching Matrix

Next Brand

$$\text{Current Brand} \quad \begin{array}{c} \\ N \\ R \\ F \end{array} \begin{array}{ccc} N & R & F \\ \left[\begin{array}{ccc} 0.4 & 0.4 & 0.2 \\ 0.2 & 0.5 & 0.3 \\ 0.1 & 0.2 & 0.7 \end{array}\right] \end{array} = B$$

a. Assume that buyers purchase a new pair of shoes every year. The following matrix shows the number of people who bought each of the three brands this year.

$$\begin{array}{ccc} N & R & F \\ [\,600 & 700 & 500\,] \end{array} = Q$$

How many people are projected to purchase each of the brands next year? Explain how to answer this question using matrix multiplication.

b. What would the brand-switching matrix look like if there is no change in the number of people buying each brand this year and next year?

c. You may recall the Associative Property of Multiplication for real numbers which guarantees that for all real numbers a, b, and c:
$a \times (b \times c) = (a \times b) \times c$.

For example, $3 \times (5 \times 2) = (3 \times 5) \times 2 = 30$.

Matrix multiplication also has this property, which can be used to project the number of people buying each brand several years into the future. From Part a, you know that

$$Q \times B = \left[\begin{array}{c} \text{the row matrix} \\ \text{showing how many} \\ \text{people will buy each} \\ \text{brand next year} \end{array}\right].$$

i. Compute and compare the results of $(Q \times B) \times B$ and $Q \times (B \times B)$. Explain why this is an example of the associative property. Explain the meaning of the resulting matrices.

ii. Describe the information obtained by computing $Q \times B^3$.

d. The following matrix shows the number of people projected to buy each brand two years from now.

$$\begin{array}{ccc} \text{N} & \text{R} & \text{F} \\ [\,378 & 653 & 769\,] \end{array}$$

Using this matrix, find a matrix showing how many people will buy each brand one year from now. Compare with results from Part a.

9 A new nutrition plan that Antonio is considering restricts his drinks to water, milk, and fruit juice. The matrix below shows the amount of protein and calories per cup for skim milk and fruit juice.

Protein and Calories (per cup)

$$\begin{array}{c} \\ \text{Protein (g)} \\ \text{Calories (g)} \end{array} \begin{array}{cc} \text{Milk} & \text{Juice} \\ \left[\begin{array}{cc} 8 & 2 \\ 85 & 120 \end{array}\right] \end{array}$$

The plan recommends that Antonio drink enough milk and juice each day to get a total of 10 grams of protein and 180 calories from those sources. He wants to know how much milk and juice he must drink to meet these recommendations.

a. Construct a column matrix showing the recommended daily totals for protein and calories. Label the rows of the matrix.

b. Let x represent the number of cups of skim milk he should drink each day and y represent the number of cups of juice he should drink. Set up a matrix equation that models this situation.

c. Write a system of equations that represents this situation.

d. How many cups of milk and juice should Antonio drink daily to meet the recommended plan?

10 Sometimes a problem that involves two variables connected by linear relationships can be solved directly with matrices, rather than by first finding a system of linear equations and then using matrices.

a. The Fairfield Hobbies and Games store sells two types of ping-pong sets. A Standard Set contains two paddles and one ball, and a Tournament Set contains four paddles and six balls. Enter this information into the matrix below.

Ping-Pong Sets

$$\begin{array}{c} \\ \text{Balls} \\ \text{Paddles} \end{array} \begin{array}{cc} \text{Tourn} & \text{Std} \\ \left[\begin{array}{cc} \_ & \_ \\ \_ & \_ \end{array}\right] \end{array}$$

b. The store receives a bulk shipment of ping-pong equipment consisting of 100 paddles and 110 balls. The owner wants to know how many of each type of ping-pong set she can make using this equipment. Let s represent the number of Standard Sets and t represent the number of Tournament Sets. Complete the following matrix equation so that it represents this situation.

$$\begin{bmatrix} 6 & 1 \\ 4 & 2 \end{bmatrix} \begin{bmatrix} - \\ - \end{bmatrix} = \begin{bmatrix} - \\ - \end{bmatrix}$$

c. Solve the matrix equation. How many sets of each type can the owner make using the balls and paddles in the bulk shipment?

d. Now, solve the problem another way. Write an equation that shows the relationship among s, t, and the total number of balls. Write another equation that shows the relationship among s, t, and the total number of paddles. Solve this system of equations. Check that you get the same answer as in Part c.

11 The owner of two restaurants in town has decided to promote business by allocating to each restaurant a certain amount of money to spend on restaurant renovation and community service projects. He has asked the manager of each restaurant to submit a proposal stating what percentage of their money they would like to spend in each of these two categories. The matrix below shows their proposals.

Funding Requests

	Rest. A	Rest. B
Renovation	70%	45%
Community Service	30%	55%

The owner has decided to allocate a total of $16,000 to renovations and $14,000 to community projects. The managers want to know how much money they will have to spend in each category.

a. Represent this situation with a matrix equation and with a system of linear equations.

b. Using a method of your choice, determine how much money each restaurant should be allocated.

Reflections

12 Think about the way matrices are added and multiplied.

a. Why should it not be surprising that matrix addition has the same properties as addition of numbers?

b. Why might it be reasonable to suspect that matrix multiplication would not have the same properties as multiplication of numbers?

13 Reproduced below is the comparison of methods for solving a linear equation and solving a matrix equation from Investigation 2 (page 140).

Solving a Linear Equation

$3x = 6$
(*inverse of 3*) $\times 3x =$ (*inverse of 3*) $\times 6$
$x =$ (*inverse of 3*) $\times 6$
$x = \frac{1}{3} \times 6$

Solving a Matrix Equation

$AX = D$
(*inverse of A*) $\times AX =$ (*inverse of A*) $\times D$
$X =$ (*inverse of A*) $\times D$
$X = A^{-1} \times D$

a. For the equation $3x = 6$, the third line of the comparison could have been written:

$x = 6 \times$ (*inverse of 3*)

Explain why the corresponding matrix equation could *not* have been written:

$X = D \times$ (*inverse of A*)

b. How do you know that D must be multiplied *on the left* by the inverse of A?

14 If you were advising a friend who is about to learn the inverse-matrix method for solving systems of linear equations, what would you tell your friend about things to watch out for, easy parts, shortcuts, or other tips?

15 You have solved systems of linear equations using graphs, tables, the substitution method, the elimination method, and the inverse-matrix method. When solving a system of linear equations, how do you decide which method to use?

Extensions

16 Matrix equations and inverse matrices can be useful for solving systems of equations involving more than two variables. For example, Isabelle is considering a nutrition plan that restricts her drinks to skim milk, orange juice, tomato juice, and water. The matrix below shows the amount of protein, carbohydrate, and calories per cup for each beverage except water.

Protein, Carbohydrate, and Calories (per cup)

	M	OJ	TJ
Protein (g)	8	2	2
Carbohydrate (g)	12	29	10
Calories	85	120	45

The plan recommends that Isabelle drink enough milk and juice each day to get a total of 15 grams of protein, 46 grams of carbohydrate, and 246 calories from these sources. She wants to know how much of each beverage she must drink to meet these recommendations exactly.

a. Set up a matrix equation that models this situation. Use multiplication by an inverse matrix to solve this equation. How much skim milk, orange juice, and tomato juice must Isabelle drink each day?

b. Write a system of three linear equations that represents this situation.

17 Consider systems of equations with more than two variables.

a. The graph of an equation in the form $ax + by + cz = d$ is a plane in three-dimensional space. Consider a system of three such equations in three variables.

$$x - y + 2z = 1$$
$$2x + y + z = 1$$
$$4x - y + 5z = 5$$

  i. In what ways can three planes intersect?

  ii. Suppose you solve such a system of three linear equations and you find a single solution, (x, y, z), with specific values of x, y, and z. How would this situation be represented by the graphs of the three equations?

b. Try using an inverse matrix to solve the linear system in Part a. Do these three planes intersect in a single point? (You may wish to check your solution by using graphing software that can graph the three planes, if available.)

c. Solve the linear system below using matrix methods.

$$12w - 5x + y + 7z = 8$$
$$8w - 3x + 2y + 7z = 3$$
$$10w - 2x + 2y + 7z = 7$$
$$13w + x + 2y + 8z = 13$$

18 There is another matrix method for solving systems of linear equations, called *row reduction*, that is actually more efficient for a computer to implement than the inverse-matrix method which you learned in this lesson, and it works more generally. You will learn more about the row reduction method in later courses. To get an idea of how this method works, consider the following system of linear equations.

$$2x - y = 7$$
$$3x + 4y = -6$$

a. Solve this system using the elimination method.

b. As you carried out the elimination method in Part a, you multiplied the equations by constants and you added or subtracted equations. When doing these operations, you were essentially just operating on the coefficients of x and y and the constants on the right side of the equations. This suggests that these operations could be done just as well on a matrix with entries that are the coefficients and constants. Such a matrix is sometimes called the **augmented matrix**. Using such an augmented matrix, you can represent the above system as follows:

$$A = \begin{bmatrix} 2 & -1 & 7 \\ 3 & 4 & -6 \end{bmatrix}$$

The first two columns contain the coefficients of the variables, and the last column contains the constants on the right side of the equations. Now, carry out the operations you may have used in the elimination method, but just apply them to the entries in the matrix A, as follows.

Step 1: Rewrite the first row of matrix A so that it represents the system with the first equation replaced by 3 times the first equation.

Step 2: Rewrite the modified matrix so that it represents the system with the second equation replaced by -2 times the second equation.

Step 3: Finally, rewrite this modified matrix so that row 2 is replaced by the sum of modified rows 1 and 2.

  c. Write the system of equations represented by the final matrix in Part b.

  d. Use the results of Part c to solve the original system of equations. Check that you get the same solution(s) as in Part a.

  e. Compare the methods in Part a and Part b. Did you use the same multipliers in Parts a and b? If not, modify your elimination method in Part a so that it more closely corresponds to the row reduction method in Part b.

19 In Extensions Task 18, you learned about the row reduction matrix method for solving a system of linear equations. This method will work when the inverse-matrix method fails. For example, consider the following system:

$$2x + y = 7$$
$$4x + 2y = 10$$

  a. Solve this system using the elimination method. What is the solution?

  b. Try to solve this system using the inverse-matrix method. What happens?

  c. Study this application of the row reduction method.

Step 1: Form the augmented matrix: $\begin{bmatrix} 2 & 1 & 7 \\ 4 & 2 & 10 \end{bmatrix}$

Step 2: Multiply the top row by 2 and replace the top row with the result: $\begin{bmatrix} 4 & 2 & 14 \\ 4 & 2 & 10 \end{bmatrix}$

Step 3: Subtract the bottom row from the top row and replace the bottom row with the result: $\begin{bmatrix} 4 & 2 & 14 \\ 0 & 0 & 4 \end{bmatrix}$

Step 4: Write the system represented by the above matrix.

Step 5: Since $0 = 4$ is a contradiction, conclude there are no solutions to this system.

Compare this solution to the solution you found using the elimination method in Part a.

d. Use the row reduction method to solve each of the following systems of equations.

    i. $-2x + y = -1$
        $3x - y = 1$

    ii. $x - y = 3$
         $4x + y = 32$

    iii. $3x + y = -2$
          $6x + 2y = 0.5$

    iv. $3x - y = 2$
         $9x - 3y = 6$

Review

20 Consider the two matrices below.

$$M = \begin{bmatrix} 3 & 4 & 5 \\ 6 & 7 & 8 \\ 9 & -4 & 16 \end{bmatrix} \qquad N = \begin{bmatrix} 1 & 0 & 2 \\ 1 & 1 & 1 \\ 3 & 5 & 10 \end{bmatrix}$$

a. Compute $M \times N$.
b. Compute $M + N$.
c. Compute $M \times 2N$.
d. Compute $2M - 3N$.

21 Find the length of the third side of each right triangle.

a.

b.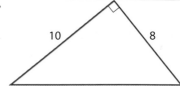

c.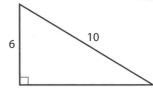

d.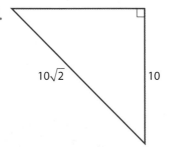

22 Identify all types of quadrilaterals (parallelogram, rhombus, rectangle, square, trapezoid) for which each property always applies.

    a. Opposite sides are parallel.

    b. There is exactly one pair of parallel sides.

    c. Diagonals are the same length.

    d. All sides are the same length.

    e. Adjacent sides are perpendicular.

    f. Opposite sides are the same length.

    g. Opposite angles are congruent.

23 Write an equation for the line that fits each set of conditions.

   a. has y-intercept of $(0, 7)$ and slope of $\frac{1}{3}$

   b. contains $(-3, -2)$ and has slope 3

   c. contains $(-2, 6)$ and $(2, 14)$

   d. contains $(6, 5)$ and $(9, -5)$

24 Draw a graph that matches each equation.

   a. $y = -\frac{2}{3}x + 6$

   b. $y = 4x - 9$

   c. $3x + 9y = 18$

   d. $5x - 3y = 30$

   e. $y = -2$

   f. $x = 5$

25 Evaluate each of the following expressions when $x = -3$, $y = 5$, and $z = -1$.

   a. $xyz - 10$

   b. $x^2 + y^2 + z^2$

   c. $2z^3 - x^2y$

   d. $y - 7x - z + xy$

   e. $(x + y)^{-1}$

   f. $-3(x + 3z)^2$

26 The box plots below give the distributions of heights (in inches) for the Denver Nuggets and the Houston Rockets at the beginning of the 2006–2007 basketball season.

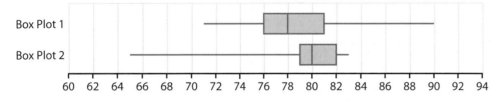

Heights of Players (in inches)

   a. The shortest player in the NBA during the 2006–2007 season played for the Denver Nuggets, and the tallest player played for the Houston Rockets. Use this information to match each box plot with the correct team.

   b. How tall was the shortest player in the NBA? How tall was the tallest player?

   c. Approximately what percentage of the Denver Nuggets players were at least 82 inches tall?

   d. Which team had the greater median height? Explain your reasoning.

   e. Does either team have a player whose height is an outlier? Explain your reasoning.

27 Complete the chart below to show equivalent fraction, decimal, and percent representations.

Fraction	Decimal	Percent
$\frac{1}{4}$		
	0.20	
		15
	1.75	

28 Solve each equation or inequality.

a. $5x + 4 = 2(x - 1)$

b. $6(2x - 10) = -4(4 - x)$

c. $2x + 8(12 + 3x) = 20x$

d. $2 - 4x \geq 20$

e. $x + 8 < 5 - 2x$

f. $15x + 120 > 0$

29 Recall that by definition, a rectangle is a parallelogram with four right angles. Use the diagram and the outline of an argument below to show that if a parallelogram has a right angle, then it is a rectangle. In the parallelogram below, $\angle D$ is a right angle. Provide a justification for each numbered statement.

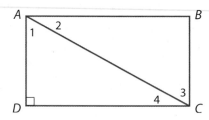

(1) $\overline{AB} \cong \overline{CD}$ and $\overline{AD} \cong \overline{CB}$.

(2) $\triangle ACD \cong \triangle CAB$

(3) $\angle B$ is a right angle.

(4) $m\angle 1 + m\angle 4 = 90°$ and $m\angle 2 + m\angle 3 = 90°$.

(5) $m\angle 1 + m\angle 2 = 90°$ and $m\angle 3 + m\angle 4 = 90°$.

(6) $\angle DAB$ is a right angle and $\angle BCD$ is a right angle.

(7) $\square ABCD$ is a rectangle.

Looking Back

In this unit, you learned about matrices, matrix operations, and properties of matrices. You learned how to use matrix methods to solve a wide variety of problems. In addition, you extended your understanding of vertex-edge graphs and systems of linear equations.

Matrices are rectangular arrays consisting of rows and columns of numbers. Operations on matrices include row sums, column sums, matrix addition and subtraction, multiplication of a matrix by a number, matrix multiplication, and finding the inverse of a matrix. Properties of matrices and matrix operations are often similar to, yet sometimes different from, the corresponding properties of real numbers and their operations. Matrices can be used to organize and display data, and to represent, analyze, and solve problems in diverse situations ranging from tracking inventories to ranking tournaments.

The tasks in this final lesson will help you review, pull together, and apply what you have learned about matrices and matrix methods.

① **Tracking Physical Fitness** The U.S. Department of Education collects data on the physical fitness of American youths. The information below shows the average time, in minutes, for students 10–11 years old to run three-quarters of a mile and for students 12–17 years old to run one mile.

Boys		
1980	1989	
6.5	7.3	10–11 year olds
8.4	9.1	12–13 year olds
7.2	8.6	14–17 year olds

Girls		
1980	1989	
7.4	8.0	
9.8	10.5	
9.6	10.7	

Source: *Youth Indicators 1991: Trends in the Well-being of American Youth.* Washington, DC: U.S. Government Printing Office, 1991.

a. Write down two trends you see in the data. Write down one fact that you find surprising and explain why it surprises you.

b. The data as given are organized in two matrices titled "Boys" and "Girls." Reorganize the data into two different matrices titled "1980" and "1989." Don't change the row labels.

c. Combine the 1980 matrix and the 1989 matrix, from Part b, to construct a single matrix that shows the change in average times from 1980 to 1989.

   i. What matrix operation did you use to combine the 1980 and 1989 matrices into this new matrix?

   ii. What are the column labels of this new matrix?

d. Think about the total time for three-person races, where the first leg of the race is three-quarters of a mile run by 10–11 year olds, the second leg is one mile run by 12–13 year olds, and the third leg is one mile run by 14–17 year olds.

   i. What do you think would have been the typical total time for such a race in 1989 if all three legs were run by girls?

   ii. What matrix operation did you use to answer this question?

e. Enrique claimed that in 1980, the 10–11 year-old boys ran faster than the 14–17 year-old boys, since 6.5 is less than 7.5. Do you agree? Explain. If you disagree, describe a method for making a more accurate comparison between the younger and older boys.

② Analyzing Flight Options The vertex-edge graph below shows the direct flights between seven cities for a major airline.

Direct Flights

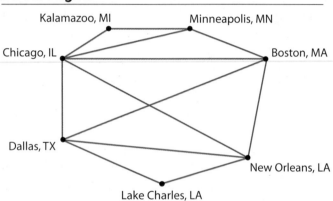

a. Construct the adjacency matrix for the graph. List the vertices in alphabetical order.

b. How many cities can be flown to directly from Chicago? What matrix operation can be used to answer this question?

c. What is the fewest number of stops at intermediate cities needed to fly from Kalamazoo to Lake Charles?

d. It is tiring and time-consuming to have more than one stopover, that is, more than two segments on a flight. Use the adjacency matrix to construct a new matrix that shows which pairs of cities can be connected with a flight of two segments or less.

   i. What matrix operations did you perform to get the new matrix?

   ii. How can you use the new matrix to identify cities connected by a flight with no more than one stopover?

e. Is there anything about the vertex-edge graph that seems unusual or out of place? Is the graph an acceptable model for this particular problem? Explain.

3 **Producing Manuals** Marcus, the owner of a small software company, is producing two manuals for a new software product. One manual is a brief start-up guide, and the other is a larger reference guide.

He contracts to have the manuals bound and shrink-wrapped at a local printer. However, he is on a deadline to ship the manuals in two days, and the machines that do the jobs are only available for a limited time. For the next two days, the binding machine is available for a total of 18 hours, and the shrink-wrap machine is available at later times for 10 hours. The matrix below shows how many seconds each machine requires for each manual.

Time Required (in seconds)

	Start-up Guide	Reference Guide
Bind	30	45
Shrink-Wrap	15	30

Marcus wants to know how many of each type of manual will be ready to ship in two days, if the printer uses all the available time on each machine for his job.

a. Represent this situation in four ways:

  i. using a system of linear equations

  ii. using a matrix equation

  iii. using graphs

  iv. using tables

b. Using one of the representations from Part a, determine how many of each type of manual will be ready to ship in two days.

c. Verify your answer using two of the other representations in Part a.

d. The next week, Marcus receives an unexpectedly large order. He needs 2,000 start-up guides and 800 reference guides bound and shrink-wrapped. Use matrix multiplication and scalar multiplication (multiplying a matrix by a number) to determine how many hours will be needed on each of the machines at the printer.

Summarize
the Mathematics

In this unit, you used matrix methods to analyze problem situations, and you examined properties of matrices.

a In order for information to be useful, it must be organized.

    i. Describe how matrices can be used to organize information.

    ii. Can the same information be displayed in a matrix in different ways? Explain.

    iii. What are some advantages of using matrices to organize and display information? What are some disadvantages?

b Sometimes a situation involves two variables that are linked by two or more conditions. These situations often can be modeled by a system of two linear equations. Describe how to use matrices to solve a system of linear equations.

c List all of the different operations on matrices that you learned about in this unit. For each operation:

    i. Give an example showing how to perform the operation using paper-and-pencil.

    ii. Describe how to perform the operation using your calculator or computer software.

    iii. Describe how the operation can be used to help you analyze some situation.

Be prepared to share your descriptions and examples with the class.

✔ Check Your Understanding

Write, in outline form, a summary of the important mathematical concepts and methods developed in this unit. Organize your summary so that it can be used as a quick reference in future units and courses.

UNIT 3

Coordinate Methods

The use of coordinates to specify locations in two dimensions, as on a map, is familiar to you. The general idea of representing points in terms of coordinates has important mathematical applications in computer modeling. It enables designers and engineers to describe geometric ideas in algebraic language. Objects such as lines, circles, and other curves in two dimensions and surfaces in three dimensions can be expressed in terms of functions and equations. Transformations such as rotation or enlargement of these shapes and surfaces can be accomplished by operations on coordinates.

In this unit, you will study coordinate methods for representing, analyzing, and transforming two-dimensional shapes. Key ideas will be developed through your work on problems in three lessons.

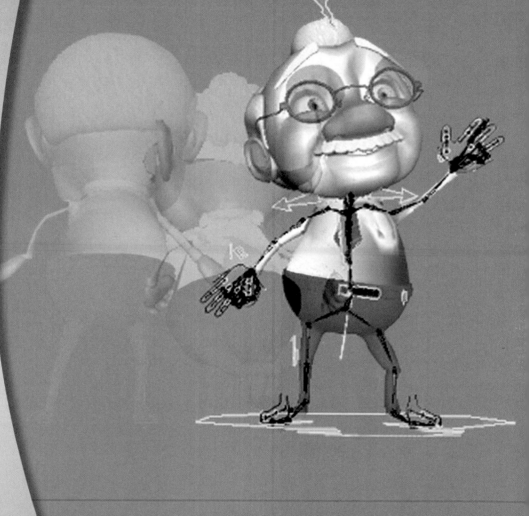

camera-1

80

Graph ...

Lessons

1 A Coordinate Model of a Plane

Use coordinates to represent points, lines, and geometric figures in a plane and on a computer or calculator screen and to analyze properties of shapes.

2 Coordinate Models of Transformations

Use coordinates to describe transformations of the plane and to investigate properties of shapes that are preserved under various types of transformations.

3 Transformations, Matrices, and Animation

Develop and use matrix representations of polygonal shapes and transformations to create computer animations.

A Coordinate Model of a Plane

Computer-generated graphics influence your world in striking ways. They are an important feature of Web sites where you search for information. They are also key elements of the video games you may play and the animated and special-effects films you enjoy.

Computer graphics are now the most commonly used tool in the design of automobiles, buildings, home interiors, and even clothing. And of course, computer or calculator graphics have been important tools in your study of mathematics. They have helped you produce, trace, and analyze graphs of data and functions. They have also helped you create geometric shapes and discover some of their properties.

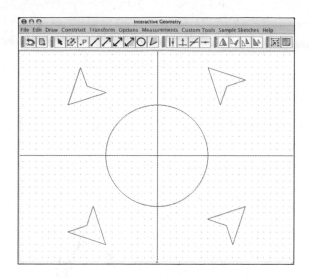

Examine the computer graphics screen shown above. The dots represent points with integer coordinates.

a How do you think the images are produced?

b How could you describe the location of the circle? Of each quadrilateral?

c How do the four quadrilaterals appear to be related? How could you check your claim?

d What equation could be used to describe the horizontal axis? A line 7 units above and parallel to the horizontal axis?

e What equation could be used to describe the vertical axis? A line 10 units to the left of and parallel to the vertical axis?

In this unit, you will study some of the mathematics, called **coordinate geometry**, that underlies computer graphics. As you complete the investigations in this first lesson, you will learn how to use coordinates to model points, lines, and geometric shapes in a plane and on a calculator or computer screen. You will also learn how to use coordinate representations of geometric ideas such as slope, distance, and midpoint to analyze properties of shapes.

Representing Geometric Ideas with Coordinates

Computer images on a screen are composed of lighted *pixels* (screen points) whose coordinates satisfy specific conditions. By specifying the conditions on pixels, CAD (computer-aided design) software and geometry drawing programs can be used to create all sorts of shapes and designs.

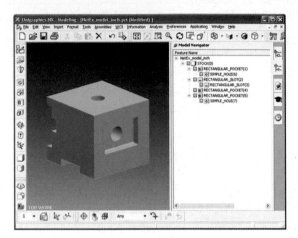

In this investigation, you will explore how a two-dimensional graphics program called *Interactive Geometry* uses coordinates in drawing and in calculating measures of geometric figures. Other software programs may work differently, but they are all based on the same mathematical ideas. As you work on the following problems, look for answers to these questions:

> *How can you create a polygon using interactive geometry software?*
>
> *What information and calculations are needed to find slopes, lengths, and midpoints of sides?*

Creating Shapes As a class or in pairs, experiment with the drawing capabilities of interactive geometry software.

1 Explore how each command in the Draw menu can be used to create examples of the objects listed.

a. Draw commands can also be implemented by selecting the appropriate tool displayed on the software toolbar.

Draw
Figure or Polygon
Point
Segment
Ray
Line
Half Plane
Circle
Angle

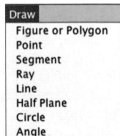

Match each icon in your software toolbar with the corresponding Draw command.

b. In a clear window, draw a circle congruent to the one on the computer screen shown on page 163. Explain how you know the two circles are congruent.

c. In a clear window, draw a quadrilateral congruent to the ones on the computer screen shown on page 163. Discuss how you know that the two quadrilaterals are congruent.

2 Now explore how to use interactive geometry software to create special quadrilaterals. First, clear the window.

a. Draw a rectangle. Record the coordinates of the vertices in the order in which you drew them. Discuss how you know that the displayed figure is a rectangle.

b. Clear the window and then find a different method to draw the same rectangle as in Part a. Describe your method.

c. Clear the window and then draw a parallelogram that is *not* a rectangle. Record the coordinates of the vertices in the order in which you drew them. Discuss how you know the displayed figure is a parallelogram.

d. By *clicking and dragging* a point, you can generate shapes for which some conditions remain the same and other conditions vary. Click and drag one of the vertices of your parallelogram in Part c. What types of shapes can you create?

Calculating Slopes and Lengths Next, explore some of the measurement capabilities of interactive geometry software.

3 Using a clear window, draw a rectangle *ABCD* with coordinates $A(2, 9)$, $B(10, 9)$, $C(10, -6)$, and $D(2, -6)$.

a. Draw the diagonals of the rectangle.

b. Use the "Slopes" command in the Measurements menu to find the slopes of the lines containing each side and the slopes of the two diagonals. Discuss why the reported slopes are reasonable.

Measurements
Coordinates
Lengths
Angles
Slopes
Perimeter/Circumference
Area
Calculation

c. How do you think the software calculates these slopes?

d. Calculate the slopes of the lines containing the diagonals without the use of technology. Compare your results to those in Part b. Explain any differences.

4 Delete the reported slopes. Then use the "Lengths" command in the Measurements menu to calculate the lengths of the sides of the rectangle you created in Problem 3.

a. How do you think the software calculates these lengths?

b. Suppose you have two points with coordinates (a, b) and (c, b) with $a < c$.

   i. How do you know that the points are on a *horizontal line*?

   ii. Write an algebraic expression for the length of the segment or distance between the points.

c. Suppose you have a *vertical line.*

    i. What is true of the coordinates of all points on the line?

    ii. Using variables, write coordinates for a point on the same vertical line as the point with coordinates (a, b).

    iii. Write an algebraic expression for the length of the segment or distance between the points.

5 Next, use the "Lengths" command to find the lengths of the diagonals of the rectangle you created in Problem 3.

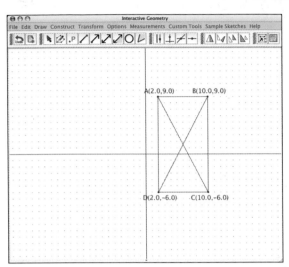

a. Explain how the software could use the coordinates of points A, C, and D to calculate the length of $\overline{AC}$. How could the software use the coordinates of points A, C, and B to calculate the length of $\overline{AC}$?

b. Test your ideas in Part a by calculating the length of $\overline{BD}$ and compare your answer to the software calculation.

c. What theorem justifies the method you used?

d. Now consider points $P(-1, 3)$ and $Q(2, 7)$ in a coordinate plane.

    i. Make a sketch on a coordinate grid showing points P and Q and $\overline{PQ}$.

    ii. Find the length of the segment $\overline{PQ}$ or distance between points P and Q. Compare your answer and method with those of your classmates. Resolve any differences.

e. Use similar reasoning to find the distance between points $S(-5, 4)$ and $T(3, -2)$.

6 To generalize the method you used for calculating distance between two points in a coordinate plane, consider general points $A(a, b)$ and $B(c, d)$ graphed below.

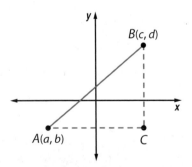

a. Make a copy of the diagram showing the coordinates of point C.

b. Write expressions for the distances AC and BC.

c. Write a formula for calculating the distance AB. Compare your formula with that of your classmates and resolve any differences.

d. When points A and B are on a horizontal or vertical line, will your formula calculate the correct distance AB? Why?

Interactive geometry software uses a method equivalent to yours to calculate the distance between two points. In order to do this, the software needs information or *input* (in this case, the coordinates of two points); instructions for *processing* the information (in this case, a formula); and then instructions on what to do with the results or *output* (in this case, it displays the distance). Specifying such instructions is called **programming**.

Before writing a program, it is helpful to prepare an algorithm that lists the main sequence of steps needed to accomplish the task. The *Distance Between Two Points Algorithm* below could be used to guide program writing for any computer or calculator. In Applications Task 3, you will analyze a calculator program that implements this algorithm.

Distance Between Two Points Algorithm

Step 1: Input the coordinates of one point.
Step 2: Input the coordinates of the other point. } input

Step 3: Use the coordinates and the formula in Problem 6 Part c to calculate the desired distance. } processing

Step 4: Display and label the distance. } output

7 Use the questions below to help write a *Slope Algorithm*, similar to the Distance Between Two Points Algorithm, that could be used to prepare a program to calculate and display the slope of a line through the points $A(a, b)$ and $B(c, d)$.

- What information would you need to input?

- What formula could be used in the processing portion?

- What information should be displayed in the output?

Calculating Midpoints You now have a method for calculating the slope of a line and a method for calculating the distance between two points. Thus, you can compute the length and the slope of a segment in a coordinate plane. Coordinates also can be used by a graphics program to calculate the **midpoint** of a segment, that is, the point on a segment that is the same distance from each endpoint.

8 Use interactive geometry software to draw a rectangle *ABCD* with vertices $A(2, 9)$, $B(10, 9)$, $C(10, -6)$, and $D(2, -6)$.

a. Use the "Midpoint" command in the Construct menu (or select the corresponding icon in the toolbar) to find the midpoint of each side of the rectangle.

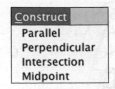

   i. Record the coordinates of the midpoints.

   ii. How do you think the software found these midpoints?

b. What should your software report as the midpoint of the segment with endpoints $P(-3, 2)$ and $Q(1, 2)$? With endpoints $S(3, 3)$ and $T(3, -2)$? Check your conjectures.

c. Suppose you have two points $P(a, b)$ and $Q(c, b)$. Write expressions for the coordinates of the midpoint of $\overline{PQ}$. Compare your expressions with others and resolve any differences.

d. Repeat Part c for the case of points $P(a, b)$ and $Q(a, c)$.

⑨ Use the "Midpoint" command to find the midpoint of diagonal $\overline{AC}$ of the rectangle in Problem 8. Of diagonal $\overline{BD}$. Record the coordinates.

a. What do you notice about the midpoints of the diagonals of the rectangle?

b. How do you think the software found these midpoints? Test your conjecture using a coordinate grid to find the midpoint of the segment joining the points $S(-3, -3)$ and $T(5, 5)$. Verify using your software.

c. Now consider a segment with general endpoints $A(a, b)$ and $B(c, d)$. Make a conjecture about the coordinates of the midpoint M of this segment.

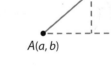

   i. Test your conjecture using a coordinate grid and the points $V(2, -4)$ and $W(6, 8)$.

   ii. Check that your calculated point is the midpoint by verifying that:

      • it is on the line containing the points V and W, and

      • it is *equidistant* from points V and W.

⑩ Use the following questions to help write a *Midpoint Algorithm* that could be used to prepare a program to calculate and display the coordinates of the midpoint of a segment with endpoints $A(a, b)$ and $B(c, d)$.

• What information would you need to input?

• How will the processing portion of your algorithm differ from the algorithms to calculate distance and slope?

• What formula or formulas could be used in the processing portion?

• What information should be displayed in the output?

Summarize
the Mathematics

In this investigation, you reviewed the idea of slope and developed methods for calculating the length and midpoint of a segment in a coordinate plane.

a Write a formula for calculating the slope m of the line containing points $P(x_1, y_1)$ and $Q(x_2, y_2)$ where $x_1 \neq x_2$. Explain in words how the formula determines slope. Why must $x_1 \neq x_2$?

b Write a formula for calculating the distance d between any two points $P(x_1, y_1)$ and $Q(x_2, y_2)$. Explain in words how the formula determines distance.

c Write a formula for the coordinates of the midpoint M of the segment with endpoints $P(x_1, y_1)$ and $Q(x_2, y_2)$. Explain in words how the formula determines coordinates of the midpoint.

Be prepared to share your formulas and thinking with the class.

✔ Check Your Understanding

In your previous studies, you saw that because triangles are rigid, they are often used in the design of complex structures such as the octagonal air control tower shown below.

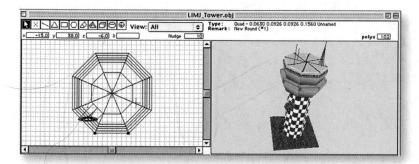

Suppose a triangular component in a CAD display of a different tower has vertices $A(0, 0)$, $B(4, -8)$, and $C(8, -4)$.

a. Make a drawing of $\triangle ABC$ on a coordinate grid. Then find the length of each side.

b. What kind of triangle is $\triangle ABC$? Explain your reasoning.

c. Find the coordinates of the midpoints of $\overline{AC}$ and $\overline{BC}$.

d. Find the slopes of $\overline{AB}$ and of the segment joining the midpoints found in Part c. How are these segments related?

e. Use the click-and-drag feature of interactive geometry software to test if the relationship you found in Part d holds in the case of other triangles, including obtuse triangles. Summarize your findings.

Reasoning with Slopes and Lengths

In the *Linear Functions* unit of Course 1, you discovered that nonvertical parallel lines have equal slopes. That is, if you know that two nonvertical lines are parallel, then you can conclude that they have equal slopes. Also, if the slopes of two lines are equal, then you can conclude that the two lines are parallel. These two facts are summarized in the statement:

> *Two nonvertical lines are parallel if and only if their slopes are equal.*

All vertical lines are, of course, parallel. This fact and the fact that all horizontal lines are parallel are helpful in reasoning about shapes. To create rectangles in Investigation 1, you also likely used the fact that the coordinate axes are perpendicular to each other. As you work on the problems in this investigation, look for answers to the following questions:

> *How can you use slopes to create and reason about figures in a coordinate plane?*

> *In general, how can you determine if two lines in a coordinate plane are perpendicular?*

1 In the *Patterns in Shape* unit of Course 1, a **parallelogram** was defined as a quadrilateral with opposite sides the same length. You have probably discovered that when working with shapes in a coordinate plane, it is easier to calculate slopes than lengths of sides. Using coordinate methods, you can show that a quadrilateral with opposite sides parallel also has opposite sides the same length—so, it is a parallelogram.

a. Examine Diagram I below.

  i. Find the coordinates of point R so that opposite sides are parallel.

  ii. Calculate and compare the lengths of $\overline{PQ}$ and $\overline{SR}$. Of $\overline{PS}$ and $\overline{QR}$.

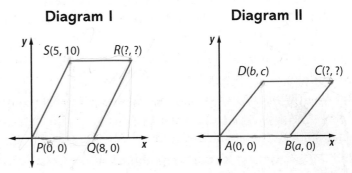

Diagram I

S(5, 10)   R(?, ?)

P(0, 0)   Q(8, 0)

Diagram II

D(b, c)   C(?, ?)

A(0, 0)   B(a, 0)

b. Now examine Diagram II that gives general coordinates.

  i. Find the coordinates of point C so that opposite sides are parallel.

  ii. Calculate and compare expressions for the lengths of $\overline{AB}$ and $\overline{DC}$. Of $\overline{AD}$ and $\overline{BC}$.

c. Based on your prior work in Course 1 and your work above, describe two ways to test if a quadrilateral is a parallelogram.

2 Consider quadrilateral *EFGH* shown below. Does quadrilateral *EFGH* appear to be a rectangle? Verify your conjecture by completing the following tasks.

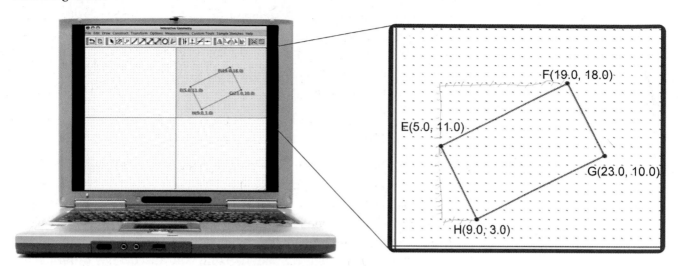

a. Use slopes to explain why quadrilateral *EFGH* is a parallelogram.

b. Based on your work in Problem 1, what can you conclude about the lengths of the sides of quadrilateral *EFGH*? Why?

c. Describe two ways you could use the distance formula to test if parallelogram *EFGH* (denoted □*EFGH*) is a rectangle.

d. Use one of the methods you described in Part c to show that □*EFGH* is a rectangle.

3 Calculating distances and using the converse of the Pythagorean Theorem is one way to show that two intersecting segments (and the lines containing them) are perpendicular. Interactive geometry software uses a less complex method to determine if two lines (segments) are perpendicular. By investigating the slopes of perpendicular segments, you can discover a simpler method. Share the work among your classmates.

a. For each set of points below, plot the points on a coordinate grid. Then verify the points represent vertices of a right triangle. In each case, identify the right angle.

   i. $A(4, 4)$, $B(8, -2)$, $C(14, 2)$ **ii.** $J(-3, 0)$, $K(0, 6)$, $L(3, -3)$

   iii. $P(8, -7)$, $Q(0, 1)$, $R(6, 7)$ **iv.** $X(2, 3)$, $Y(6, 19)$, $Z(10, 18)$

b. Identify the line segments that determine a right angle in each of the triangles in Part a. Then find the slopes of those segments. Write your slopes in reduced fraction form.

   i. What appears to be true about the slopes of perpendicular segments?

   ii. What appears to be true about the product of the slopes of perpendicular segments?

c. Check your observations in Part b using the slopes of the sides of rectangle *EFGH* in Problem 2.

4 Now, using a coordinate grid, draw and label two intersecting segments with the indicated conditions. Check if the segments are perpendicular by measuring the angle formed with the square corner of a sheet of paper.

  a. slope of $\overline{AB}$ is -2; slope of $\overline{BC}$ is $\frac{1}{2}$

  b. slope of $\overline{DE}$ is $\frac{3}{4}$; slope of $\overline{DF}$ is $-\frac{4}{3}$

  c. slope of $\overline{GH}$ is $-\frac{5}{6}$; slope of $\overline{HI}$ is $\frac{6}{5}$

  d. slope of $\overline{JK}$ is $\frac{3}{2}$; slope of $\overline{JL}$ is $-\frac{2}{3}$

5 Look back at your results in Problem 4.

  a. What appears to be true about pairs of nonvertical lines (or intersecting segments) with slopes whose product is -1? That is, with slopes that are **opposite reciprocals**?

  b. Check your conjecture with another pair of intersecting segments whose slopes are opposite reciprocals.

  c. Write a statement summarizing your discoveries in Problems 3 and 4 about perpendicular lines. Write your statement in an "if-and-only-if" form like the statement for parallel lines on page 170.

6 Computer-aided design makes extensive use of simple polygons in creating meshes that outline more complex shapes.

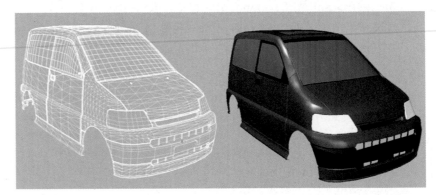

Coordinates of consecutive vertices of different triangles and quadrilaterals are given below. In each case, carefully draw the figure on a coordinate grid and answer as precisely as possible the following questions.

  • If it is a triangle, is it a right triangle, an isosceles triangle, or an equilateral triangle?

  • If it is a quadrilateral, is it a square, rectangle, rhombus, parallelogram, or kite?

Then describe the properties you used to determine your classifications. *Your analysis of at least three of the cases should be done without the use of interactive geometry software or calculator programs.*

  a. $A(-2, 2)$, $B(8, 6)$, $C(4, -4)$

  b. $D(6, 3)$, $E(-3, 9)$, $F(-6, 3)$, $G(3, -3)$

  c. $J(6, -3)$, $K(3, 9)$, $L(-6, -6)$

  d. $P(-4, 0)$, $Q(8, 0)$, $R(4, 8)$, $S(-4, 12)$

  e. $T(-5, 6)$, $U(-1, 8)$, $V(3, 0)$, $W(-1, -2)$

7 Examine the measurements of side lengths and angles of quadrilateral *TUVW* from Part e of Problem 6 drawn with interactive geometry software. The information was produced by selecting the polygon and using the "Coordinates," "Lengths," and "Angles" commands in the Measurements menu.

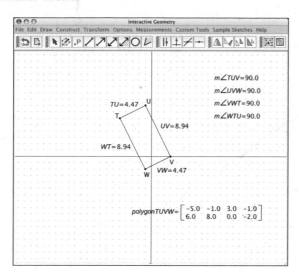

In addition to providing length and angle measure information, the window provides the matrix representation of quadrilateral *TUVW*.

$$TUVW = \begin{array}{cccc} T & U & V & W \\ \left[\begin{array}{cccc} -5 & -1 & 3 & -1 \\ 6 & 8 & 0 & -2 \end{array}\right] \end{array}$$

Matrices provide an efficient way for the computer to store and keep track of the vertices of polygons.

a. In the same manner, represent two of the other polygons in Problem 6 with a matrix.

b. Quadrilateral *WXYZ* is to be a rectangle.

$$WXYZ = \begin{bmatrix} 1 & 6 & 10 & ? \\ 2 & 0 & 10 & ? \end{bmatrix}$$

   i. Find the coordinates of the fourth vertex. Describe how you found the coordinates.

   ii. Draw quadrilateral *WXYZ* on a coordinate grid. Does *WXYZ* appear to be a rectangle?

   iii. Verify that *WXYZ* is a rectangle by giving evidence related to its sides and angles.

8 In Problem 5 of Investigation 1, you may have observed that the diagonals of rectangle *ABCD* were the same length.

a. Use graph paper or interactive geometry software to draw several other rectangles and calculate the lengths of the diagonals in each case. What did you find?

b. You can use general coordinates to justify that the two diagonals of *any* rectangle are the same length.

   i. Begin by placing a rectangle in a coordinate plane so that general coordinates are easy to work with as shown below. What are the coordinates of point *C*?

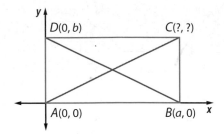

   ii. Show that $AC = BD$.

c. Without using coordinates, how could you use ideas of congruent triangles from Course 1 to justify that $AC = BD$?

d. In Problem 9 (page 168) of Investigation 1, you found that the diagonals of the rectangle intersected at their midpoints. That is, the diagonals of the rectangle *bisected* each other. Use the diagram and general coordinates above to justify that the diagonals of any rectangle bisect each other.

e. Explain why the diagonals of any square are congruent and bisect each other.

Summarize
the Mathematics

Calculating lengths and slopes of segments and using properties of parallel and perpendicular lines can help to determine the nature of geometric figures in a coordinate plane.

ⓐ Suppose a line ℓ in a coordinate plane has slope $\dfrac{p}{q}$.

   i. What is the slope of a line parallel to ℓ? Why must this be the case?

   ii. What is the slope of a line perpendicular to ℓ? Why does this seem reasonable?

ⓑ Consider quadrilateral *QUAD* with vertex matrix

$$QUAD = \begin{bmatrix} x_1 & x_2 & x_3 & x_4 \\ y_1 & y_2 & y_3 & y_4 \end{bmatrix}.$$

   i. How would you determine if *QUAD* is a parallelogram?

   ii. How would you determine if *QUAD* is a square?

ⓒ How could you use coordinates to justify that the diagonals of any square are perpendicular?

Be prepared to discuss your ideas and reasoning with the class.

✓Check Your Understanding

Consider quadrilateral *PQRS* with vertex matrix

$$PQRS = \begin{bmatrix} 8 & 28 & 24 & 4 \\ 4 & 12 & 28 & 20 \end{bmatrix}.$$

a. Draw quadrilateral *PQRS* on a coordinate grid.

b. What special kind of quadrilateral is *PQRS*? Use coordinates to justify your answer.

Investigation 3 Representing and Reasoning with Circles

In Investigations 1 and 2, you learned how to represent and analyze polygons in a coordinate plane. You can describe their sides using linear equations and study their properties using ideas of distance and slope. Polygons, particularly triangles and quadrilaterals, are the building blocks for architectural designs. Industrial, automotive, and aerospace designs often require that shapes have circular components.

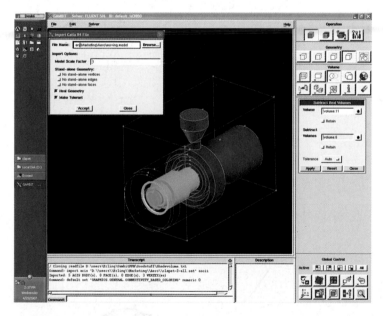

Your work on the problems in this investigation will help you answer these questions:

What information is needed to create a circle in a coordinate plane?

How can you represent circles in a coordinate plane with equations?

How can you use general coordinates of points to reason about special properties of circles?

1 As a class, explore how interactive geometry software could be used to create the design shown at the right.

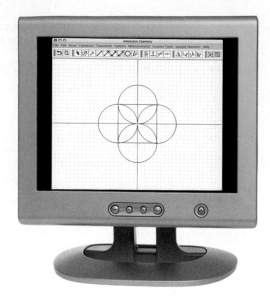

a. What information was needed by the software to draw each circle? Why do you think that information is sufficient?

b. Clear the window and redraw the square, centered at the origin, with side length 10 units.

c. Draw a circle **inscribed in the square**, that is, a circle that touches each side of the square at one point. Describe the points of contact of the circle and square.

d. Draw a circle **circumscribed about the square**, that is, a circle that passes through each vertex of the square.

e. What is the radius of each circle in Parts c and d?

2 Here are two circles with center at the origin O and radius 10 drawn in a coordinate plane.

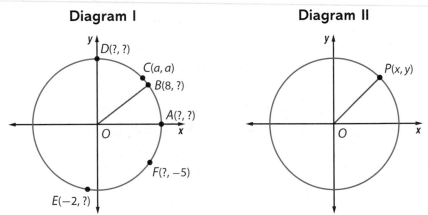

a. What must be true about the distance from point O to any other point on the circle?

b. Without the help of software, find the missing coordinate(s) of points A through F on the circle in Diagram I.

c. Suppose $P(x, y)$ is any point on the circle in Diagram II.

  i. What must be true about the distance OP?

  ii. Write an equation showing the relationship between x, y, and the radius of the circle.

d. Write an equation for a circle with its center at the origin and with radius 7. With radius $\sqrt{3}$. With radius r.

3 A calculator-produced circle is shown below. The **Zsquare** window has a scale on both axes of 1 unit.

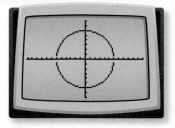

   a. What is the radius of the circle?

   b. Write an equation for this circle.

   c. What expressions could be placed in the [Y=] menu to produce the circle? Do your expressions produce a circle with the same radius?

   d. Use your calculator to produce a copy of the circle shown in the computer display on page 163.

4 Some of the circles you created in Problem 1 did not have their centers at the origin. However, you can use reasoning similar to that in Problem 2 to find equations for these circles.

   a. What is the center and radius of the circle whose center is on the positive *x*-axis?

   i. Suppose $P(x, y)$ is any point on that circle. Explain why it must be the case that $\sqrt{(x - 5)^2 + y^2} = 5$.

   ii. Use that information to write an equation for the circle that does not involve a radical symbol.

   b. Write similar equations for:

   i. the circle whose center is on the positive *y*-axis.

   ii. the circle whose center is on the negative *x*-axis.

   iii. the circle whose center is on the negative *y*-axis.

   c. Verify that the coordinates of the vertices of the square satisfy your equations of the four circles that contain those vertices. Share the workload with your classmates.

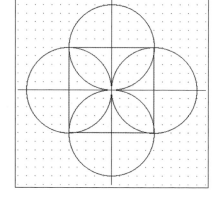

5 Now try to generalize your work in Problems 2–4 to a circle whose center is not on an axis.

   a. Use reasoning similar to that in Problem 4 to find the equation of a circle with center $C(h, k)$ and radius r.

   b. Compare your equation with those of your classmates. Resolve any differences.

   c. Rewrite your equation in Part b for the case when $C(h, k)$ is the origin. What do you notice?

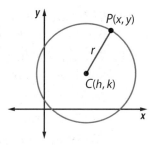

6 Without using technology, determine which of the following equations describe a circle in a coordinate plane. For each equation that represents a circle, determine the center, the radius, and one point on the circle. For each equation that does *not* represent a circle, explain why not.

a. $x^2 + y^2 = 25$

b. $x^2 + y = 16$

c. $3x^2 + 3y^2 = 108$

d. $(x - 5)^2 + (y - 1)^2 = 81$

e. $3x^2 + y^2 = 9$

f. $x^2 + (y + 5)^2 = 1$

7 Coordinates as employed by interactive geometry software open new windows to geometry by allowing you to easily create figures and search for patterns in them. Complete Parts a–c using your software. You can create the figures yourself or use the "Explore Angles in Circles" custom tool.

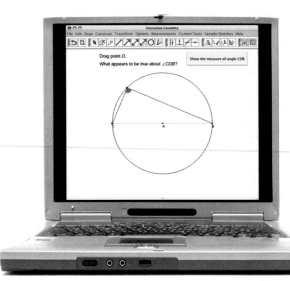

a. Draw a circle with center *A* and diameter with endpoint *B*. Label the other endpoint *C*.

b. Construct a new point *D* on the circle. Then draw $\overline{BD}$ and $\overline{CD}$.

c. Click and drag point *D* along the circumference of the circle.

　i. What appears to be true about $\angle CDB$ in all cases?

　ii. How is your conjecture supported by calculations from the Measurements menu?

d. State your conjecture in the form:
An angle inscribed in a semicircle ... .
Compare your conjecture with your classmates and resolve any differences.

8 As you saw in Investigation 2, coordinates can provide a powerful way to justify conjectures you make about geometric figures. The key is to position the figure in a coordinate plane so that general coordinates are easy to work with. A circle with center at the origin and radius *r* is shown below. Point *A(a, b)* is a general point on the circle, different from points *P* and *Q* which are endpoints of a diameter on the *x*-axis.

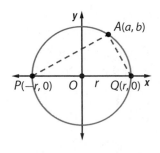

Use these general coordinates and the following questions to help justify the conjecture you made in Problem 7:

An angle inscribed in a semicircle is a right angle.

a. What are some possible methods you could use to justify that $\angle PAQ$ is a right angle?

b. What are the coordinates of points P and Q?

c. Since point $A(a, b)$ is on the circle, what must be true about the distance OA? How is that distance related to the coordinates a and b?

d. Study Jack's argument below. He shows that $\triangle PAQ$ is a right triangle, and so $\angle PAQ$ is a right angle. Check the correctness of Jack's reasoning and give reason(s) justifying each step. If there are any errors in Jack's reasoning, correct them.

Jack's argument

The length of $\overline{PA} = \sqrt{(a + r)^2 + b^2}$, so $(PA)^2 = (a + r)^2 + b^2$.   (1)

The length of $\overline{AQ} = \sqrt{(r - a)^2 + b^2}$, so $(AQ)^2 = (r - a)^2 + b^2$.   (2)

The length of $\overline{PQ} = 2r$, so $(PQ)^2 = 4r^2$.   (3)

$$(PA)^2 + (AQ)^2 = (a + r)^2 + b^2 + (r - a)^2 + b^2 \qquad (4)$$
$$= (a^2 + 2ar + r^2 + b^2) + (r^2 - 2ar + a^2 + b^2) \qquad (5)$$
$$= 2a^2 + 2r^2 + 2b^2 \qquad (6)$$
$$= 2(a^2 + b^2) + 2r^2 \qquad (7)$$
$$= 2r^2 + 2r^2 \qquad (8)$$
$$= 4r^2 \qquad (9)$$
$$= (PQ)^2 \qquad (10)$$

Therefore, $\triangle PAQ$ is a right triangle with $\angle PAQ$ a right angle.   (11)

e. Now examine Malaya's argument justifying the conjecture that $\angle PAQ$ is a right angle. Check the correctness of Malaya's reasoning and give reason(s) justifying each step. Correct any errors in Malaya's reasoning.

Malaya's argument

The slope of $\overline{PA}$ is $\dfrac{b}{a + r}$.   (1)

The slope of $\overline{QA}$ is $\dfrac{b}{a - r}$.   (2)

The product of the slopes is $\left(\dfrac{b}{a + r}\right)\left(\dfrac{b}{a - r}\right) = \dfrac{-b^2}{a^2 - r^2}$.   (3)

Since $a^2 + b^2 = r^2$, it follows that $a^2 - r^2 = -b^2$.   (4)

This means that the product of the slopes is $\dfrac{b^2}{-b^2} = -1$.   (5)

So, $\overline{PA} \perp \overline{AQ}$ and $\angle PAQ$ is a right angle.   (6)

Summarize
the Mathematics

In this investigation, you discovered how to write equations for circles in a coordinate plane and used coordinates to make general arguments.

a What is the equation of a circle with center at the origin and radius r?

b What is the equation of a circle with center at (h, k) and radius r?

c What formula was the key to deriving the equation of a circle?

d How can you tell by looking at an equation whether its graph is a circle?

e Why are general coordinates such as (a, b) used in reasoning about geometric properties?

Be prepared to share your equations and thinking with the class.

✔Check Your Understanding

A circle with center at $(3, -4)$ is drawn so that it is **tangent** to the x-axis. That is, the circle touches the x-axis at only one point called the *point of tangency*.

a. Draw the circle on a coordinate plane.

b. What are the coordinates of the point of tangency?

c. Write an equation for the circle.

d. Write an equation for a congruent circle with center at the origin.

e. Graph the circle in Part d on your graphing calculator.

Applications

1 Use graph paper, the **Line(** command from the DRAW menu of your graphing calculator, or interactive geometry software to draw a model of a kite with vertices $A(5, -6)$, $B(7, -2)$, $C(5, 2)$, and $D(-9, -2)$.

   a. Does your drawing appear to be that of a kite? Use careful reasoning with the coordinates to justify that $ABCD$ is a kite, that is, a quadrilateral with exactly two pairs of congruent adjacent sides.

   b. Draw the cross braces of the kite and find their lengths using coordinates.

   c. Use coordinates to find the midpoints of $\overline{AC}$ and $\overline{BD}$.

   d. Justify that the midpoint of $\overline{AC}$ is on $\overline{BD}$. How is this fact seen in your drawing?

2 Use graph paper, the **Line(** command from the DRAW menu of your graphing calculator, or interactive geometry software to draw a model of a school crossing sign. Assume the height is the same as the width of the base, and the length of the vertical edges is half that of the base. Locate the shape on the coordinate axes so that one side of the shape is on the x-axis and the y-axis is a line of symmetry.

   a. Give the coordinates of each vertex.

   b. Determine the length of each side using coordinates. Which pairs of sides are the same length?

   c. Use coordinates to find the height of your model sign.

   d. Find the area of your model sign.

3 The following program, designed for one type of graphing calculator, computes the distance between two points in a coordinate plane. The left-hand column is the program; the right-hand column describes the function of the commands.

DIST Program

Program	Function in Program
ClrHome	Clears display screen
Input "X COORD",A Input "Y COORD",B Input "X COORD",C Input "Y COORD",D	Enters x- and y-coordinates of two points
$\sqrt{((A-C)^2+(B-D)^2)}\rightarrow L$	Calculates distance and stores value in memory location L
Disp "DISTANCE IS",L	Outputs calculated distance with label

---

a. Describe how this program uses the Distance Between Two Points Algorithm on page 167.

b. What does the program call the coordinates of the two points?

c. Explain how the processing portion actually calculates the distance.

d. Enter the program DIST in your calculator (modified as necessary for your particular calculator). Check your program for accuracy by testing several pairs of points.

4 Modify the program in Applications Task 3 so that it will compute the slope of a nonvertical segment determined by two points. Call your new program SLOPE. Check your program for accuracy by testing it with several points.

5 In Investigation 1, Problem 10 (page 168), you wrote a midpoint algorithm for calculating the coordinates of the midpoint of a segment. A program for a graphing calculator that will compute the midpoint of a segment is shown below.

MIDPT Program

Program		Function in Program
ClrHome		1. Clears display screen
Input "X COORD",A		2. _____
Input "Y COORD",B		3. _____
Input "X COORD",C		4. _____
Input "Y COORD",D		5. _____
(A+C)/2→X		6. _____
(B+D)/2→Y		7. _____
Disp "MIDPOINT COORDS"		8. Displays words, MIDPOINT COORDS
Disp X		9. _____
Disp Y		10. _____
Stop		11. _____

---

a. Analyze this program and explain the purpose of each command line as was done for lines 1 and 8.

b. Enter the program MIDPT in your calculator. (Depending on your calculator, you may need to modify the commands slightly.) Test the program on pairs of points of your choosing.

6 Drilling teams from oil companies search around the world for new sites to place oil wells. Increasingly, oil reserves are being discovered in offshore waters. The Gulf Oil Company has drilled two high-capacity wells in the Gulf of Mexico 5 km and 9 km from shore, as shown in the diagram on page 183. The 20 km of shoreline is nearly straight, and the company wants to build a refinery on shore between the two wells. Since pipe and labor cost money, the company wants to find the location that will serve both wells and uses the least amount of pipe when it is laid in straight lines from each well to the refinery.

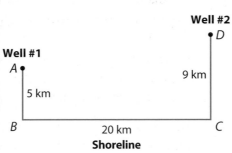

Well #1

A •

5 km

B

20 km

Shoreline

Well #2

• D

9 km

C

a. How can coordinates be used to model this situation?

b. What distance(s) should you try to minimize to use the least amount of pipe?

c. Do you think the refinery should be closer to *B*, to *C*, or at the midpoint of the shoreline? Make a conjecture.

d. Determine your best estimate for the location of the refinery. About how much pipe will be required?

e. There are several methods for solving this problem, including:

- Analyze tables or graphs of a function relating total length of pipe to distance of refinery from point *B*.

- Use point *D*, its reflection across $\overleftrightarrow{BC}$, and congruent triangles.

- Use the click-and-drag feature of interactive geometry software.

Select a method different from what you used in Part d and use that method to solve this problem. Compare your answer with that found in Part d.

7 A CAD face-view drawing of a building includes a quadrilateral *PQRS* whose vertices are given by the matrix:

$$\begin{array}{cccc} P & Q & R & S \end{array}$$
$$\begin{bmatrix} 4 & 8 & 14 & 10 \\ 4 & -2 & 2 & 8 \end{bmatrix}$$

a. Sketch the quadrilateral on a coordinate grid.

b. What kind of quadrilateral is *PQRS*? Give reasons to support your response.

8 $\overline{AB}$ has endpoints $A(-5, 0)$ and $B(4, 3)$. $\overline{CD}$ has endpoints $C(-3, 9)$ and $D(1, -3)$. The equations of the lines containing $\overline{AB}$ and $\overline{CD}$ are $x - 3y = -5$ and $3x + y = 0$, respectively.

a. How could you quickly check that these equations are correct?

b. Verify that the lines are perpendicular.

c. Find the point of intersection of $\overline{AB}$ and $\overline{CD}$ by solving the system of equations.

d. Find the midpoints of $\overline{AB}$ and $\overline{CD}$. Compare your results with Part c.

e. What kind of quadrilateral is *ACBD*? Explain your reasoning.

9 In Check Your Understanding Part d (page 169), you discovered that the line segment connecting the midpoints of two sides of a particular triangle was parallel to the third side. You may have also noticed that the length of the *midsegment* was half the length of the third side. With coordinates, you can verify this is true for any triangle.

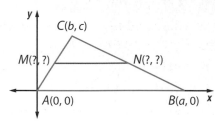

Using the above placement of △*ABC* in a coordinate plane:

a. Find the coordinates of the midpoint *M* of $\overline{AC}$. Of the midpoint *N* of $\overline{BC}$.

b. Use coordinates to explain why $\overline{MN} \parallel \overline{AB}$.

c. Show that $MN = \frac{1}{2}AB$.

10 Quadrilateral *ABCD* is a rhombus with general coordinates.

a. Determine the coordinates of point *C*.

b. Show that $\overline{AC} \perp \overline{BD}$.

c. Show that $\overline{AC}$ bisects $\overline{BD}$ and that $\overline{BD}$ bisects $\overline{AC}$.

d. Write a statement that summarizes this general property of rhombuses.

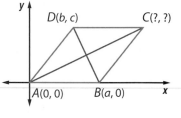

11 The circle shown below was produced from the DRAW menu of a graphing calculator using the command **Circle(0,0,6)**. It is displayed in the standard **ZSquare** window.

a. Write an equation for the displayed circle.

   i. Use your equation to find two points on the circle whose *x*-coordinate is 3.

   ii. What two expressions could be used in the ⬛Y=⬛ menu to produce a graph of the circle?

b. Use the **Circle(** command to produce a circle with center at (2, 4) and radius 10. What might be a good window to use to display the circle?

   i. Write an equation for the circle.

   ii. Use your equation to find two points on the circle whose *x*-coordinate is 5.

c. Use the **Circle(** command to produce the circle defined by $(x + 5)^2 + (y - 8)^2 = 84$.

   i. Write an equation for a circle that has the same center and is tangent to the *x*-axis.

   ii. Write an equation for a circle that has the same center and is tangent to the *y*-axis.

   iii. Write an equation for a circle that is tangent to both the *x*- and *y*-axes and is congruent to the given circle that you graphed. How many circles are possible? How are their centers related?

Connections

12 In Course 1, you may have conducted an experiment in which you placed several equal weights at various positions on a yardstick and found the balance point or *center of gravity*. The balance point corresponded to the mean of the distances from zero on the yardstick.

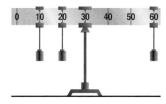

Test a similar idea for two-dimensional shapes.

a. Cut out a triangle from a sheet of cardboard or tag board that is about the size of a $\frac{1}{2}$-sheet of notepaper.

- Experiment with the cutout to try to find a point at which it will balance on the top of your finger or a pencil.

- Now place the cutout on a coordinate grid and record the coordinates of its vertices.

- Compute the mean of the *x*-coordinates and the mean of the *y*-coordinates. Locate this point on your coordinate grid and on the cardboard cutout.

- Verify by balancing that the point you found is the center of gravity.

b. Repeat Part a for a rectangle. For a parallelogram that is not a rectangle. What do you notice?

c. Repeat Part a for a quadrilateral that is not a parallelogram. What do you notice?

13 An engineering school offers a special reading and writing course for all entering students. Students are assigned to one of two sections based on performance on a placement test. Section A emphasizes reading skills; Section B stresses writing skills. The mean test scores for Section A are 64.2 (reading) and 73.8 (writing). For Section B, the mean reading and writing scores are 74.4 and 57.6, respectively. Placement test scores of five students are shown below.

Placement Test Scores

	Reading Score	Writing Score
Jim	68	64
Emily	67	67
Anne	70	62
Miguel	66	69
Gloria	60	60

a. Represent the reading and writing scores of each student listed in the table as a point on a coordinate grid. Label the points. On the same grid, also plot and label the points corresponding to the mean scores for Sections A and B.

b. Using only the visual display, assign students to Section A or B. What influenced your choices?

c. Suppose students are assigned to the section whose mean point is closest to their point.

   i. Assign each student to a section. Compare your assignments to that in Part b.

   ii. Why would this assignment criterion make sense for placement of students?

d. Are there any students for whom neither section appears appropriate? Explain your response.

14 How do you use the concept of mean in your procedure to calculate the coordinates of the midpoint of a segment?

15 Suppose (x_1, y_1) and (x_2, y_2) are given points with $x_2 > x_1$. Then the x-coordinate of the midpoint is half the distance from x_1 to x_2 added to x_1.

a. Show that the above statement is true when $x_1 = 6$ and $x_2 = 11$.

b. Explain why the above statement makes sense.

c. Rewrite the expression $x_1 + \dfrac{x_2 - x_1}{2}$ in a simpler equivalent form. Is this the x-coordinate of the midpoint?

16 So far, you have been drawing polygons in a coordinate plane by plotting and connecting vertices in order. You can also think of polygons as being *enveloped* by a family of lines. Examine the lines below and the quadrilateral that is enveloped by them. The scale on both axes is 1 unit.

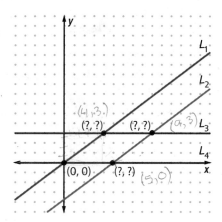

a. Without use of technology, match each equation given below with the line through the corresponding side of the quadrilateral. Describe clues you used to determine the matches.

Equation I $y = 0$ **Equation II** $3x - 4y = 15$ $(5,0)$ $(0, -1/4)$

Equation III $y = 3$ **Equation IV** $3x - 4y = 0$

b. Determine the coordinates of the vertices of the quadrilateral.

c. Explain as precisely as possible how you can verify the quadrilateral is a rhombus.

d. The equations in Part a describe lines that contain the sides of the quadrilateral. The equations will describe only the points on the sides if you restrict the input values for x and y.

   i. In the case of the equation for the side determined by the vertices $(0, 0)$ and $(4, 3)$, explain why $0 \leq x \leq 4$ ($x \geq 0$ and $x \leq 4$) and $0 \leq y \leq 3$.

   ii. For each of the remaining equations in Part a, describe the restrictions on x and y so that the equation describes just the side of the rhombus.

17 A circle with radius 3 and center at the origin is shown at the right.

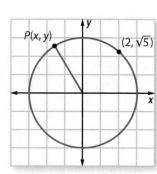

a. Justify that the point with coordinates $(2, \sqrt{5})$ is on the circle.

b. Use symmetry to name the coordinates of three other points on the circle.

c. Without any calculations, identify the coordinates of four other points on the circle.

18 In this lesson, you saw how geometric figures and distance measurement in a plane can be represented using coordinates. Complete a table like the one below, which summarizes key features of a two-dimensional **coordinate model** of geometry. Then give a specific example of each idea from the coordinate model.

Geometric Idea	Coordinate Model	Example
Point	Ordered pair (x, y) of real numbers	
Plane	All possible ordered pairs (x, y) of real numbers	(No example needed)
Line	All ordered pairs (x, y) satisfying a linear equation $ax + by = c$	$2x + y = 8$
Segment length		
Midpoint		
Circle		
Parallel lines		
Perpendicular lines		

19 State Police radios can transmit up to 25 miles. Officer Jacobs patrols Highway 20 which runs straight north from the state police post to Driftwood, located 50 miles from the police post. Officer Kelley patrols Highway 45, which runs straight west from the police post to the state line, 50 miles west of the police post.

a. This situation can be represented with a coordinate model as shown below.

   i. What does the 50×50 grid represent?

   ii. What does point O_1 represent? Point O_2? Point P?

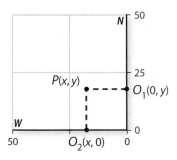

b. Find the probability that the officers can talk to each other if calls can be relayed through the police post.

c. Find the probability that the officers can talk to each other if calls are *not* relayed through the police post.

Reflections

20 How can you show that two triangles are congruent by calculating with the coordinates of the vertices?

21 In the first investigation, you invented formulas for calculating the distance between two points in a coordinate plane and finding coordinates of the midpoint of the segment determined by those points. You were asked to write your formulas for general points (x_1, y_1) and (x_2, y_2), that is, using *subscript notation*. What advantages or disadvantages do you see in using subscript notation in these cases?

22 In the definition of the midpoint of a segment on page 167, the phrase "on a segment" is included. Why is that phrase needed?

23 For points $P(x_1, y_1)$ and $Q(x_2, y_2)$ in a coordinate plane:

- the slope of line PQ is $\dfrac{\Delta y}{\Delta x}$ or $\dfrac{y_1 - y_2}{x_1 - x_2}$.
- the distance PQ is $\sqrt{(\Delta x)^2 + (\Delta y)^2}$ or $\sqrt{(x_1 - x_2)^2 + (y_1 - y_2)^2}$.

In each case, the differences of coordinates are calculated. When calculating the slope or calculating the distance, does the order in which you subtract the coordinates make any difference? Illustrate and explain your reasoning.

24 An equilateral triangle has two vertices at $(0, 0)$ and $(0, 8)$. What are the possible coordinates of the third vertex? Illustrate and explain your reasoning.

25 Quadrilateral *ABCD* has vertex *A* at the origin and adjacent sides of length 13 and 5 units. For each part below, illustrate and explain your reasoning.

a. Find coordinates for *B*, *C*, and *D* so that quadrilateral *ABCD* is a rectangle.

b. Find coordinates for *B*, *C*, and *D* so that quadrilateral *ABCD* is a parallelogram (but not a rectangle).

26 Suppose you are given the coordinates of three points in a plane. How can you use calculations on the coordinates:

a. to test whether or not the three points lie on a line?

b. to test whether or not the three points are the vertices of a right triangle?

c. to test whether or not one of the points is on the perpendicular bisector of the segment with the other two points as endpoints?

d. to test whether or not one of three points is the center of a circle containing the other two points.

Extensions

27 In addition to the graphics window, interactive geometry software may also include a programming window that can be used to create both simple and complex figures by entering program commands. In Lesson 3, you will use the programming window to create animations.

To become familiar with the programming window, use the "Design by Robot" custom tool to complete the following tasks.

a. Enter the commands below in the programming window of your software. "Right 90" means "rotate the robot 90° clockwise" from its "home" position.

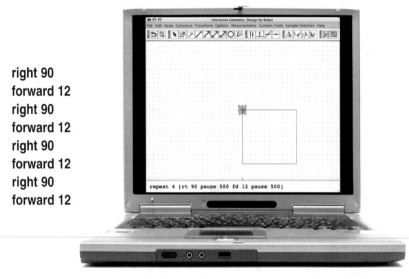

right 90
forward 12
right 90
forward 12
right 90
forward 12
right 90
forward 12

i. Compare the figure you created with that shown on the screen above.

ii. Compare the program you entered with that shown on the above screen. How are they similar and how are they different in execution? In structure?

iii. Use the clearscreen (**cs**) and **home** commands to clear the window and position the robot at the origin of the coordinate system. Then modify one of the programs so that the robot draws a rectangle that is *not* a square.

b. As you saw in Part a, blocks of the same instructions that are used in succession can be put into a single "repeat" statement. Predict the shape created by this repeat statement:

repeat 3 [fd 10 rt 120]

Check your prediction.

c. Use the programming window to direct the robot to draw each of the following shapes. Before beginning each part, use the clearscreen (**cs**) and **home** commands to clear the window and position the robot at the origin of the coordinate system.

  i. an isosceles right triangle with two sides of length 10

  ii. a parallelogram that is not a rectangle

  iii. a regular hexagon with side length of 8

d. Compare the shapes created by each repeat statement.

Shape I   repeat 20 [fd 2 rt 18]

Shape II   repeat 180 [fd 0.2 rt 2]

e. Enter the following program. Describe the design produced and how it was created.

```
cs home
repeat 75 [fd 12 pause 250 rt 125 pause 250]
```

28 In Investigation 1, you discovered a formula for the midpoint of a segment in a coordinate plane. If you know the coordinates of the midpoint of a segment and the coordinates of one endpoint, how can you find the coordinates of the other endpoint? Illustrate your idea for the segment $\overline{AB}$ where A has coordinates $(8, -2)$ and the coordinates of the midpoint are $(11.5, 2)$.

29 The four lines with equations $x + y = 2$, $y = 1$, $x + y = -2$, and $x - 3y = 2$ envelop a quadrilateral.

a. Find the coordinates of its vertices.

b. Sketch the quadrilateral.

  i. For each equation, what restrictions on the input values for x and y are needed so the equation describes only the side of the figure?

  ii. What kind of quadrilateral is formed? How do you know?

c. Find equations for the lines containing the diagonals of the quadrilateral. What restrictions on the input values for x and y are needed for the lines to describe only the diagonals?

d. What are the coordinates of the point of intersection of the diagonals?

e. Verify your answer in Part d using a method different from the one you used to find that answer.

30 Look back at Applications Task 10. Write the converse of the statement you wrote in Part d.

a. Do you think the converse statement is always true? Test your conjecture using graph paper or interactive geometry software.

b. To prove your conjecture, would it be easier to use general coordinates or the idea of congruent triangles? Explain your reasoning and provide a sample proof.

c. Write the statement from Applications Task 10, Part d and its converse as a single "if and only if" statement.

31 Streets in a city or neighborhood are often built in a rectangular grid. The street layout may be represented by a rectangular coordinate system. In this situation, distances can be measured along streets, as a car would drive, not diagonally across blocks. (Of course, there are no one-way streets!) The shortest street distance between two locations is called the **taxi-distance**. For example, on the following coordinate grid, the taxi-distance between points P and Q is 5.

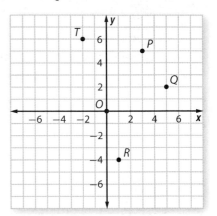

a. Find the taxi-distance between the given points O and P, and between points T and R.

b. A Dial-a-Ride dispatcher receives a request for a pickup at point $X(-8, 20)$. Available vans are stationed at point $A(12, 8)$ and at point $B(-11, -7)$. Which van should make the pickup? Why?

c. Write a formula for computing the taxi-distance TD between points $P(a, b)$ and $Q(c, d)$.

d. Draw a graph of all points in the plane (not just points with integer coordinates) whose taxi-distance from $(0, 0)$ is 2. Do the same for all points whose taxi-distance from $(0, 0)$ is 4.

e. What would be a reasonable name for the figures you graphed in Part d?

f. In *taxi geometry*, what would be a reasonable value for π?

32 Use interactive geometry software to construct a triangle, $\triangle ABC$, segment by segment.

a. Using the "Midpoint" and "Perpendicular" commands, construct the lines that are the perpendicular bisectors of the sides of your triangle. What appears to be true about the three perpendicular bisectors of the sides?

b. Test your conjecture in Part a with several different types of triangles including right, obtuse, and acute by clicking and dragging a vertex of $\triangle ABC$. Write a summary of your findings.

c. Suppose P is the point where the perpendicular bisectors of the sides of $\triangle ABC$ meet. What seems to always be true about the lengths PA, PB, and PC? Explain carefully why that must be the case.

d. Given any triangle in the plane, is it possible to draw a circle that contains the three vertices? Explain.

Review

33 Use your understanding of square roots to solve each of the following equations.

  a. $5 = \sqrt{x}$

  b. $9 = \sqrt{t - 100}$

  c. $2\sqrt{y} = 14$

34 Suppose the strip pattern shown below extends indefinitely in both directions.

  a. Describe all of the symmetries of the pattern that also preserve color.

  b. Describe all of the symmetries of the pattern if color is ignored.

35 Write an equation for the line passing through the indicated pair of points.

  a. $D(3, 2)$ and $E(9, 4)$

  b. $S(3, 1)$ and $T(5, -2)$

36 Consider the lines $y = x$ and $y = -x$.

  a. On the same coordinate grid, draw graphs of these lines.

  b. What is the measure of the angle formed by the positive x-axis and the line $y = x$? Explain your reasoning as carefully as possible.

  c. Are these two lines perpendicular? Explain your reasoning.

37 Expand each of the following products to an equivalent expression in standard quadratic form.

  a. $(x + 3)^2$

  b. $(5 + y)^2$

  c. $(t - 8)^2$

  d. $(p - 6)(p + 6)$

38 Suppose that Lauren randomly surveyed 150 students in her school and asked if they attended the school musical last year. Sixty of the students indicated that they had attended the school musical last year. Based upon this data, approximately how many of the 1,150 students in the school do you expect will attend the musical this year?

39 Solve each equation without using calculator graphs or tables of values.

    a. $12 + 2(5x - 8) = 0$

    b. $(x + 4)(7 - x) = 0$

    c. $(3x + 9) - (5x - 8) = 0$

    d. $6(2^x) - 48 = 0$

40 The cost of painting a wall is directly proportional to the area of the wall. A wall that is 22 feet long and 10 feet high costs $30 to paint.

    a. How much will it cost to paint a wall that is 28 feet long and 12 feet high?

    b. Find the constant of proportionality for this situation and explain what it means in terms of the context.

41 Examine the linear graphs shown below.

    a. Match each equation with its graph.

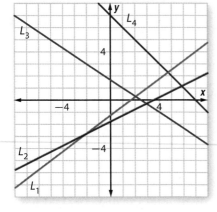

Equation I $y + x = 7$

Equation II $5 - 2x = 3y$

Equation III $4y - 2x = -7$

Equation IV $3x - 4y = 5$

    b. Did you match equations to graphs by finding x- and y-intercepts from the equations? If not, try it.

    c. Did you match equations to graphs by rewriting equations in the form $y = a + bx$ and finding slopes? If not, try it.

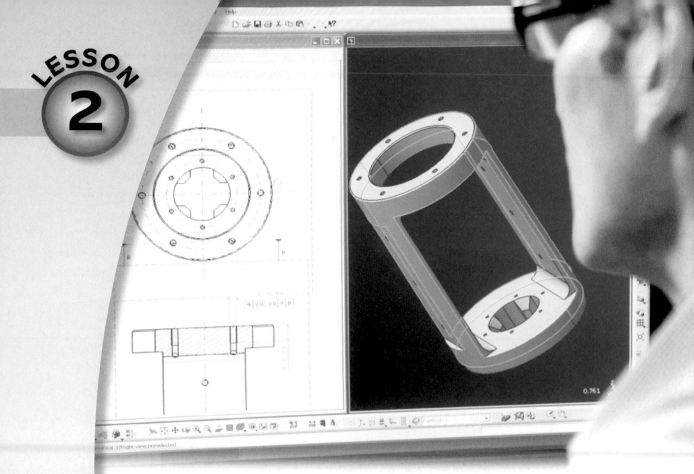

Coordinate Models of Transformations

In Lesson 1, you modeled polygons by specifying coordinates of consecutive vertices and connecting them in order with segments. You modeled circles with equations that specified conditions on coordinates of points on the circles. Coordinate representations of figures permit you to easily reposition and resize shapes in a coordinate plane or on a computer or calculator screen.

Examine the computer graphics screen on the next page.

Think About This Situation

The computer graphics display above shows some of the ways in which a figure in a coordinate plane can be transformed in terms of position and/or size.

a How could you create the humanoid in the second quadrant using interactive geometry software? How could you describe the figure using coordinates?

b What transformation of the figure in Quadrant II would produce the image in Quadrant I? The image in Quadrant III? The image in Quadrant IV?

c How do you think coordinates might be used to describe these transformations?

In this lesson, you will learn how coordinates can be used to transform shapes. You will investigate coordinate representations of **rigid transformations** that provide a way to reposition figures in a plane without changing the shape or the size of the figures. You will also investigate coordinate representations of **similarity transformations** that can be used to resize figures while maintaining their shapes. Finally, you will explore some of the matrix algebra of transformations—how matrices are used to represent transformations and how they are combined to form new transformations.

Computer graphics enable designers to model two- and three-dimensional figures and to also easily manipulate those figures. For example, interior design software permits users to select images of various types and sizes of furniture and position them at different places in a room layout by sliding and rotating the shapes. Automotive design software permits users to create symmetric components of vehicles by reflecting and/or rotating basic design elements. Complete models of vehicles can be rotated and viewed from different angles.

In this investigation, you will explore how coordinates are used in computer graphics software to transform the position of shapes in a plane. As you work on the following problems, look for answers to these questions:

How can coordinates be used to describe a sliding motion or translation?

How can coordinates be used to describe a turning motion or rotation?

How can coordinates be used to describe a mirror or line reflection?

1 Interactive geometry software provides tools to reposition a shape by translation, rotation about a point, and reflection across a line. Other software will have similar commands. As a class or in pairs, experiment with the first three commands in the Transform menu and the corresponding functions in the menu bar.

Transform
Reflect
Rotate
Translate
Scale
Reflect Using
Rotate By
Translate By
Scale By

a. Begin by exploring how the Translate command can be used to transform shapes in a plane. Draw a shape or select the "Humanoid" from Sample

Sketches. Translate the shape to a different position. Observe how the original shape and its *image* appear to be related. Repeat for at least three other translations, including: one that slides the shape horizontally; one that slides the shape vertically; and one that slides the shape in a slanted direction.

b. Next, explore how the Rotate command can be used to transform shapes in a plane. Draw a figure or select the "Humanoid" shape from Sample Sketches. Rotate the figure counterclockwise about the origin. Observe how the original figure and its image appear to be related. Repeat for at least three other counterclockwise rotations about the origin, including: a 90° rotation; a 180° rotation; and a 45° rotation.

c. Now, explore how the Reflect command can be used to transform a shape in a plane. Draw a shape or select the "Humanoid" shape from Sample Sketches. Reflect the shape across a line of your choice. Observe how the original shape and its image appear to be related. Repeat for at least three other line reflections, including: a reflection across the *x*-axis; a reflection across the *y*-axis; and a reflection across the line $y = x$. In each case, first clear the window and redraw your shape.

d. How do you think the software determines the position of the translated image of a shape? The rotated image of a shape? The reflected image of a shape?

Translating Shapes A **translation,** or sliding motion, is determined by distance and direction. By looking carefully at a simple shape and its translated image, you can discover patterns relating the coordinates of the shape and the coordinates of its image.

 On the screen below, a flag *ABCDE* and its translated image *A′B′C′D′E′* are shown.

Horizontal Translation

a. Describe the translation as precisely as you can.

b. Explain how the translated image of the flag could be produced using only the translated images of points *A*, *B*, *C*, *D*, and *E*.

c. Under this translation, what would be the image of $(0, 0)$? Of $(1, -5)$? Of $(-5, -4)$? Of (a, b)?

d. Write a rule you can use to obtain the image of any point (x, y) in the coordinate plane under this translation. State your rule in words and in symbolic form $(x, y) \rightarrow (\underline{\quad}, \underline{\quad})$.

3 The screens below show a flag *ABCDE* and its image under two other translations.

Vertical Translation

Oblique Translation

a. Describe the vertical translation as precisely as you can. The diagonal (oblique) translation.

b. Under the vertical translation, what would be the image of (0, 0)? Of (2, 5)? Of (4.1, −2)? Of (*a*, *b*)?

c. Write a rule you can use to obtain the image of any point (*x*, *y*) under the vertical translation. State your rule in words and in symbolic form (*x*, *y*) → (__, __).

d. Under the oblique translation, what would be the image of (0, 0)? Of (2, 5)? Of (4.1, −2)? Of (*a*, *b*)?

e. Write a rule you can use to obtain the image of any point (*x*, *y*) under the oblique translation. State your rule in words and in symbolic form.

4 Compare the transformation rules you developed for Part d of Problem 2 and for Parts c and e of Problem 3. Write a general rule that tells how to take any point (*x*, *y*) and find its translated image if the preimage is moved horizontally *h* units and vertically *k* units. Compare your rule with others and resolve any differences.

You now have a rule you can use to find the translated image of any point when you know the **components of the translation**—the horizontal and vertical distances and directions the point is moved (left or right, up or down). This is exactly the information a calculator or computer graphics program needs in order to display a set of points and their translated images.

5 Use the following questions to help write an algorithm that would guide a programmer in the development of a translation program that displays the original figure (called the **preimage**) and its translated image and connects corresponding vertices of the two figures.

• What information would you need to input?

• What formula or formulas could be used in the processing portion?

• What information should be displayed in the output?

Rotating Shapes Rotations about the origin have similar coordinate models. A **rotation,** or turning motion, is determined by a point called the *center of the rotation* and a *directed angle of rotation*. A flag *ABCDE* and its images under counterclockwise rotations of 90°, 180°, and 270° about the origin are shown below.

Rotations About the Origin

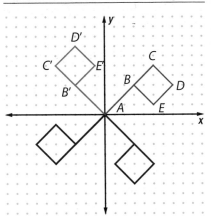

6 Consider flag *ABCDE* above and its image under a 90° counterclockwise rotation about the origin.

a. On a copy of the table below, record the coordinates of the images of the five points on the flag under a 90° counterclockwise rotation about the origin. Explain why the rotated image of the flag could be produced using only the rotated images of points *A, B, C, D,* and *E*.

Preimage	90° Counterclockwise Rotated Image
A(0, 0)	A'(,)
B(3, 3)	B'(,)
C(5, 5)	C'(,)
D(7, 3)	D'(,)
E(5, 1)	E'(,)

b. Use any patterns you see between preimage and image points in your completed table to help plot the points $(-2, -5)$, $(-4, 1)$, $(5, -3)$, and their images under a 90° counterclockwise rotation about the origin on a new coordinate grid.

   i. For each preimage point, use dashed segments to connect the preimage to the origin and the origin to the image.

   ii. Connect each preimage segment to its image segment with a "turn" arrow that shows the directed angle of rotation.

c. Write a rule relating the coordinates of any preimage point (x, y) and its image point under a 90° counterclockwise rotation about the origin. State your rule in words and in symbolic form.

d. According to your rule, what is the image of (0, 0)? Why does this image make sense?

e. How should the slope of the line through a preimage point and the origin be related to the slope of the line through the origin and the image point? Verify your idea by computing and comparing slopes.

f. Write an algorithm to guide the development of a program for a 90° counterclockwise rotation about (0, 0) that displays the preimage and image figures.

7 As you probably expect, counterclockwise rotations of 180° and 270° about the origin also have predictable coordinate patterns. Use a copy of the screen at the top of page 200 showing flag *ABCDE* and its images to explore these patterns.

a. Investigate patterns in the coordinates of the preimage and image pairs when points are rotated 180° about the origin.

b. Write a rule relating the coordinates of any preimage point (*x, y*) and its image point under a 180° rotation about the origin. State your rule in words and in symbols.

c. How is the slope of the line through two preimage points related to the slope of the line through the images of those points? What does this tell about a line and its image under a 180° rotation?

d. Similarly, search for patterns in the coordinates of the preimage and image pairs when points are rotated 270° counterclockwise about the origin.

e. Write a rule relating the coordinates of any preimage point (*x, y*) and its image point under a 270° counterclockwise rotation about the origin. State your rule in words and in symbols.

f. Describe how you could modify the algorithm in Part f of Problem 6 so that it would guide development of a program to rotate a point 180° or 270° counterclockwise about the origin instead of 90° counterclockwise.

Reflecting Shapes

Line reflections can also be expressed using coordinates. A **line reflection** is determined by a "mirror line" (or line of reflection) that is the perpendicular bisector of the segment connecting a point and its reflected image. A point on the line of reflection is its own image. In the following problems, you will build coordinate models for reflections across vertical and horizontal lines, as well as across the lines $y = x$ and $y = -x$.

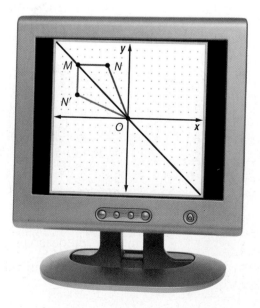

8 A flag *ABCDE* and its reflected image across the *y*-axis are shown on the screen below.

Reflected Across the *y*-axis

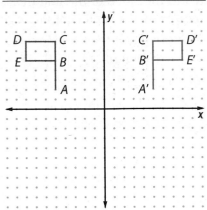

a. Investigate patterns in the coordinates of preimage and image pairs when points are reflected across the *y*-axis.

b. Explain why the reflected image of the flag could be produced using only the reflected images of points *A*, *B*, *C*, *D*, and *E*.

c. Write a rule which tells how to take any point (*x*, *y*) and find its reflected image across the *y*-axis. State your rule in words and in symbols.

d. On a copy of the diagram, use dashed segments to connect point *A* to point *A'* and point *D* to point *D'*. Use coordinates to verify that the *y*-axis is the perpendicular bisector of $\overline{AA'}$ and $\overline{DD'}$.

9 The table below shows coordinates of six preimage points and coordinates (*a*, *b*) of a general point. Plot each of the six points and its reflected image across the *x*-axis.

a. Record the coordinates of the image points in a table like the one below.

Preimage	Reflected Image Across *x*-axis
(−4, 1)	(−4, −1)
(3, −2)	
(−2, −5)	
(4, 5)	
(0, 1)	
(−3, 0)	
(*a*, *b*)	

b. What pattern relating coordinates of preimage points to image points do you observe? Use the pattern to give the coordinates of the image of (*a*, *b*).

c. Write a rule that tells how to take any point (x, y) and find its reflected image across the x-axis. State your rule in words and in symbols.

d. How is the x-axis related to the segment determined by any point (a, b) not on the x-axis and its reflected image? Justify your answer using coordinates.

10 Draw the graph of $y = x$. Plot each preimage point in the table below and its reflected image across that line. Connect each preimage/image pair with a dashed segment.

a. Record the coordinates of the image points in a copy of the table below.

Preimage	Reflected Image Across $y = x$
$(-4, 1)$	$(1, -4)$
$(3, -2)$	
$(-2, -5)$	
$(4, 5)$	
$(0, 1)$	
$(-3, 0)$	
(a, b)	

b. Describe a pattern relating coordinates of preimage points to image points.

c. Write a rule relating the coordinates of any preimage point (x, y) to its reflected image across the line $y = x$. State your rule in words and in symbols.

d. How is the line of reflection, $y = x$, related to the segment determined by any point (a, b) not on the line and its image? Justify your answer.

11 Next, investigate patterns in the coordinates of the preimage and image pairs when points are reflected across the line $y = -x$.

a. Draw the graph of $y = -x$. Then plot the six preimage points in the table in Problem 10 and their reflected images across the line.

b. Describe a pattern relating coordinates of preimage points to coordinates of image points.

c. Write a rule relating the coordinates of any preimage point (x, y) and its reflected image across the line $y = -x$. State your rule in words and in symbols.

d. How is the segment determined by a point and its reflected image related to the line $y = -x$?

12 You now have coordinate models for the following line reflections.

- reflection across the x-axis
- reflection across the y-axis
- reflection across the line $y = x$
- reflection across the line $y = -x$

Sharing the workload among your classmates, develop planning algorithms that would guide a programmer in the development of line reflection programs for each of these four line reflections. Identify the input, processing, and output portions of each of your algorithms.

Summarize
the Mathematics

In this investigation, you developed coordinate rules relating points and their images under different rigid transformations: translations, rotations about the origin, and line reflections.

a A translation is determined by a single point and its image.

    i. Suppose a translation slides the point $O(0, 0)$ to the point $A(a, b)$. Write a symbolic rule $(x, y) \rightarrow (\_\_, \_\_)$ that describes this translation.

    ii. Suppose a translation slides the point $A(a, b)$ to the point $B(c, d)$. Write a symbolic rule $(x, y) \rightarrow (\_\_, \_\_)$ that describes this translation.

b Summarize the coordinate rules for these rotations about the origin.

    i. For a rotation of $90°$ counterclockwise: $(x, y) \rightarrow (\_\_, \_\_)$

    ii. For a rotation of $180°$ counterclockwise: $(x, y) \rightarrow (\_\_, \_\_)$

    iii. For a rotation of $270°$ counterclockwise: $(x, y) \rightarrow (\_\_, \_\_)$

    iv. For a rotation of $270°$ clockwise: $(x, y) \rightarrow (\_\_, \_\_)$

c Summarize the coordinate rules for line reflections:

    i. Across the x-axis: $(x, y) \rightarrow (\_\_, \_\_)$

    ii. Across the y-axis: $(x, y) \rightarrow (\_\_, \_\_)$

    iii. Across the line $y = x$: $(x, y) \rightarrow (\_\_, \_\_)$

    iv. Across the line $y = -x$: $(x, y) \rightarrow (\_\_, \_\_)$

Be prepared to explain your coordinate rules and strategies you could use to remember or redevelop them.

✓Check Your Understanding

Consider the following matrix representation of △ABC.

$$\triangle ABC = \begin{bmatrix} -1 & 4 & 3 \\ 2 & -3 & 5 \end{bmatrix}$$

a. On separate grids, sketch and label △ABC and its image under each of the following transformations.

   i. Reflection across the y-axis

   ii. Translation with horizontal component −3 and vertical component 2

   iii. Reflection across the line $y = x$

   iv. Rotation of 180° about the origin

   v. Rotation of 90° counterclockwise about the origin

b. For one of the transformations in Part a, use coordinates to show that △ABC and its transformed image are congruent.

Investigation 2 Modeling Size Transformations

In the previous investigation, you found patterns in the coordinates of preimage/image pairs for transformations with which you were familiar. For those transformations, the distance between any pair of preimage points was the same as the distance between their images. As a result, under these rigid transformations, a polygon and its image had the same size and shape—they were congruent.

   In this investigation, you will reverse the procedure. You will start with a rule relating coordinates of any preimage and its image, and you will explore how the transformation affects familiar shapes.

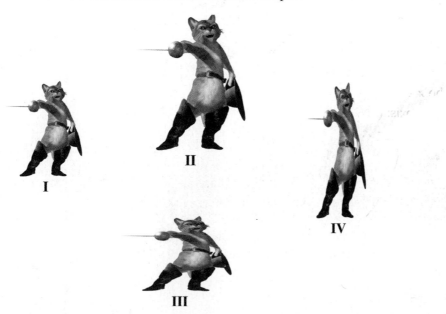

As you complete the problems in this investigation, look for an answer to this question:

   How can coordinates be used to rescale or resize a shape?

① Consider first the transformation defined by the following rule:

$$\begin{array}{ccc} \text{preimage} & & \text{image} \\ (x,\ y) & \longrightarrow & (3x,\ y) \end{array}$$

This rule is read "the x-coordinate of the image is 3 times the x-coordinate of the preimage; the y-coordinate of the image is the same as the y-coordinate of the preimage."

a. Which of Figures II, III, or IV on the previous page appears to be the image of Figure I under this transformation? Explain your reasoning.

b. On a coordinate grid, plot the points $X(1, 1)$, $Y(5, 1)$, and $Z(5, 5)$. Draw $\triangle XYZ$ and its image under this transformation.

c. Examine your preimage and image shapes. What characteristics of $\triangle XYZ$ are also characteristics of its image? How do the shapes differ?

d. How do you think the perimeter of $\triangle XYZ$ will compare to the perimeter of its image? How do you think the area of $\triangle XYZ$ will compare to the area of its image? Test your conjectures.

e. Which of Figures II, III, or IV on the previous page could be the image of Figure I when transformed by the rule: $(x, y) \rightarrow (x, 3y)$? What clue(s) did you use?

Your work on Problem 1 has shown that even a simple transformation might not preserve all characteristics of the preimage shape. By modifying the transformation rule slightly, you can create a transformation which has many interesting and useful characteristics.

② A **size transformation** (or **dilation**) of magnitude 3 centered at the origin is defined by the following rule:

$$\begin{array}{ccc} \text{preimage} & & \text{image} \\ (x,\ y) & \longrightarrow & (3x,\ 3y) \end{array}$$

a. On a copy of the diagram shown here, draw the image of quadrilateral *ABCD* under this size transformation. Label image vertices A', B', C', and D'.

b. Examine your preimage and image shapes. Make a list of all of the properties of quadrilateral *ABCD* that seem to also be properties of quadrilateral $A'B'C'D'$. Also, describe how the two shapes seem to differ.

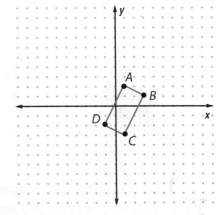

3 Making visual comparisons, as you did in Problem 2, is useful; but such comparisons should be made with some skepticism. You should always seek additional evidence to support or refute your visual conjectures. This is where coordinate representations and formulas for distance and slope can be very helpful. Use these ideas to examine more carefully quadrilateral *ABCD* and its image quadrilateral *A'B'C'D'* that you drew in Problem 2.

a. Compare the length of $\overline{AB}$ with the length of $\overline{A'B'}$. Does the same relation hold for other preimage/image pairs of segments? Explain.

b. How does $\overleftrightarrow{AB}$ appear to be related to $\overleftrightarrow{AD}$? Does the same relationship hold for their images? Give evidence to support your claim.

c. How do the perimeters of quadrilateral *ABCD* and quadrilateral *A'B'C'D'* compare?

d. How does $\overleftrightarrow{BC}$ appear to be related to $\overleftrightarrow{AD}$? Is this relationship true for their images? Justify your conclusion.

e. What kind of quadrilateral is *ABCD*? Is the image quadrilateral *A'B'C'D'* the same kind of quadrilateral? Explain your reasoning.

f. How do the areas of quadrilaterals *ABCD* and *A'B'C'D'* appear to be related? State a conjecture. Test your conjecture. How does the magnitude of the size transformation come into play here?

4 Refer back to your drawing of quadrilateral *ABCD* and its image quadrilateral *A'B'C'D'* under the size transformation of magnitude 3.

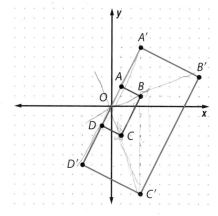

a. Use a ruler to draw lines through *A* and *A'*, *B* and *B'*, *C* and *C'*, and *D* and *D'*. Extend the lines to intersect the axes. What do you notice about the intersection of these four lines?

b. Find the equations of the four lines in Part a. Use these equations to verify your observation in Part a.

c. The size transformation has its **center** at the origin since the lines in Part a intersect at (0, 0). What is the image of the center (0, 0) under this size transformation?

d. Compare the distances from the center *O* to a point and to the image of that point. State a conjecture.

  i. Find the distance from *O* to *A* and from *O* to *A'*. From *O* to *B* and from *O* to *B'*. Do these distances confirm your conjecture?

  ii. Make similar comparisons for distances from point *O* to points *C* and *D* and to their images. What seems to be true? Modify your original conjecture, if necessary, based on the evidence.

e. Now try to generalize your finding. How should the distances from *O*(0, 0) to *P*(*a*, *b*) and from *O*(0, 0) to *P'*(3*a*, 3*b*) be related? Show why this must be the case by calculating the distances *OP* and *OP'*.

f. Complete the following statement:

> *If O is the center of a size transformation with magnitude k and the image of P is P', then OP'* = _____ *and* $\dfrac{OP'}{OP}$ = _____.

Compare your general statement with that of others. Resolve any differences.

5 Next, consider a size transformation with magnitude 0.5 and center at the origin.

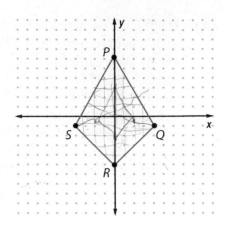

a. Write a rule for this size transformation.

b. On a copy of the diagram shown here, plot and label the image of quadrilateral *PQRS* under this size transformation. How do you think quadrilateral *PQRS* and its image are related in terms of shape? In terms of size?

c. Compare segment lengths in the image with corresponding lengths in quadrilateral *PQRS*. How does the magnitude 0.5 affect the relation between lengths? Between perimeters?

d. Find the area of the image quadrilateral. Compare it to the area of quadrilateral *PQRS*. How does the magnitude 0.5 affect the relation between areas?

6 Now, consider $\triangle PQR = \begin{bmatrix} 3 & -3 & -2 \\ 4 & 2 & -1 \end{bmatrix}$.

a. Sketch $\triangle PQR$ on a coordinate grid.

b. Sketch the image, $\triangle P'Q'R'$, resulting from transforming $\triangle PQR$ with a size transformation of magnitude 2.5 and center at the origin.

c. Compare lengths of corresponding preimage and image sides.

d. How are $\overleftrightarrow{PQ}$ and $\overleftrightarrow{QR}$ related? How are $\overleftrightarrow{P'Q'}$ and $\overleftrightarrow{Q'R'}$ related? Give evidence to support your claim.

e. Use the information in Parts c and d to help you determine the area of $\triangle PQR$ and $\triangle P'Q'R'$. Compare the areas and relate them to the magnitude 2.5.

7 Size transformations are used by interactive geometry software to resize figures in a coordinate plane. As a class or in pairs, experiment with the "Scale" (or similar) command.

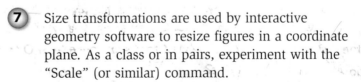

Transform
Reflect
Rotate
Translate
Scale
Reflect Using
Rotate By
Translate By
Scale By

a. Draw $\triangle XYZ = \begin{bmatrix} 8 & 4 & -3 \\ 6 & -7 & -4 \end{bmatrix}$.

b. Find images of $\triangle XYZ$ when transformed with magnitudes 3, 1.5, and 2.5. In each case, compare side lengths and areas of the preimage and image triangles. Are the results of your comparisons consistent with what you would have predicted? Explain.

c. Select the "Humanoid" from Sample Sketches. Experiment with commands in the Transform menu to produce a series of images that suggests the figure is walking toward the front of the screen.

d. Clear the window. Experiment with commands in the Transform menu to produce a series of images that suggests the figure is walking toward the back of the screen.

Summarize
the Mathematics

In this investigation, you explored properties of size transformations.

a Explain why the transformation in Problem 1 $(x, y) \rightarrow (3x, y)$ is or is not a size transformation.

b Suppose a size transformation with magnitude $k > 0$ and center at the origin O maps A onto A', B onto B', and C onto C'.

  i. Write a rule that can be used to obtain the image of any point (x, y) in the coordinate plane under this size transformation. State your rule in words and in symbolic form.

  ii. How is the length $A'B'$ related to the length AB?

  iii. If $\triangle ABC$ has an area of 25 square units, what is the area of $\triangle A'B'C'$? Why does this make sense in terms of the formula for the area of a triangle?

  iv. How is the distance from O to C' related to the distance from O to C?

  v. Where do $\overleftrightarrow{AA'}$ and $\overleftrightarrow{CC'}$ intersect? Does $\overleftrightarrow{BB'}$ intersect there too?

c How are size transformations similar to rigid transformations? How are they different?

d What is the magnitude of a size transformation that is also a rigid transformation? Describe such a transformation.

Be prepared to explain your conclusions to the entire class.

✔ Check Your Understanding

A size transformation with magnitude 3.5 and center at the origin is applied to a right triangle with legs of length 4 and 5 units.

a. What are the lengths of the three sides of the image triangle?

b. What is the area of the given triangle and of its image?

c. Write an algorithm for a program that will accept the coordinates of a point and the magnitude of this or any other size transformation (with center at the origin). The program should display the point, its image, and their coordinates.

Investigation 3 · Combining Transformations

Computer graphics software includes the creation, storage, and manipulation of figures that can be simple or complex. Translations, rotations, and size transformations are essential to many graphics applications to change the position, tilt, and size of shapes. You now have the basic tools for creating computer graphics images because you know the coordinate rules that define some of these key transformations.

What kinds of transformations do you think were needed to create the characters' walk from the back of the landscape to the front?

As you work on the problems of this investigation, look for answers to these questions:

How can rigid transformations and/or size transformations be combined to form new transformations?

When combining two transformations, how can you predict what the new transformation will be and how it will affect the preimage?

Use interactive geometry software or graph paper to complete the following problems. You should focus your attention on (1) preimage/image point coordinate patterns and (2) how the shape of the preimage and its image are related.

1 This first problem will help you experiment with combinations of transformations.

7,15 *13,6* *7,2*

-15,7 *-6,13* *-2,7*

 a. Draw $\triangle ABC = \begin{bmatrix} 7 & 13 & 7 \\ 15 & 6 & 2 \end{bmatrix}$ in a coordinate plane.

   i. Draw the image of $\triangle ABC$ reflected across the y-axis. Find the coordinates of B', the image of B. Similarly, find the coordinates of A' and C'.

   ii. Next, draw the image of $\triangle A'B'C'$ rotated 90° counterclockwise about the origin. Find the coordinates of B'', the image of B'. What are the coordinates of A'' and C''?

 b. Now, examine $\triangle ABC$, $\triangle A'B'C'$, and $\triangle A''B''C''$.

   i. Are they congruent? Explain your reasoning.

   ii. What is the measure of $\angle B$? How do you know? What is the measure of $\angle B''$?

   iii. Compare the **orientation** (the clockwise or counterclockwise labeling of the vertices) of $\triangle ABC$ to the orientation of $\triangle A''B''C''$. What do you notice? Why does your observation make sense?

 c. If you first rotate $\triangle ABC$ 90° counterclockwise about the origin and then reflect that image across the y-axis, do you think you will get the same *final* image, $\triangle A''B''C''$? Try it! Explain what happens.

 d. Is the final image $\triangle A''B''C''$ in Part c related to $\triangle ABC$ by a single transformation? Is so, describe the transformation.

In the following problems, you will investigate more systematically the effects of various combinations of transformations.

2 **Combining Two Translations** Examine the effects of applying one translation followed by another. This two-step transformation is called a **composition of translations**.

 a. Draw $\triangle ABC$ with vertex coordinates $A(1, 5)$, $B(6, 2)$, $C(8, 11)$.

 b. Translate $\triangle ABC$ using the translation with horizontal component 6 and vertical component -10. Then apply the translation with horizontal component -8 and vertical component 2 to the image of $\triangle ABC$. What were the net horizontal and vertical distances that you translated the figure and in which directions?

 c. Compare the size, shape, and orientation of $\triangle ABC$ and the final image.

 d. Compare the coordinates of the vertices of the final image with those of $\triangle ABC$. What pattern do you observe?

e. Could you apply a *single* transformation to △ABC and obtain the same final image? If so, describe that transformation as completely as possible.

f. Repeat Parts a–e for a new figure and two new translations of your choice. Keep a record of the coordinates of the vertices of your figure and its final image. Also, record the components of your translations.

g. Look back at your work in Parts a–f. Suppose the following two translations are applied in succession.

$$(x, y) \rightarrow (x + a, y + b)$$
$$(x, y) \rightarrow (x + h, y + k)$$

Write a symbolic rule $(x, y) \rightarrow (\_\_, \_\_)$ which describes the new combined transformation. Compare your rule with that of others and resolve any differences.

3 **Combining Two Size Transformations** Next, investigate the effects of successively applying two size transformations with center at the origin.

a. Draw △ABC with vertex coordinates $A(-5, 1)$, $B(0, -4)$, $C(2, 5)$.

b. Apply a size transformation of magnitude 2 to △ABC. Then apply a size transformation of magnitude 1.5 to the image of △ABC.

c. Compare the size, shape, and orientation of △ABC to the final image.

d. Compare the coordinates of the vertices of the final image with those of △ABC. What pattern do you observe?

e. Make a conjecture about the effects of applying two size transformations in succession, both centered at the origin. Test your conjecture using a different figure and two new size transformations with center at the origin; one should have magnitude k, $0 < k < 1$. Keep a record of the coordinates of the vertices of your figure and its image. Also, record the magnitudes of your size transformations.

f. Look back at your work in Parts a–e. Suppose the following two size transformations with magnitudes $k > 0$ and $m > 0$ are applied in succession.

$$(x, y) \rightarrow (kx, ky)$$
$$(x, y) \rightarrow (mx, my)$$

Write a symbolic rule that describes the new combined transformation. Compare your rule with that of others and resolve any differences.

The process of successively applying two transformations is called **composing** the transformations. The transformation that maps the *original* preimage to the *final* image is called the **composite transformation**.

4 **Combining Two Rotations** Use methods similar to your work in Problems 2 and 3 to investigate the effects of composing two rotations about the origin.

a. Draw $\triangle ABC$ with vertex coordinates $A(1, 4)$, $B(7, 1)$, $C(12, 8)$.

b. Rotate $\triangle ABC$ 90° counterclockwise about the origin. Then rotate the image of $\triangle ABC$ 180° about the origin. Compare the coordinates, size, shape, and orientation of $\triangle ABC$ to the final image.

c. Make a conjecture about the effects of composing two rotations about the origin. Test your conjecture using a different figure and two new counterclockwise rotations about the origin. Keep a record of the degree measures of your rotations.

d. Look back at your work in Parts a–c. Write a statement summarizing your findings.

5 **Combining Two Reflections** Examine the effects of composing two line reflections that you have studied.

a. Draw a triangle or a quadrilateral of your choice.

b. Reflect the figure across the *x*-axis. Then reflect the image across the line $y = x$.

   i. Compare the size, shape, and orientation of the preimage and image under the composite transformation.

   ii. What kind of transformation does the composition of the two line reflections appear to be? Be as specific as you can.

   iii. Write a symbolic rule $(x, y) \rightarrow (\_\_, \_\_)$ which describes the composite transformation.

   iv. Does the order in which you compose the two line reflections lead to the same final image?

c. Now reflect the figure across the *x*-axis and then reflect the image across the *y*-axis.

   i. Compare the size, shape, and orientation of the preimage and image under the composite transformation.

   ii. What kind of transformation does the composition of the two line reflections appear to be?

   iii. Write a symbolic rule $(x, y) \rightarrow (\_\_, \_\_)$ which describes the composite transformation.

d. What is the effect of reflecting across the same line twice?

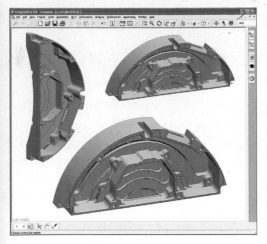

When composing transformations, the two transformations do not have to be two of the same kind. In the problems that follow, you will explore compositions of size transformations and rigid transformations (translations, rotations, and reflections). Such composite transformations allow shapes to be rotated and enlarged or reduced in computer graphics applications.

6 **Combining a Rotation and a Size Transformation**

Investigate the effects of composing a size transformation and a counterclockwise rotation with centers at the origin.

a. Suppose you rotate a triangle 90° counterclockwise about the origin and then apply a size transformation of magnitude 2 to the image triangle.

   i. Predict how the coordinates of the preimage will be related to those of the image.

   ii. Predict how the lengths of corresponding sides will be related.

   iii. Predict how the measures of corresponding angles will be related.

   iv. Predict how the areas of the preimage and image will be related.

b. Check your predictions by applying the composite transformation to △ABC with vertex coordinates A(1, 4), B(7, 1), and C(12, 8).

c. Reverse the order in which the transformations are applied—apply the size transformation first, then the rotation. How are the preimage and the image related this time?

d. On the basis of Parts a–c, what would you predict would happen if you used a different rotation with center at the origin but the same size transformation? The same rotation but a different size transformation? Would the order in which you applied the transformations lead to different final images? Test your conjectures.

7 **Combining a Size Transformation and a Translation**

Make a conjecture about the effects on shapes of composing a size transformation with center at the origin and a translation.

a. Test your conjecture by drawing △ABC with vertex coordinates A(2, 8), B(14, 4), and C(18, 10). Then find the image of △ABC under the composite transformation: size transformation of magnitude $\frac{1}{2}$ with center at the origin followed by a translation with components 3 and −10. What are the coordinates of the vertices of the image of △ABC under this composite transformation?

b. How are segment lengths affected by composing a size transformation with a translation?

c. How are angle measures affected by composing a size transformation with a translation?

d. How are areas affected by composing a size transformation with a translation?

e. Does the order in which you apply the transformations lead to different final images? If it does, are the effects on segment lengths, angle measures, and areas different also? Give evidence supporting your claims.

Figures that are related by a size transformation, or by a composite of a size transformation with a rigid transformation, are called **similar**. The composite transformation is called a **similarity transformation**. In Problems 6 and 7, each of the preimage/image pairs of shapes are examples of similar shapes. The magnitude of the similarity transformation is the **scale factor**. It is the multiplier you use to convert lengths in the original figure to those in the similar image.

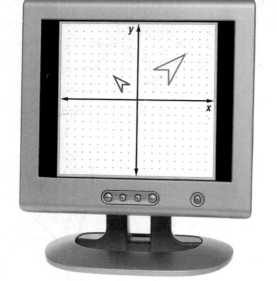

8 Examine the computer graphics screen to the right.

a. How would you check to see if the two dart shapes are similar?

b. Describe in words a similarity transformation that maps the smaller dart onto the larger dart.

c. Write a coordinate rule $(x, y) \rightarrow (\_\_, \_\_)$ that describes this similarity transformation.

d. Write a coordinate rule $(x, y) \rightarrow (\_\_, \_\_)$ that maps the larger dart onto the smaller dart. What is the scale factor?

9 Consider the similarity transformation that is the composition of these two transformations in the order given:

$$\textbf{Transformation I} \qquad (x, y) \rightarrow \left(\tfrac{2}{3}x, \tfrac{2}{3}y\right)$$
$$\textbf{Transformation II} \qquad (x, y) \rightarrow (x, -y)$$

a. Describe each of the transformations as completely as possible.

b. Suppose $\triangle PQR = \begin{bmatrix} 6 & 18 & 12 \\ 12 & 0 & -6 \end{bmatrix}$. What is the image of $\triangle PQR$ under the similarity transformation?

c. Write a rule $(x, y) \rightarrow (\_\_, \_\_)$ which describes the similarity transformation. Compare your rule to that of others and resolve any differences.

d. What is the scale factor of the similarity transformation?

e. How would your answers to Parts b–d change if Transformation II was applied first to $\triangle PQR$?

Summarize
the Mathematics

In this investigation, you explored compositions of rigid transformations and size transformations.

a What kind of transformation is formed by composing the transformations given in each case below? Be as specific as you can.

   i. Two translations

   ii. Two counterclockwise rotations about the origin

   iii. Two line reflections involving lines intersecting at the origin

   iv. Two size transformations with center at the origin

b What kind of transformation is formed by composing a rigid transformation and a size transformation? Under such a transformation:

   i. how are corresponding segments of the preimage and image shapes related?

   ii. how are corresponding angles related?

   iii. how are corresponding areas related?

c Does the order in which you compose two transformations make a difference? Explain.

d If you are given symbolic rules for two transformations, how can you find a symbolic rule for the composite transformation?

Be prepared to share your ideas with the class.

✔ Check Your Understanding

A figure is congruent to its image under the composition of two rigid transformations. The figure is similar to its image under the composition of a size transformation and a rigid transformation.

a. Refer to this coordinate grid. The scale on both axes is 1. For each pair of triangles given, determine if they are congruent or similar. In each case, describe in words a transformation or a composition of transformations that will map the first triangle onto the second.

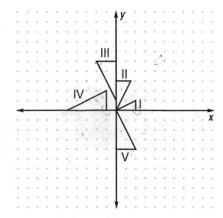

   i. △I and △IV

   ii. △II and △V

   iii. △III and △V

   iv. △V and △I

b. For each transformation you identified in Part a, write a symbolic rule that describes the transformation.

Applications

1 Copy each polygon below on a separate coordinate grid. Draw and label the transformed image according to the given rule. Identify as precisely as you can the type of transformation.

a. $(x, y) \rightarrow (x + 2, y - 3)$

b. $(x, y) \rightarrow (-y, x)$

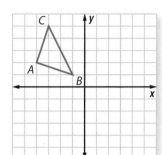
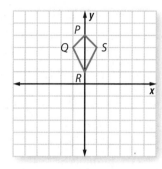

c. $(x, y) \rightarrow (-x, y)$

d. $(x, y) \rightarrow (-x, -y)$

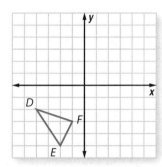
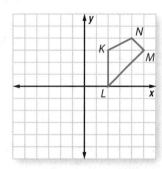

2 $\triangle ABC$ has vertices as follows: $A(1, 2)$, $B(4, 4)$, and $C(3, 6)$.

a. Draw $\triangle ABC$ on a coordinate grid. Then draw and label the image of $\triangle ABC$ under each of the following transformations.

   i. Translation with horizontal component 5 and vertical component -4

  ii. Horizontal translation 7 units to the left

 iii. Translation that maps the origin to the point $(-3, -6)$

b. Choose one of the image triangles in Part a and verify that it is congruent to $\triangle ABC$.

③ △*PQR* has vertices as follows: *P*(3, −2), *Q*(6, −1), and *R*(4, 3).

a. On separate coordinate grids, draw △*PQR* and its image under each of the following transformations. Label the vertices of the images.

   i. Rotation of 180° about the origin

   ii. Rotation of 90° counterclockwise about the origin

   iii. Rotation of 90° clockwise about the origin

b. Choose one of the image triangles in Part a and verify that it is congruent to △*PQR*.

④ Consider □*ABCD* = $\begin{bmatrix} 1 & 2 & 6 & 5 \\ -1 & 2 & 2 & -1 \end{bmatrix}$.

a. On separate coordinate grids, draw □*ABCD* and its image under each of the following transformations. Label the vertices of the images.

   i. Reflection across the *x*-axis

   ii. Reflection across the line *y* = *x*

   iii. Reflection across the *y*-axis

b. Choose one of the image quadrilaterals in Part a and verify that it is a parallelogram.

c. What is the perimeter of □*ABCD*? How do you know that each of the image parallelograms in Part a will have the same perimeter?

⑤ Consider △*PQR* = $\begin{bmatrix} -2 & 2 & 0 \\ -1 & 1 & 5 \end{bmatrix}$.

a. What kind of triangle is △*PQR*? How do you know?

b. What is the area of △*PQR*?

c. On separate coordinate grids, draw △*PQR* and its image under each of the following transformations. Label the vertices of the images.

   i. Translation that maps the origin to the point (−2, −2)

   ii. Counterclockwise rotation of 270° about the origin

   iii. Reflection across the line *y* = −*x*

d. What kind of triangle is each of the three image triangles in Part c? How do you know?

e. Find and compare the areas of the three image triangles in Part c.

6 A triangle translation program that implements the translation planning algorithm in Investigation 1, Problem 5 (page 199) is given below.

a. Analyze this program and explain the purpose of each command line not already described.

b. Enter the program in your calculator and test the program with a sample triangle.

TRANSL Program

Program	Function in Program
ClrHome	1. Clears the home screen.
Input "X1 COORD-PRE",A	2. Requests input for x-coordinate of one vertex. Stores the value in variable named A.
Input "Y1 COORD-PRE",B	3. _____
Input "X2 COORD-PRE",C	4. _____
Input "Y2 COORD-PRE",D	5. _____
Input "X3 COORD-PRE",E	6. _____
Input "Y3 COORD-PRE",F	7. _____
Input "X COMP-TRANS",H	8. _____
Input "Y COMP-TRANS",K	9. Requests input for the vertical component of the translation. Stores the value in variable K.
ClrHome	10. _____
Disp "PREIMAGE"	11. _____
Pause	12. Pause stops a program from continuing until [ENTER] is pressed.
ClrDraw	13. Clears all drawings.
Line(A,B,C,D)	14. Draws a line segment from vertex (A, B) to vertex (C, D).
Line(C,D,E,F)	15. _____
Line(E,F,A,B)	16. _____
Pause	17. _____
ClrHome	18. _____
Disp "IMAGE"	19. _____
Pause	20. _____
Line(A+H,B+K,C+H,D+K)	21. _____
Line(C+H,D+K,E+H,F+K)	22. _____
Line(E+H,F+K,A+H,B+K)	23. _____

7 Look back at the TRANSL program in Applications Task 6.

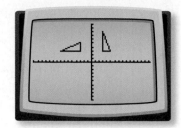

   a. How would you modify the program to display a triangle and its image under a 90° counterclockwise rotation about the origin?

   b. Enter the modified program, name it ROT90 in your calculator, and test the program with a sample triangle.

8 Copy each polygon below on a separate coordinate grid. Draw and label the transformed image according to the given rule. Identify as precisely as you can the type of transformation.

   a. $(x, y) \rightarrow (3x, 3y)$

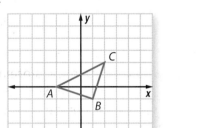

   b. $(x, y) \rightarrow (-x + 8, y)$

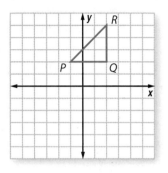

   c. $(x, y) \rightarrow \left(\frac{1}{2}x, \frac{1}{2}y\right)$

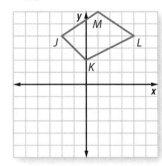

   d. $(x, y) \rightarrow (x, -y - 4)$

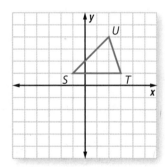

9 A picture is to be placed in a brochure, and the designer wants it positioned in a 2" × 3" frame. If the original picture is 6" × 9", what size transformation should be applied to the picture so it will fit the frame?

10 Consider quadrilateral $ABCD = \begin{bmatrix} 6 & -4 & -3 & 7 \\ 2 & 0 & -5 & -3 \end{bmatrix}$.

a. Draw quadrilateral *ABCD* on a coordinate grid.

b. Draw the image quadrilateral, *A'B'C'D'*, resulting from transforming *ABCD* with a size transformation of magnitude 2.5 and center at the origin.

c. How do $\overleftrightarrow{AB}$ and $\overleftrightarrow{BC}$ appear to be related? How do $\overleftrightarrow{A'B'}$ and $\overleftrightarrow{B'C'}$ appear to be related? Verify your conjectures using coordinates.

d. How do $\overleftrightarrow{AB}$ and $\overleftrightarrow{A'B'}$ appear to be related? How do $\overleftrightarrow{BC}$ and $\overleftrightarrow{B'C'}$ appear to be related? Verify your conjectures using coordinates.

e. Find the area of quadrilateral *ABCD*.

　　i. Predict the area of quadrilateral *A'B'C'D'*.

　　ii. Check your prediction.

f. Connect each preimage point and its image with a line. What is true about the lines?

11 Preimage and image pairs of a figure under certain transformations are shown below. The image figure is darker blue. In each case, identify as precisely as you can the type of transformation. Then write a coordinate rule for the transformation.

a.

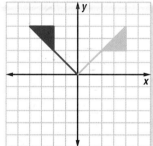

$(x, y) \rightarrow (\_\_, \_\_)$

b.

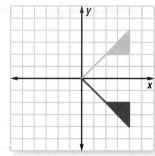

$(x, y) \rightarrow (\_\_, \_\_)$

c.

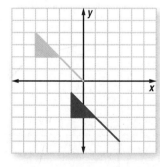

$(x, y) \rightarrow (\_\_, \_\_)$

d.

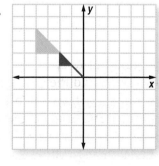

$(x, y) \rightarrow (\_\_, \_\_)$

12 Size transformations have their center at the origin. However, in producing graphics displays, it is sometimes useful to enlarge or reduce a figure using a center different from the origin.

a. Consider the following procedure for applying a size transformation to a figure when the center of the size transformation is $A(2, 1)$ and the magnitude is 3.

Draw $\triangle PQR$ on a coordinate grid. Plot point A.

$$\triangle PQR = \begin{bmatrix} 3 & 4 & 4 \\ 3 & 3 & 5 \end{bmatrix}$$

Step 1: Determine the horizontal and vertical components of the translation that will translate $A(2, 1)$ to the origin. Find the image of $\triangle PQR$ under that translation. Label as $\triangle 1$ the image of $\triangle PQR$.

Step 2: Apply a size transformation to $\triangle 1$ using the origin as center and 3 as the magnitude. Label the new image $\triangle 2$.

Step 3: Find the components of the translation that maps the origin back to $A(2, 1)$. Then find the image of $\triangle 2$ under that transformation. Label the final image $\triangle P'Q'R'$.

b. Examine $\triangle PQR$ and $\triangle P'Q'R'$. Does this procedure produce the desired result—that is, a size transformation of magnitude 3 with center $A(2, 1)$? Explain.

c. Write a symbolic rule $(x, y) \rightarrow (\_\_, \_\_)$ that describes this composite transformation.

13 Modify the "translate-transform-translate back" procedure outlined in Applications Task 12 to create a procedure for rotating $\triangle PQR$ 90° counterclockwise about the point $A(2, 1)$. Then write a symbolic rule $(x, y) \rightarrow (\_\_, \_\_)$ that describes this composite transformation.

14 For each of the following pairs of triangles, describe a composition of two or more transformations that will map the first triangle to the second triangle. Write a symbolic rule $(x, y) \rightarrow (\_\_, \_\_)$ that describes the composite transformation.

a. $\triangle$I onto $\triangle$II

b. $\triangle$I onto $\triangle$III

c. $\triangle$II onto $\triangle$IV

d. $\triangle$I onto $\triangle$IV

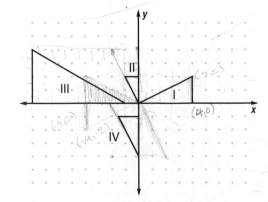

Connections

15 Suppose a size transformation with magnitude k and center at the origin maps points $A(x_1, y_1)$ and $B(x_2, y_2)$ onto points A' and B', respectively.

   a. What are the coordinates of points A' and B'?

   b. Ivan provided the following general argument to show that distance $A'B' = k \cdot$ (distance AB).

   Check the correctness of Ivan's work by giving a reason justifying each step. Correct any errors that you find.

$$\text{distance } A'B' = \sqrt{(kx_1 - kx_2)^2 + (ky_1 - ky_2)^2} \quad (1)$$

$$\text{distance } A'B' = \sqrt{k^2(x_1 - x_2)^2 + k^2(y_1 - y_2)^2} \quad (2)$$

$$\text{distance } A'B' = k\sqrt{(x_1 - x_2)^2 + (y_1 - y_2)^2} \quad (3)$$

$$\text{distance } A'B' = k \cdot (\text{distance } AB) \quad (4)$$

   c. Use the definition of slope and reasoning with the general coordinates of points A, B, A', and B' to show that $\overleftrightarrow{A'B'}$ is parallel to $\overleftrightarrow{AB}$.

   d. State in words the property of size transformations you justified in Part c.

16 In Investigation 1, you developed a rule for a reflection across the y-axis. You can use a "translate-transform-translate back" method (see Applications Tasks 12 and 13) to develop rules for reflections across other vertical lines.

   a. Draw the vertical line $x = 5$ on a coordinate grid. To find the coordinates of the reflected image of point $A(a, b)$ across this line:

   i. translate the line $x = 5$ so that it coincides with the y-axis. What are the horizontal and vertical components of that translation? What are the coordinates of A_1, the image of $A(a, b)$ under the translation?

   ii. What are the coordinates of the image of A_1 when reflected across the y-axis? Label the image point A_2.

   iii. What are the components of the translation that maps the y-axis (back) to the line $x = 5$? What are the coordinates of A', the image of A_2 under that translation?

   iv. Write a symbolic rule $(x, y) \rightarrow (\_\_, \_\_)$ that describes the composite transformation.

   b. The rule $(x, y) \rightarrow (-x + 2h, y)$ gives the coordinates of the reflected image of any point (x, y) across the vertical line $x = h$.

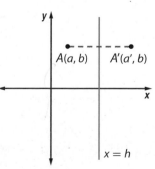

   i. Verify the rule for the point $P(2, -6)$ reflected across the line $x = -3$.

   ii. Use the translate-transform-translate back connection to show that the general rule is correct.

17 In Investigation 1, you developed a rule for a reflection across the x-axis. Symbolic rules can also be developed for reflections across other horizontal lines. Use reasoning similar to that in Connections Task 16 to write a rule that gives the coordinates of the image of any point (x, y) when reflected across the horizontal line y = k.

18 Draw and label a triangle or quadrilateral of your choice. Reflect the figure across the x-axis and then reflect the image across the line y = -6.

a. Compare the size, shape, and orientation of the preimage and image under the composite transformation.

b. What kind of transformation does the composite transformation appear to be? Write a symbolic rule (x, y) → (__, __) that describes the composite transformation.

c. Does the order in which you apply the two line reflections lead to the same final image?

19 Investigate the effects of composing a reflection across a line with a translation in a direction parallel to the line.

a. On a coordinate grid, draw a triangle △ABC near the x-axis to represent a duck foot. Record the coordinates of its vertices.

b. Reflect △ABC across the x-axis, then translate the image horizontally 8 units. Label the final image △A′B′C′.

c. How are the coordinates of △A′B′C′ related to those of △ABC? Write a coordinate rule for this composite transformation.

d. Apply the same combination of the two transformations to △ABC but in the opposite order. Does the order in which you apply the translation and reflection matter?

e. Now apply the coordinate rule you gave in Part c at least three more times to △A′B′C′. Describe how alternate images such as images one and three or two and four are related.

f. The combination of a reflection across a line and a translation in a direction parallel to the line is called a **glide reflection**.

   i. Start with a new triangle. Then apply a glide reflection in which the reflection line is the y-axis.

   ii. Write a coordinate rule for this glide reflection.

20 The order in which two transformations are composed is often important.

 a. In Investigation 3, you found that the order in which you applied a size transformation and a rotation about the origin made no difference. Use the symbolic representations for a size transformation $(x, y) \rightarrow (kx, ky)$ and a 90° counterclockwise rotation about the origin $(x, y) \rightarrow (-y, x)$ to justify this observation.

 b. Order makes a difference in composing size transformations and translations. Use symbolic representations of these transformations to explain why and how the positions of the images differ.

 c. Order makes a difference in composing two different line reflections. Use symbolic representations of a reflection across the x-axis and a reflection across the line $y = x$ to explain why and how the positions of the images differ.

21 Refer to the table you completed for Connections Task 18 in Lesson 1 (page 188). Extend that table to include the coordinate models of rigid transformations you developed in this lesson.

Geometric Transformation	Coordinate Model	Example
Translation	$(x, y) \rightarrow (x + h, y + k)$	
Reflection across y-axis		
Reflection across line $x = h$	See Connections Task 16	
Reflection across x-axis		
Reflection across line $y = k$	See Connections Task 17	
Reflection across line $y = x$		
Reflection across line $y = -x$		
90° counterclockwise rotation about origin		
180° rotation about origin		
270° counterclockwise rotation about origin		
Size transformation magnitude k and center at the origin		
Similarity transformation		

22 If you reflect a point $P(x, y)$ across the y-axis, its image is $P'(-x, y)$. If you then reflect the image point $P'(-x, y)$ across the y-axis, the final image is $P(x, y)$, your original point.

  a. For each rigid transformation (translation, line reflection, rotation, and glide reflection (see Connections Task 19)), describe a transformation with which it can be composed so that for any point, the preimage and the final image are the same.

  b. For a size transformation with magnitude k and center at the origin, describe a transformation with which it can be composed so that for any point, the preimage and the final image are the same.

  c. How is your work in Parts a and b related to the concept of *multiplicative inverses* for real numbers and for matrices?

Reflections

23 What clockwise rotation will be the same as a 270° counterclockwise rotation? Why? What clockwise rotation will be the same as a 90° counterclockwise rotation? Why?

24 In Investigations 1 and 2, you and your classmates developed symbolic rules for describing various transformations of a coordinate plane. If you forget one of these rules, how would you go about reconstructing it? Illustrate with a rule for a specific transformation.

25 How is the **Zoom** feature on your graphing calculator or computer software like a size transformation?

  a. What determines the center?

  b. What determines the magnitude?

26 The coordinate models of size transformations that you investigated had their centers at the origin of a coordinate plane. Think about how you could enlarge a figure *without* using a coordinate grid. Draw a triangle PQR on a sheet of plain paper. Mark a point C on the paper. How could you use a ruler to find the image of $\triangle PQR$ under a size transformation with the given point as center and magnitude 3? Write an explanation of your method that could be used by a classmate.

27 For all the rigid transformations and size transformations you examined in Lesson 2, the image of a line is a line. For some of the transformations, the image of a line is *always* parallel to the preimage line. For which transformations is the image of a line parallel to the preimage? Verify your choices by showing that the image of the line containing points $(-1, 3)$ and $(2, 5)$ is parallel to the preimage line for each transformation.

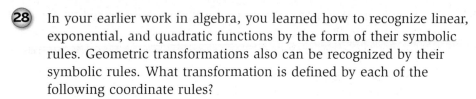

28 In your earlier work in algebra, you learned how to recognize linear, exponential, and quadratic functions by the form of their symbolic rules. Geometric transformations also can be recognized by their symbolic rules. What transformation is defined by each of the following coordinate rules?

a. $(x, y) \rightarrow (5y, 5x)$

b. $(x, y) \rightarrow \left(-\frac{1}{2}x, -\frac{1}{2}y\right)$

c. $(x, y) \rightarrow (4x - 12, 4y + 8)$

Extensions

29 A line ℓ contains points $A(a, b)$ and $B(c, d)$. Use these general coordinates to justify each of the following statements:

a. Under a 180° rotation about the origin, the image of a line is a line parallel to the preimage line.

b. Under a translation with components h and k, the image of a line is a line parallel to the preimage line.

30 In Investigation 3, Problem 5 (page 213), you found that the composition of reflections across two intersecting lines was a rotation about their point of intersection. In Connections Task 18 (page 224), you found that the composition of reflections across two parallel lines was a translation.

a. Use interactive geometry software to explore the composition of reflections across any two intersecting lines. Is the composite transformation always a rotation? If so, what is the center of the rotation? How are the direction and amount of rotation related to the two intersecting lines?

b. Use interactive geometry software to explore the composition of reflections across any two parallel lines. Is the composite transformation always a translation? If so, how are the direction and magnitude related to the two parallel lines?

31 This task will provide you additional experiences using the programming window of your interactive geometry software and in analyzing programs for the "Design by Robot" custom tool.

a. Predict the shape created by the command:

fd 3 rt 90 fd 4 rt 143 fd 5

Check your prediction.

b. The **fd** and **rt** instructions translate and rotate the position of the robot.

i. For each translation instruction in Part a, describe the magnitude and direction of the translation.

ii. For each rotation instruction in Part a, describe the center and angle of rotation (clockwise/counterclockwise).

c. How is the idea of composition of transformations used in the command in Part a?

d. Enter the *procedure* below in your software programming window.

```
program triangle
parameter [k]
fd 3*k rt 90
fd 4*k rt 143
fd 5*k
home
end
```

    i. Explore what happens when you type each of these commands in the command window (after closing the pop-up window).

- triangle 1
- triangle 2
- triangle 4

    ii. How does the "triangle" procedure use the idea of transformation of shape?

e. Predict the figure created by the following program. Then check your prediction. Explain any differences between your prediction and the figure created.

```
cs home
repeat 5 [repeat 4 [fd 10 rt 90] rt 72]
```

32 The word **ATTENTION** is to be illustrated in a space that is 10" high by 30" wide. The letters are to take up the whole space. The font available produces a word 10" high by 24" wide. Find a transformation that will scale the letters in a horizontal direction to fill the space.

33 Investigate the effects of the transformation described by the following rule.

$$(x, y) \rightarrow (2x, 3y)$$

a. Describe the image of square $ABCD$ with vertices $A(0, 0)$, $B(0, 2)$, $C(2, 2)$, $D(2, 0)$ under this transformation.

b. Are any points in the coordinate plane their own images?

c. Consider points on a line. Are their images also on a line?

d. Consider midpoints of segments. Are their images the midpoints of the image segments?

e. What is the effect of this transformation on the length of line segments?

f. What is the effect of this transformation on areas?

34 Investigate the effects of the transformation described below.

$$(x, y) \rightarrow (x + y, x)$$

   a. Describe the image of square $ABCD$ with vertices $A(0, 0)$, $B(0, 2)$, $C(2, 2)$, $D(2, 0)$ under this transformation.

   b. Are any points in the coordinate plane their own images?

   c. When you transform points on a line, are the images also on a line?

   d. When you transform midpoints of segments, are the images also midpoints of the image segments?

   e. What is the effect of this transformation on the length of line segments?

   f. What is the effect of this transformation on areas?

35 Look back at your work in Applications Tasks 12 and 13 with an eye to generalizing the results. Use the translate-transform-translate back method to develop symbolic rules $(x, y) \rightarrow (\_\_, \_\_)$ for each of the following transformations:

   a. Size transformation with center $C(a, b)$ and magnitude k

   b. 90° counterclockwise rotation about the point $C(a, b)$

   c. **Half-turn** (180° rotation) about the point $C(a, b)$

Review

36 Ava put $1,500 into a bank account that earns 5% interest compounded yearly. She does not take any money out of the account.

   a. How much money will Ava have after 5 years?

   b. How long will it take for the balance in the account to reach $3,000?

37 Find the values of x and y for each matrix equation.

   a. $\begin{bmatrix} x & 3 \\ 4 & 5 \end{bmatrix}\begin{bmatrix} 6 & -2 \\ 0 & y \end{bmatrix} = \begin{bmatrix} -12 & 16 \\ 24 & 12 \end{bmatrix}$

   b. $\begin{bmatrix} 4 & y \\ x & 1 \end{bmatrix}\begin{bmatrix} 3 & -2 \\ 10 & -5 \end{bmatrix} = \begin{bmatrix} 42 & -23 \\ 31 & -19 \end{bmatrix}$

38 Rewrite each equation so that y is expressed as a function of x.

   a. $4x + 7y = 10$

   b. $3(2x + y) = 2$

   c. $\dfrac{12y - 6x}{4} = 20$

   d. $\dfrac{x}{6} + 2y = 32$

39 The measure of ∠ABC is 30°. Find the measures of each indicated angle.

a. ∠CBD

b. ∠DBA

c. ∠EBF

d. ∠CBE

e. ∠FBA

f. ∠FBD

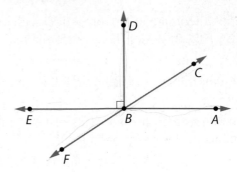

40 Use properties of numbers and mental computation to evaluate the following.

a. 4(360) + 4(140)

b. 4 · 18 · 25

c. 25 + 569 + 75

d. $(25^2)(4^2)$

41 A jar contains red, blue, and green marbles. There are 3 red marbles, 5 blue marbles, and 8 green marbles in the jar.

a. Lucy places 10 yellow marbles in the jar. For each color, find the probability of drawing a marble of that color from the jar.

b. How many more yellow marbles should Omar add to the jar if he wants the probability of drawing a yellow marble to be 0.60?

42 For each of the isosceles right triangles shown below, find the lengths of the remaining sides and the measures of the remaining angles.

a.

b.

43 Each day on her way to school, Madison stops and picks up her friend Clara. For the last two weeks, Madison has kept track of how long it takes her to get to school each day. The times, in minutes, are listed below:

15, 12, 11, 16, 10, 18, 16, 12, 14, 14

a. Find the mean and median times that it takes Madison to pick up Clara and get to school.

b. Madison kept track of the times for another week and computed the mean and median times for all 15 days. The median time for the three weeks was the same as the median time for the first two weeks, but the new mean was greater. Find five possible times that would give these results. Explain the reasoning that you used to find your times.

Transformations, Matrices, and Animation

In the first two lessons of this unit, you used coordinates to model geometric shapes such as polygons and circles, and to investigate geometric relationships such as perpendicularity. Coordinates provided ways to describe and reason about familiar and not-so-familiar ideas numerically and algebraically. You also explored how coordinate representations can be used to transform shapes in computer and calculator graphics displays. In this lesson, you will investigate how graphics displays can be linked to create animated effects.

Consider the following sequence of images in a mock animation of a space shuttle. These images were created using a two-dimensional photo image of the shuttle superimposed on an aerial photo of Earth.

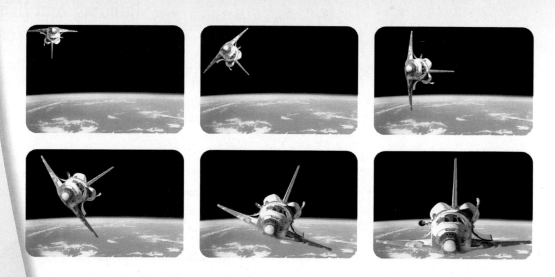

Examine how the sequence of images changes from frame to frame.

a Where do you think the origin of a coordinate system was placed in creating this animation?

b What point(s) on the shuttle image would you use in determining how each image was transformed?

c Describe the types of transformations that appear to have been used in creating the animation.

d Computer animations are frequently used in movies and video games. Are there other applications of computer animation with which you are familiar?

In the investigations of this lesson, you will learn how to use matrices to perform transformations of two-dimensional shapes and create simple animations. The tools that you develop have straightforward extensions to work in three dimensions and the methods that are typically used in computer animation.

Investigation 1 — Building and Using Rotation Matrices

For the purpose of this lesson, you can simplify the space shuttle animation by representing the space shuttle and sequence of images with two-dimensional figures similar to the ones shown below. Such simple representations are used when a "storyboard," or outline, of an animation is developed.

As a class, study the animation created by the interactive geometry custom tool "Animate Shuttle." In that animation, the space shuttle performs a rollover maneuver as if in preparation for re-entry.

As you work on the problems of this investigation, look for answers to the following questions:

How can a rotation with center at the origin be represented by a matrix?

How can rotation matrices be used to animate the rotation of two-dimensional shapes?

1 One possible coordinate model of the shuttle is shown below.

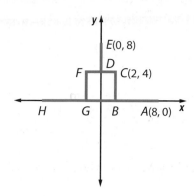

a. What are the coordinates for points *F* and *H*? For points *B* and *G*?

b. Find the coordinates of the image of the shuttle model when rotated 180° about the origin.

c. Write a symbolic rule $(x, y) \rightarrow$ (__, __) that gives the coordinates of the image of any point $P(x, y)$ under a 180° rotation.

d. What is a symbolic rule that gives the coordinates of the image of any point $P(x, y)$ under a 90° counterclockwise rotation about the origin?

e. How would you modify your rule in Part d so that it describes a 90° *clockwise* rotation about the origin?

2 Matrix multiplication can be used to express each of the rotations in Problem 1. To do this, coordinates of points (x, y) need to be represented as one-column matrices, $\begin{bmatrix} x \\ y \end{bmatrix}$. For example, the one-column or **point matrix** for $(-2, 4)$ is $\begin{bmatrix} -2 \\ 4 \end{bmatrix}$.

a. Look back at the symbolic rule for a 180° rotation that you found in Problem 1 Part c. To build a 2 × 2 matrix representation for the 180° rotation, find numbers a, b, c, and d that make this matrix equation true.

180° Rotation Matrix		General Point Matrix		Image Point Matrix
$\begin{bmatrix} a & b \\ c & d \end{bmatrix}$	×	$\begin{bmatrix} x \\ y \end{bmatrix}$	=	$\begin{bmatrix} -x \\ -y \end{bmatrix}$

i. Test your rotation matrix by using it to find the rotation images of points $A(8, 0)$ and $C(2, 4)$. Compare your image points with those found in Problem 1 Part b.

ii. Why is the general point matrix placed to the right of the rotation matrix?

b. Determine the matrix for a 90° counterclockwise rotation about the origin.

$$\begin{bmatrix} a & b \\ c & d \end{bmatrix} \begin{bmatrix} x \\ y \end{bmatrix} = \begin{bmatrix} -y \\ x \end{bmatrix}$$

   i. Check your rotation matrix by using it to find the images of points $B(2, 0)$ and $C(2, 4)$ in Problem 1.

   ii. Do these image points make sense?

c. Geometrically, you know that a 90° counterclockwise rotation followed by another 90° counterclockwise rotation gives a 180° rotation. See if multiplying the matrix for the 90° counterclockwise rotation by itself yields the matrix for the 180° rotation.

d. What do you notice about the entries of the matrices used to express these rotations?

3 One advantage of a matrix representation of a transformation is that you can use it quickly to transform an entire shape. Consider

$$\triangle AEH = \begin{bmatrix} 8 & 0 & -8 \\ 0 & 8 & 0 \end{bmatrix}$$ determined by the tips of the shuttle model.

a. Multiply the matrix representation of $\triangle AEH$ by the 90° counterclockwise rotation matrix. Using the coordinate rule for the 90° counterclockwise rotation, verify that the result of your calculation is the image triangle, $\triangle A'E'H'$.

b. When transforming an n-sided polygon using matrices, why should the coordinate matrix of the polygon be the factor on the right?

4 Designing animations often requires use of rotations through many different angles, in addition to those that are multiples of 90°. When building matrix representations for rotations and other transformations, it is very useful to know what happens to the points $(1, 0)$ and $(0, 1)$. Diagram I below shows the images of points $P(1, 0)$ and $Q(0, 1)$ under a 45° counterclockwise rotation about the origin.

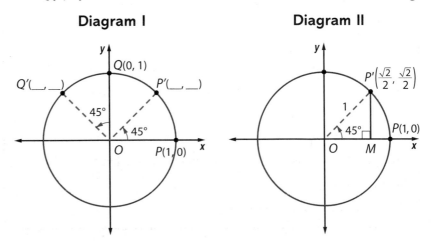

a. Explain why the image of point P and the image of point Q will be on a circle of radius 1 with center at the origin.

b. Using Diagram II, explain as precisely as you can why the image of point P under the 45° rotation has coordinates $\left(\frac{\sqrt{2}}{2}, \frac{\sqrt{2}}{2}\right)$.

c. Find the coordinates of point Q'.

d. Now find the entries of the 45° counterclockwise rotation matrix $R_{45°} = \begin{bmatrix} a & b \\ c & d \end{bmatrix}$ by solving the two matrix equations below. Begin by entering the coordinates of point P' and point Q' in the appropriate column matrices.

  i. $\begin{bmatrix} a & b \\ c & d \end{bmatrix} \begin{bmatrix} 1 \\ 0 \end{bmatrix} = \begin{bmatrix} \underline{} \\ \underline{} \end{bmatrix}$

  ii. $\begin{bmatrix} a & b \\ c & d \end{bmatrix} \begin{bmatrix} 0 \\ 1 \end{bmatrix} = \begin{bmatrix} \underline{} \\ \underline{} \end{bmatrix}$

  iii. So, $R_{45°} = \begin{bmatrix} \underline{} & \underline{} \\ \underline{} & \underline{} \end{bmatrix}$.

e. Check that multiplying the 45° counterclockwise rotation matrix by itself (with entries expressed in radical form) gives the matrix for a 90° counterclockwise rotation about the origin that you found in Problem 2 Part b.

5 Look back at the entries for the 45° counterclockwise rotation matrix $R_{45°}$ and how they were calculated.

a. How are the entries of matrix $R_{45°}$ related to the rotation images of $P(1, 0)$ and $Q(0, 1)$?

b. Does the pattern hold for the 180° and 90° rotation matrices you found in Problem 2? Explain.

6 A computer or calculator program can be written that will rotate the space shuttle model counterclockwise about the origin using steps of 45°. Study the Roll Over Algorithm given below.

Roll Over Algorithm

Step 1. Set up the coordinate matrix representing the space shuttle.

Step 2. Set up the 45° counterclockwise rotation matrix.

Step 3. Draw the shuttle.

Step 4. Compute and store the coordinates of the shuttle rotated 45°.

Step 5. Clear the old shuttle and draw the rotated image.

Step 6. Pause.

Step 7. Repeat Steps 4–6 as needed.

---

a. Identify the input, processing, and output parts of the Roll Over Algorithm.

b. Step 7 is a *control* command. It controls the action of the algorithm. To make the shuttle rotate all the way around once, how many times should Steps 4–6 be performed?

(7) The roll over portion of the animation can be created using commands such as those below. Note how assigning names to the shuttle coordinate matrix and the rotation matrix simplifies the programming.

Roll Over Program

```
let shuttle = [[8,0][2,0][2,4][0,4][0,8][0,4][–2,4][–2,0][–8,0][8,0]]
let rotmatrix = [[0.7071,0.7071][–0.7071,0.7071]]
draw shuttle
repeat 8 [draw [let shuttle = [rotmatrix*shuttle]] pause 500]
```

a. Discuss with your classmates how the commands in this program match corresponding steps in the Roll Over Algorithm.

b. Test the program by entering it in the Command window of your interactive geometry software.

c. Predict the animation that will be produced by replacing the last *three* lines of the program by these *two* lines:

```
draw shuttle
repeat 8 [draw [let shuttle = [rotate shuttle 45]] pause 500]
```

Run the program to test your prediction. Make notes of any misunderstandings of programming commands.

Summarize the Mathematics

In this investigation, you explored how to find matrix representations for certain rotations and how matrices can be used to create an animation.

a Explain how to use the coordinate rule for a 270° counterclockwise rotation about the origin to find the matrix representation for that rotation. Find the matrix.

b How would you modify the Roll Over program so that it will rotate the space shuttle *clockwise* about the origin in steps of 45°?

c Describe a systematic way of determining the entries of a rotation matrix.

d The matrix $\begin{bmatrix} 0.5 & 0.866 \\ -0.866 & 0.5 \end{bmatrix}$ is the matrix for a 60° clockwise rotation about the origin. Explain how to use the matrix to find a coordinate rule for the image of a point (x, y) under this rotation.

Be prepared to share your ideas and reasoning with the class.

✓ Check Your Understanding

Build a matrix that represents a 135° counterclockwise rotation about the origin.

a. Use the matrix to find the rotation image of the point $(-1, 5)$.

b. Use the matrix to find the rotation image of $\triangle HJK = \begin{bmatrix} -1 & 4 & 3 \\ 2 & -3 & 5 \end{bmatrix}$.

c. Sketch $\triangle HJK$ and its rotation image on a coordinate grid.

Investigation 2 · Building and Using Size Transformation Matrices

Consider again the sequence of images in the mock animation of a space shuttle.

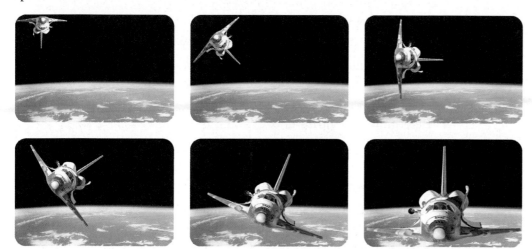

Note how the shuttle image increases in size from frame to frame as it rolls over and moves forward. This simulation was accomplished by using size transformations, translations, and rotations. In this investigation, all size transformations will be centered at the origin.

As you work on problems in this investigation, look for answers to the following questions:

How can size transformations be represented using matrices?

How can you animate size change and translation of two-dimensional shapes?

How can size transformations and translations be combined with rotations to create more complex animations?

1 Begin by examining the space shuttle model shown below.

   a. What is the coordinate matrix for a similar shuttle model (in the same position) whose sides are twice the length of those in the given model? Half the length?

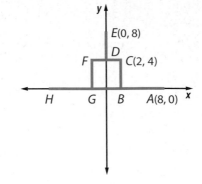

   b. Write two coordinate rules $(x, y) \rightarrow$ (__, __) that would resize the shuttle model as described in Part a.

   c. What is the coordinate matrix for the image of the given shuttle model when translated:

   i. 5 units to the right?

   ii. 3 units down?

   iii. 5 units to the right and 3 units down?

   d. Write coordinate rules for these three transformations in the form $(x, y) \rightarrow$ (__, __).

2 Matrix representations for size transformations with center at the origin can be found using the same method you used to find rotation matrices.

   a. Determine the entries a, b, c, and d of the matrix for a size transformation with magnitude 2.

$$\begin{bmatrix} a & b \\ c & d \end{bmatrix}\begin{bmatrix} x \\ y \end{bmatrix} = \begin{bmatrix} 2x \\ 2y \end{bmatrix}$$

   b. What should be the image of the point $F(-2, 4)$ under this transformation? Multiply the transformation matrix by the one-column matrix for point F and check to see if you get the correct image point.

   c. Multiply the size transformation matrix you found in Part a by the matrix for $\triangle AEH$, where A, E, and H are the wing tips of the shuttle model.

   i. Compare the coordinates of the image $\triangle A'E'H'$ with those found using the appropriate coordinate rule for a size transformation with magnitude 2.

   ii. Compare the lengths of $\overline{EH}$ and $\overline{E'H'}$. Why does that relationship make sense?

   d. Find the matrix for a size transformation with magnitude $\frac{1}{2}$. With magnitude 5.

3 How could you use the idea of multiplying a matrix by a real number to find the image of a point or a polygon under a size transformation of magnitude 3 with center at the origin? Of magnitude $\frac{1}{4}$? Compare methods with others and resolve any differences.

4 Use the Roll Over Algorithm on page 235 and the following questions to help you develop an algorithm for a program that will repeatedly scale the space shuttle model by a factor of 1.5 using a size transformation with center at the origin.

- What information would you need to input?

- What processing would the program need to complete?

- What information should the calculator or computer output?

Resizing Algorithm

Step 1. Set up the coordinate matrix representing the shuttle. (input)

Step 2. _____ (_____)

Step 3. _____ (_____)

⋮

---

5 The animation described in Problem 4 can be created using commands such as:

Resizing Program
```
let shuttle = [[8,0][2,0][2,4][0,4][0,8][0,4][-2,4][-2,0][-8,0][8,0]]
let sizematrix = [[1.5,0][0,1.5]]
draw shuttle
repeat 4 [draw [let shuttle = [sizematrix*shuttle]] pause 500]
```

a. Test the program by entering it in the Command window of your software.

b. Suppose the last *three* lines of the program were replaced by these *two* lines:

```
draw shuttle
repeat 4 [draw [let shuttle = [scale shuttle 1.5]] pause 500]
```

What animation do you think will be produced by the modified program? Check your conjecture by running the modified program.

c. Write a series of commands that could be used to create an animation that repeatedly scales the shuttle by a factor of 4 with center at the origin.

d. Test your program by entering it in the Command window. Revise commands as necessary.

6 Unlike in the cases of rotations and size transformations, translations cannot be represented with 2 × 2 matrices. (See Extensions Task 18.) Instead of using matrix multiplication, the coordinate rule form is frequently used to describe translations. For example, you can translate the shuttle and display the image with the commands shown below.

```
let shuttle = [[8,0][2,0][2,4][0,4][0,8][0,4][-2,4][-2,0][-8,0][8,0]]
let shuttle = [translate shuttle [5,-3]]
draw shuttle
```

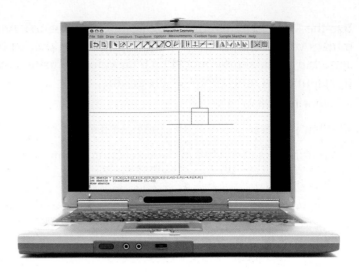

a. Write a series of commands that could be used to create an animation that repeatedly translates the shuttle 2 units to the right and 3 units up.

b. Test your program. Revise it as necessary.

In the following problems, you will explore ways in which transformations can be combined to create more complex animations.

7 Suppose you want to create an animation that starts with the shuttle wingspan repositioned so that its center is at (−4, 5) and is rescaled to 75% the original size.

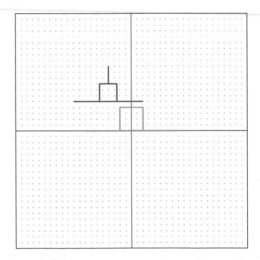

a. Describe the horizontal and vertical components of the translation and scale factor of the size transformation that will produce this image when drawn.

b. Will the order in which you choose to apply the translation and size transformation matter? Explain.

c. Write a series of programming commands that could be used to create this initial image of the shuttle.

d. Test your program and revise it as necessary.

8 By repeatedly combining a size transformation, rotation, and a translation, you can create an animation where the shuttle model appears to move from being far away to flying by the observer.

   a. Study the two algorithms for gradually growing the size of the shuttle model.

<table>
<tr><td>

Grow Algorithm 1

Step 1. Set up the coordinate matrix representing the shuttle.

Step 2. Set up a 0.25 scale factor.

Step 3. Compute the image of the original shuttle under the size transformation.

Step 4. Draw the image.

Step 5. Increase the scale factor by 0.05.

Step 6. Pause.

Step 7. Repeat Steps 3–6 as needed.

</td><td>

Grow Algorithm 2

Step 1. Set up the coordinate matrix representing the shuttle.

Step 2. Set up a 0.25 scale factor.

Step 3. Compute the image of the original shuttle under the size transformation.

Step 4. Draw the image.

Step 5. Multiply the scale factor by 1.2.

Step 6. Pause.

Step 7. Repeat Steps 3–6 as needed.

</td></tr>
</table>

   i. How does the size of the shuttle grow for each algorithm?

   ii. For each algorithm, how many times must the steps be repeated before the shuttle image reaches its original size?

   b. Suppose to start an animation sequence, the center of the shuttle wingspan needs to be translated to $(-20, 20)$ and the shuttle scaled to 25%. In which order should you perform the translation and size transformation?

   c. Study the program below that creates an animation of the space shuttle similar to that produced by the "Animate Shuttle" custom tool. Describe as precisely as you can the effect of each line.

Shuttle Animation Program

```
 1. pgm animateShuttle
 2. gridstyle grid off axes off
 3. let shuttle=[shape [[10,0][2.5,0][2.5,5][0,5][0,10][0,5][–2.5,5][–2.5,0][–10,0][10,0]]]
 4. style shuttle filled on fillcolor 0 255 0 visible off label off
 5. let currentCenter=[–18,18]
 6. let currentAngle=180
 7. let currentScale=0.2
 8. let shuttleImage=[translate [scale [rotate shuttle currentAngle] currentScale] currentCenter]
 9. style shuttleImage filled on fillcolor 200 0 200 label off visible on
10. pause 500
11. repeat 18 [let currentAngle=currentAngle+10 let currentScale=currentScale+0.05
    let currentCenter=currentCenter+[1,–1] let shuttleImage=[translate [scale [rotate shuttle currentAngle]
    currentScale] currentCenter] draw shuttleImage pause 100]
12. clear currentCenter currentAngle currentScale shuttleImage
13. gridstyle grid on axes on
14. draw shuttle
15. end
```

   d. Compare your description of the program with the execution of the "Animate Shuttle" custom tool.

Summarize
the Mathematics

In this investigation, you extended your work with matrices and animation to include size transformations with center at the origin which were then combined with translations.

a What are two different ways of representing a size transformation with magnitude *k* using matrices? Why would you choose one form over the other?

b Explain how each form in Part a can be used to find the image of a point and the image of a polygon under a size transformation.

c How could you use general coordinate rules, matrices, or drawing on a coordinate grid to show that order is important when combining size transformations and translations?

Be prepared to share your ideas and reasoning with the entire class.

✔ Check Your Understanding

Build a coordinate matrix for flag *PQRS* shown below. Then do the following:

a. Write the matrix for a size transformation of magnitude 3 centered at the origin.

b. Use the matrix to find the image of flag *PQRS* under the transformation.

c. Sketch the original flag and its image on a coordinate grid.

d. Write a series of commands for an animation program that:

- begins with the flag as shown above.

- uses composition of size transformations of magnitude 1.1 to show the flag growing and moving away from the origin until its pole is beyond (20, 0).

e. Test your program and revise it as necessary.

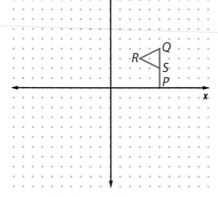

Applications

1 In Investigation 1, you determined the matrix representations for counterclockwise rotations of 90° and 45° about the origin. You also determined the matrix for a 180° rotation about the origin.

a. Find the entries of the matrix for a 90° *clockwise* rotation about the origin.

b. Write the matrix for a 45° *clockwise* rotation about the origin. Explain how its entries can be found by using Diagram II on page 234 and symmetry of the circle.

c. What matrix should you get if you multiply the matrix in Part b by itself? Verify that is the case.

d. Describe two ways to find the entries of the matrix for a 135° *clockwise* rotation about the origin. Use one of the ways to find the matrix.

2 Computer animations are used to design and test choreographing movements such as the motions of a flag typically used by Color Guards or the motion of the people themselves. Consider the two basic flag motions of a twirl and a flip. In a twirl, the entire flag is rotated 360°. In a flip, the handle of the flag is turned so that a flag pointing left then points right and vice-versa.

a. What type of transformation can be used to create the flip effect in a coordinate plane?

b. Consider a reflection across the y-axis. What is the image of $A(2, 3)$ under this reflection? The image of $B(3, -4)$?

c. Write a coordinate rule for reflection across the y-axis.

d. Use the coordinate rule to find the matrix representation of a reflection across the y-axis.

e. Write a Flag Animation Algorithm that meets the following specifications.

- Begin with the vertical flag *PQRS* with coordinate
  matrix $\begin{bmatrix} 0 & 0 & -3 & 0 \\ 0 & 8 & 6 & 4 \end{bmatrix}$.

- Using matrix multiplication, twirl the banner counterclockwise twice about the origin, showing progressive images rotated 45°.

- Flip the flag using matrix multiplication and then twirl the banner *clockwise* twice, again showing progressive images rotated 45°.

f. What are the coordinates of the final flag image after completion of the animation outlined in Part e?

3. In addition to reflection across the y-axis as investigated in Applications Task 2, two other important line reflections are reflection across the x-axis and reflection across the line $y = x$.

   a. Find the matrix representation of a reflection across the x-axis.

   b. Find the matrix representation of a reflection across the line $y = x$.

   c. Use these two matrix representations to model composition of the two line reflections, first reflecting over the x-axis. Write a matrix representation for the composite transformation.

   d. What special transformation is represented by the matrix in Part c? Explain your reasoning.

4. Transformations are used by graphic design artists to produce letters of different sizes and orientation and to position them in a layout.

   a. For each of your first and last initials, identify key points on a coordinate grid that define the shape and size of the letters. Then give a matrix that could be used to draw each letter.

   b. Placement of shapes in relation to the origin is a very useful starting point when creating graphics.

      i. Describe how you used the origin to determine the coordinates of your initials.

      ii. Imagining both letters anchored at the origin or both appearing next to each other would lead to different coordinate representations. Why might a graphic designer prefer one placement over the other?

   c. Determine the matrices for your initials rotated clockwise 90° about the origin.

   d. *Optional:* Design and test a program that will animate your initials across the screen.

5. Consider rectangle $PQRS = \begin{bmatrix} 1 & 1 & 6 & 6 \\ 2 & 5 & 5 & 2 \end{bmatrix}$.

   a. Illustrate two ways of using matrices to find the image of the rectangle under a size transformation with center at the origin and magnitude 5.

   b. Find the coordinate matrix for the image of rectangle $PQRS$ under the composition of first a rotation of 180° about the origin and then a size transformation of magnitude 3.

6 Build a coordinate matrix for a model of a rocket similar to the one shown at the right. Then do the following:

a. Write an algorithm for animating the launch of the rocket.

b. Explain your methods for performing translations and size transformations on the rocket in your algorithm.

c. Give the coordinate matrix for the rocket half way through the animation.

Connections

7 In Investigation 1, you developed matrix representations for counterclockwise rotations about the origin through angles of 45°, 90°, 180°, and 270°. In this task, you will build the matrix for a 60° counterclockwise rotation about the origin.

a. Use the diagram below of an equilateral triangle to help you determine the coordinates of points $P(1, 0)$ and $Q(0, 1)$ under a 60° counterclockwise rotation about the origin.

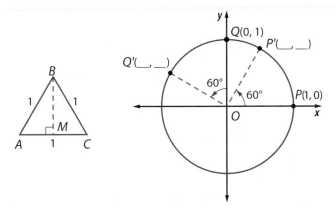

b. Use your results in Part a to help determine the entries for the 60° counterclockwise rotation matrix $R_{60°} = \begin{bmatrix} a & b \\ c & d \end{bmatrix}$. Express matrix entries in radical form, not as decimal approximations.

c. Write a coordinate rule $(x, y) \rightarrow (\_\_, \_\_)$ for the 60° counterclockwise rotation about the origin.

8 Use reasoning similar to that in Connections Task 7 to help determine the entries of the matrix $R_{30°}$ for a 30° counterclockwise rotation about the origin.

a. Verify that the product of the matrices $R_{60°}$ and $R_{30°}$ is the matrix for a 90° counterclockwise rotation about the origin.

b. Write a coordinate rule $(x, y) \rightarrow (\_\_, \_\_)$ for the 30° counterclockwise rotation about the origin.

9 In the *Patterns in Shape* unit of Course 1, you analyzed tessellations and frieze patterns in terms of their symmetries. Those symmetries were described in terms of transformations—reflections, rotations, and translations. Those same transformations can be used to create repeating patterns such as the one shown below.

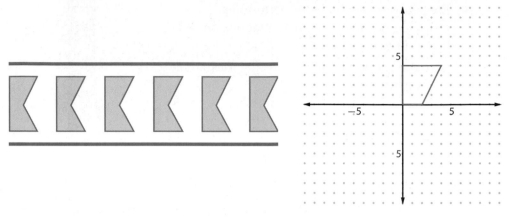

Write an algorithm that inputs the coordinates of the figure at the right and produces the frieze pattern shown at the left.

10 In Unit 2, *Matrix Methods*, you found many useful interpretations of powers of matrices. When matrices are used to represent transformations, their powers have a special interpretation. Recall that the matrix

$$R_{45°} = \begin{bmatrix} \dfrac{\sqrt{2}}{2} & -\dfrac{\sqrt{2}}{2} \\ \dfrac{\sqrt{2}}{2} & \dfrac{\sqrt{2}}{2} \end{bmatrix}$$

is the matrix representation for a 45° counterclockwise rotation about the origin.

a. Compute $R_{45°}^2$ and $R_{45°}^4$, and compare your results to the other matrices you constructed in Investigation 1.

b. What rotation do you think $R_{45°}^6$ represents? $R_{45°}^3$? Explain your reasoning.

c. Represent a 270° counterclockwise rotation about the origin as a power of the matrix for a 90° counterclockwise rotation.

11 In the *Matrix Methods* unit, you compared operations on matrices with operations on real numbers. Consider the case of square roots. The real number 1, which is the multiplicative identity, has two square roots, 1 and −1. The multiplicative identity for 2 × 2 matrices is $I = \begin{bmatrix} 1 & 0 \\ 0 & 1 \end{bmatrix}$.

a. Verify that the matrix for a line reflection across the *x*-axis is a "square root" of *I*.

b. Find two other "square roots" of *I*.

c. How many "square roots" do you think *I* has? Explain.

Reflections

12 Suppose $R = \begin{bmatrix} a & b \\ c & d \end{bmatrix}$ is the matrix for a rotation about the origin, and

$I = \begin{bmatrix} 1 & 0 \\ 0 & 1 \end{bmatrix}$ is the identity matrix.

a. What must be true about $R \times I$?

b. How are the entries of any rotation matrix related to the entries of the identity matrix?

13 Look back at the Roll Over Algorithm in Investigation 1 (page 235). If *NOW* represents the matrix of the current shuttle image and *NEXT* represents the matrix of the next image in the animation sequence, write a rule relating *NOW* and *NEXT*.

14 Look back at your work on Connections Tasks 7 and 8.

a. How could you use the matrix $R_{60°}$ to find the matrix for a 120° counterclockwise rotation about the origin?

b. How could you use the matrix $R_{30°}$ to find the matrix for a 120° counterclockwise rotation about the origin?

c. How could you use the symmetry of a circle to help write the matrix for each of the following transformations?

   i. 30° clockwise rotation about the origin

   ii. 60° clockwise rotation about the origin

d. How could you find the matrix for a 150° clockwise rotation about the origin?

15 You have been able to use geometric reasoning to find rotation matrices for special angles. In the *Trigonometric Methods* unit, you will learn how to find rotation matrices for any angle. Investigate how you could use the reporting capabilities of interactive geometry software to find rotation matrices for other angles. Find the matrix representation for each of the following transformations.

a. 15° counterclockwise rotation about the origin

b. 75° clockwise rotation about the origin

Extensions

16 View the "Animate Person" and "Animate Humanoid" Sample Sketches in your interactive geometry software. Select one of the animations.

a. Describe the transformations involved in creating that animation.

b. Describe a possible algorithm for producing the animation.

17 All of the rotations and size transformations that you have worked with in this lesson have been centered at the origin. To spin a flag around a point P other than the origin, you can use the translate-transform-translate back method (see Applications Task 13 page 222). Use this method to write a program, using matrices, that will rotate a flag about the point $P(5, 0)$ in steps of $45°$.

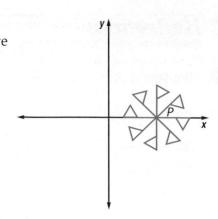

18 By combining creativity with an understanding of matrix representations of transformations, many other interesting animations are possible. For example, consider the image shown below created by successive applications of a size transformation followed by a rotation of a square:

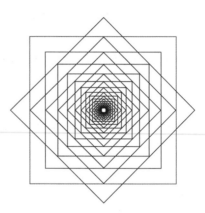

a. Estimate the scale factor of the size transformation and the angle of rotation.

b. Write a calculator program that produces a spiral-type animation, according to these specifications:

- Start with the $\triangle ABC = \begin{bmatrix} 0 & 2 & 2 \\ 0 & 0.5 & -0.5 \end{bmatrix}$.

- Successively transform $\triangle ABC$ using the composition of a size transformation of magnitude 1.1, followed by a $45°$ counterclockwise rotation. Both transformations should be centered at the origin.

- Set up the viewing window so that you can see at least 30 steps.

19 The **homogeneous coordinates** of a point (x, y) are $(x, y, 1)$; this represents the point in three-dimensional space, in a plane parallel to the x-y plane, and 1 unit above it. When plotting the point $(x, y, 1)$ in the x-y plane, the "1" is ignored. But representing points with homogeneous coordinates allows matrix multiplication to be used to translate points.

a. Let $A = \begin{bmatrix} 1 & 0 & 2 \\ 0 & 1 & 3 \\ 0 & 0 & 1 \end{bmatrix}$. What is the effect of multiplying $\begin{bmatrix} x \\ y \\ 1 \end{bmatrix}$ on the left by A?

b. Build a matrix that will translate a point 5 units to the right and 3 units down.

c. Use homogeneous coordinates to build the matrix for a translation that has horizontal component h and vertical component k.

d. How could you represent a 90° counterclockwise rotation about the origin with a 3 × 3 matrix?

e. Write an algorithm for a flag animation using homogeneous coordinates and 3 × 3 matrices to represent a rotation and a translation.

20 Consider the transformation of the plane represented by this matrix:

$$S = \begin{bmatrix} 1 & 0 \\ 0 & 1 \end{bmatrix}$$

a. Represent the rectangle shown as a matrix. Then find the image of the rectangle under the transformation represented by S.

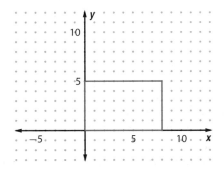

b. Sketch the rectangle and its image on a coordinate grid.

c. Describe as precisely as you can the transformation that S represents. This transformation is called a **horizontal shear**.

d. Find the coordinate rule for the transformation.

e. How does the transformation affect the perimeter of the rectangle? The area? Explain your answers.

f. Compare S to the shear transformation $T = \begin{bmatrix} 1 & 0.5 \\ 0 & 1 \end{bmatrix}$.

g. Use S or T to transform your initials in Applications Task 4 to italic font.

21 In addition to rigid transformations, size transformations, and shear transformations (Extensions Task 20), there are many other **linear transformations** of the plane. These are transformations whose coordinate rules $(x, y) \rightarrow (x', y')$ give x' and y' as linear expressions in x and y. Consider the transformation that has this coordinate rule:

$$(x, y) \rightarrow (2x + y, x + 2y)$$

a. Find the matrix representation of this transformation.

b. Write a 2×4 matrix that represents the rectangle shown below.

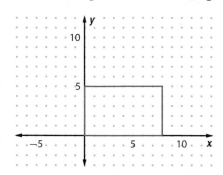

c. Multiply your matrix in Part b by the matrix for the transformation to find the transformed image of the rectangle.

d. Sketch the rectangle and its image on a coordinate grid.

e. Which of the following statements are true about the rectangle and its transformed image?

   i. Lengths do not change.

   ii. Angle sizes do not change.

   iii. Pairs of parallel sides are transformed into pairs of parallel sides.

   iv. Area does not change.

Review

22 Ten students in Mr. Malone's class measured the circumference of their wrists and their necks. Their measurements, in centimeters, are provided in the table below.

Wrist	14	16.5	15.3	18.1	20.5	23.1	21.2	17.1	19.3	17.8
Neck	27.1	32	31.5	37.3	42.1	45.2	41.6	33.8	37.5	35.2

a. Find the mean, median, and standard deviation of the wrist circumferences.

b. There are 2.54 centimeters in an inch. What would the mean, median, and standard deviation be if the students had measured their wrist circumferences using inches instead of centimeters?

c. Madison wants a rule for predicting someone's neck circumference if she knows his or her wrist circumference. Find such a rule for Madison. Explain how you found your rule.

d. Sam, a sixth-grader, has a wrist circumference of 12.3 cm. Use your rule to predict the circumference of Sam's neck.

23 Consider the line segment with endpoints $A(0, 12)$ and $B(6, 8)$.

a. Find the length of $\overline{AB}$.

b. Find the midpoint of $\overline{AB}$.

c. Find an equation of the line containing $\overline{AB}$.

d. Find an equation of the line that contains the point $A(0, 12)$ and is perpendicular to $\overline{AB}$.

e. Find an equation of a line that would be parallel to $\overleftrightarrow{AB}$.

24 Determine the equations of the lines with the indicated characteristics.

a. Has slope of $-\frac{3}{2}$ and contains the point $(0, 9)$

b. Has slope of $\frac{4}{3}$ and contains the point $(-6, -10)$

c. Has graph as shown below

d. Has graph as shown below

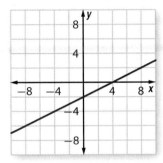 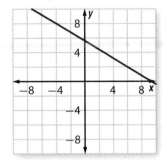

25 Raul's take-home pay for each week in the last month is represented by x_i where $x_1 = \$112.32$, $x_2 = \$89.17$, $x_3 = \$95.91$, and $x_4 = \$78.64$.

a. Compute Σx and then indicate what that number tells you about Raul's earnings.

b. Compute $\frac{\Sigma x}{4}$. What is this number called?

c. Each week, Raul has to give $25 to his parents to help pay for his car insurance. Which of the following expressions correctly represents the total amount of money Raul had left to spend during this four-week period?

I $(\Sigma x) - 25$ II $\Sigma 25x$ III $\Sigma(x - 25)$

d. Recall that $\bar{x}$ is the symbol for the mean of a set of values. Compute $\Sigma(x - \bar{x})$ for these values.

Looking Back

I n this unit, you learned how to use coordinates and matrices to represent figures in a coordinate plane and on a computer or calculator screen. Coordinate representations of slope and distance were helpful in creating and analyzing figures, particularly those with parallel or perpendicular sides. You developed methods to reposition and resize shapes using transformations represented by coordinate rules and by matrices.

You learned how the key ideas of congruence and similarity are related to transformations. The image of any figure under a rigid transformation—a translation, rotation, line reflection, or glide reflection—is congruent to the original figure, or preimage. The image of a figure under a size transformation or a similarity transformation—the composite of a size transformation and a rigid transformation—is always similar to the original figure. The connection between composition of transformations and matrix multiplication was seen to be particularly useful in creating animations of shapes.

The tasks in this final lesson will help you review and organize your thinking about coordinate methods and their connections with matrices.

① **Designing Shapes** Examine this simplified image of an insect created by a graphic artist for a cell phone screen saver.

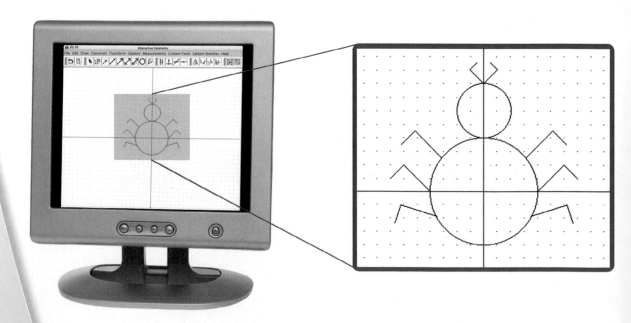

a. What equation could be used to represent the main body (not including the legs)? The head (not including the antennae)?

b. Identify the coordinates of 5 points on the main body. Identify the coordinates of 2 other points without doing any calculations.

c. Use coordinates of key points on the middle leg and foot to verify that the foot is positioned at a right angle to the leg.

d. In creating the image, the artist drew the legs and antenna on the right side of the insect and then used a transformation to complete the image. What transformation was used and how could that transformation be described using coordinates?

e. Suppose the artist wanted to reposition the insect image using a translation with horizontal and vertical components 10 and −6, respectively.

   i. Write equations that could be used to represent the main body and head of the new figure.

   ii. What would be the perimeter of the main body of the new figure?

   iii. What would be the area of the main body of the new figure?

f. Describe a single transformation that the artist could use to reposition the insect image so that it is pointing downward. Describe a different single transformation that could be used. Write a coordinate rule for each of those transformations.

g. Suppose the artist wanted to resize the insect image using a size transformation of magnitude 2 with center at the origin.

   i. Write equations that could be used to represent the main body and head of the new figure.

   ii. What would be the perimeter of the main body of the new figure?

   iii. What would be the area of the main body of the new figure?

② **Analyzing Shapes** A quadrilateral $ABCD$ has vertices with coordinates $A(4, 9)$, $B(7, 5)$, $C(-2, -3)$, and $D(-1, 9)$.

a. Sketch this quadrilateral on a coordinate grid or display it on your calculator or computer.

b. Use coordinate methods to justify that quadrilateral $ABCD$ is a kite.

c. How are the diagonals of kite $ABCD$ related? Use coordinate methods to justify your answer.

d. Find the midpoints of each side and connect them in order with line segments. What kind of polygon is formed? (Be as specific as possible.) Justify your answer.

3 **Relating Shapes** Examine the screen below. The scale on each axis is 1.

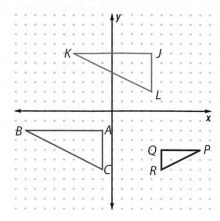

a. Explain why $\triangle ABC \cong \triangle JKL$.

b. Describe a transformation or sequence of transformations that maps $\triangle ABC$ onto $\triangle JKL$. Write a coordinate rule $(x, y) \rightarrow (\_\_, \_\_)$ that relates preimage and image points.

c. Explain why $\triangle ABC$ is similar to $\triangle QPR$.

d. Describe a transformation or sequence of transformations that maps $\triangle ABC$ onto $\triangle QPR$. Write a coordinate rule $(x, y) \rightarrow (\_\_, \_\_)$ that relates preimage and image points.

e. Describe a transformation or sequence of transformations that maps $\triangle JKL$ onto $\triangle QPR$. Write a coordinate rule $(x, y) \rightarrow (\_\_, \_\_)$ for the transformation.

f. Describe a sequence of transformations that repositions $\triangle ABC$ so that it meets both of the following specifications.

- The image is a right triangle of the same size.

- The longest leg of the image triangle lies along the positive y-axis with point A at the origin.

g. Write a coordinate rule for the composite transformation in Part f.

4 **Repositioning and Resizing a Shape** Use grid paper to display $\triangle ABC$, where

$$\triangle ABC = \begin{bmatrix} 1 & 9 & 1 \\ 7 & 2 & 2 \end{bmatrix}.$$

Use matrices to find the image of $\triangle ABC$ under each transformation below. Draw and label the vertices of the image and record the matrix representation.

a. 90° counterclockwise rotation about the origin

b. Reflection across the line $y = x$

c. Size transformation of magnitude 3 with center at the origin

⑤ **Reshaping a Shape** Consider the transformation that stretches figures vertically by a factor of 3 but does not change them in any other way.

a. Investigate how this transformation affects shapes.

    i. Begin by sketching the rectangle below and its image under the transformation.

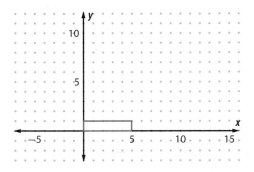

    ii. Write 2 × 4 matrices that represent the rectangle and its image.

b. Write a symbolic rule that relates the coordinates of any point (x, y) and its image under this transformation.

c. Construct a matrix representation of the transformation. Call the matrix T.

d. When you multiply a matrix of points that represents a polygon and a transformation matrix, on which side should the transformation matrix appear? Why?

    i. Multiply the coordinate matrix for the rectangle in Part a by the matrix T.

    ii. Compare the image points to those you found in Part a.

e. Consider powers of the matrix T.

    i. Using matrix multiplication, find the image of the rectangle in Part a under the transformation represented by T^2. Sketch the rectangle and its image under this composite transformation.

    ii. Describe the transformation represented by T^2.

    iii. Without computing, describe the transformation represented by T^3.

f. Consider the matrix $S = \begin{bmatrix} 2 & 0 \\ 0 & 1 \end{bmatrix}$.

    i. Make a conjecture about the effect on shapes of the transformation represented by S. Defend your conjecture.

    ii. Make a conjecture about the effect of the transformation represented by $T \times S$. Defend your conjecture.

⑥ **Animating a Shape** Design and test an animation program that launches the rocket model shown into a circular orbit of radius 15, displays 20 orbits of the rocket, and returns the rocket to its base. Use steps of 45°.

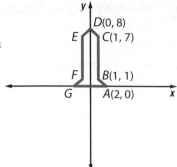

Summarize
the Mathematics

In this unit, you investigated how coordinates and matrices can be used to model geometric shapes, represent transformations, and create animations.

ⓐ How can polygons and circles be represented using coordinates? Illustrate with examples.

ⓑ How can coordinates be used to analyze properties of a polygon or to draw a polygon with special properties? Illustrate with examples.

ⓒ What transformations can be used to reposition a shape in a coordinate plane without changing its size? Illustrate with examples.

ⓓ How can you resize a shape in a coordinate plane? Illustrate with an example.

ⓔ What strategies are helpful in creating a symbolic rule for a transformation described in words? In creating a matrix representation of the transformation? Illustrate with an example.

ⓕ Describe how two transformations can be composed.

   i. How can you find a coordinate rule for a composite transformation?

   ii. How can you find a matrix representation for a transformation that is the composition of two transformations represented as matrices?

ⓖ What is a similarity transformation and how does it affect shapes? Give an example of a similarity transformation that is not a size transformation or a rigid transformation.

ⓗ How can animation effects be produced using coordinate methods?

Be prepared to share your descriptions, illustrations, and summaries with the class.

✔Check Your Understanding

Write, in outline form, a summary of the important mathematical concepts and methods developed in this unit. Organize your summary so that it can be used as a quick reference in future units and courses.

REGRESSION AND CORRELATION

Some things just seem to go together. For example, there is an association between each of the following pairs of variables: time spent studying for an exam and score on the exam, age of a car and its value, age and height of a child, duration of eruption of a geyser and time until the next eruption, and playing time and points scored in basketball. In each case, an increase in the value of one variable tends to be associated with an increase (or decrease) in the value of the second variable. Detecting, measuring, and explaining patterns of association between pairs of variables help in making decisions and predictions.

In previous units, you have learned to make scatterplots and find the least squares regression line using technology. In this unit, you will learn more about this line and how to compute a correlation, which indicates how closely the points cluster about the regression line. Key ideas will be developed through your work on two lessons.

Lessons

1 *Bivariate Relationships*

Compute and interpret rank correlation, describe shapes and characteristics of scatterplots, and identify types of association.

2 *Least Squares Regression and Correlation*

Use a regression line to make predictions, interpret the coefficients of a regression equation, and predict the effect of influential points. Compute and interpret the correlation, predict the effect of influential points, and distinguish between correlation and cause-and-effect.

Bivariate Relationships

In your previous work in *Core-Plus Mathematics*, you have used scatterplots to examine possible relationships between two quantitative (numerical) variables. In this unit, you will develop further methods for detecting and describing patterns in bivariate data.

The points on the scatterplot below show (*height of husband, height of wife*) data for a randomly selected sample of two hundred married couples.

Husbands' and Wives' Heights

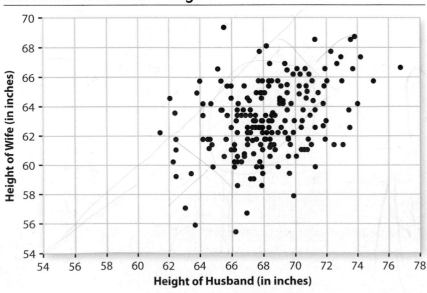

Height of Husband (in inches)

Source: D.J. Hand, *et al.* (Eds.). *A Handbook of Small Data Sets.* Chapman and Hall: 1994.

Think about the pattern in the plot.

a How would you describe the relationship between the heights of husbands and wives? How many wives are taller than their husbands? What line could you draw to help you answer this question?

b How might you describe the shape of the cloud of points on this scatterplot?

c Would you feel more confident predicting the height of a wife if her husband was 64 inches tall or 74 inches tall? Why?

d Find the couple that lies farthest from the general trend. How does this couple differ from couples that follow the general trend?

In this lesson, you will learn to describe and summarize the pattern in a scatterplot.

Investigation 1 · Rank Correlation

In the United States, we seem to be "rank happy." Sports teams are ranked on games won; motion pictures are ranked by viewer preference or gross revenue; DVDs are ranked on the number of rentals; automobiles are ranked by safety; and colleges are ranked on quality. As you work on the problems in this investigation, look for answers to this question:

How can you measure how closely the rankings of two people agree?

1 Consider the following music categories:

Rock	Hip-Hop/Rap	Classical	Pop
Latin	Dance/Electronic	R&B/Soul	Country

a. Rank your favorite type of music from the above choices with a 1. Continue ranking with a 2 for your second favorite and an 8 for your least favorite. Ties are not allowed!

b. Working with a partner, display your rankings on a scatterplot that has scales and labels on the axes. Plot one point (*one partner's rank, other partner's rank*) for each of the eight types of music. For example, in the figure at the right, Doris ranked Hip-Hop/Rap music seventh and Kuong ranked it second. Using a full sheet of paper, make your scatterplot as large as possible, with big dots, and then display it on the wall of your classroom.

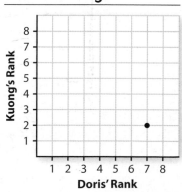

Music Rankings

c. Does there seem to be an association between your ranking and your partner's ranking? Explain your reasoning.

d. Examine the various scatterplots from pairs of students posted around the classroom. Identify the plots that show **strong positive association**, that is, the ranks tend to be similar. Describe these plots.

e. Identify the plots that show **strong negative association**, that is, the ranks tend to be opposite. Describe these plots.

f. Which plots show **weak association** or **no association**?

g. From which plots would you feel confident in predicting the ranks of one person if you know the ranks of the other?

2 By looking at the scatterplots, it is fairly easy to make a decision about the *direction* (positive, negative, or none) of the association of two variables. But, as is often the case, it is helpful to have a numerical measure to aid your visual perception of the *strength* of the association.

a. Use the music rankings to brainstorm about ways to assign a number to each scatterplot that indicates the direction and the strength of the association.

   i. Use the method you prefer to assign a number to three of the scatterplots. Do the results make sense? If not, revise your method.

   ii. Describe your method to the rest of the class.

   iii. Where everyone can see it, post a list of the methods your class has suggested.

b. One way to test a method is to use it on extreme cases. What number does each method give for a **perfect positive association**—two rankings that are identical? For a **perfect negative association**—two rankings that are opposite of each other? As a class, decide on a method that seems to make the most sense.

c. Use the following pairs of ranks prepared by Tmeka and Aida to test the method your class selected.

$$(1, 4), (2, 5), (3, 3), (4, 7), (5, 2), (6, 6), (7, 8), (8, 1)$$

Music Rankings

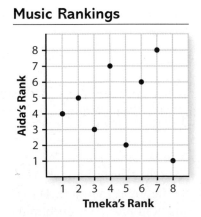

   i. Is there an obvious positive or negative association between the two rankings?

ii. Use your class' method from Part b to assign a number to this association.

iii. How does your number indicate the direction and strength of this association?

British statistician Charles Spearman (1863–1945) invented a simple measure for the strength of the association between two rankings. The measure is called a rank correlation. **Spearman's rank correlation**, r_s, is given by the formula:

$$r_s = 1 - \frac{6\Sigma d^2}{n(n^2 - 1)}$$

Here, n represents the number of items ranked, and Σd^2 represents the sum of the squared differences between the ranks.

Charles Spearman

3 Working with your partner, make a table, like the one below, to help you compute Spearman's rank correlation for your music rankings from Problem 1.

Type of Music	Your Rank	Partner's Rank	Difference of Ranks (d)	Squared Difference (d^2)
Rock				
Hip-Hop/Rap				
Classical				
Pop				
Latin				
Dance/Electronic				
R&B/Soul				
Country				

$$\Sigma d^2 =$$

a. Compute r_s for your ranking and your partner's ranking. Round your rank correlation to the nearest thousandth. Write this value on your scatterplot.

b. How could you use the lists on your calculator to help you compute r_s?

c. Exchange rankings with another pair of students. Check their work by computing r_s for their rankings.

4 Study the scatterplots and rank correlations displayed in your classroom.

    a. Which scatterplot has a rank correlation closest to 1? What can you say about the taste in music of those two classmates?

    b. Which scatterplot has a rank correlation closest to -1? What can you say about the taste in music of those two classmates?

    c. Which scatterplot has a rank correlation closest to 0? What can you say about the taste in music of those two classmates?

    d. Suppose you had ranked the types of music in exactly the same way as your partner did.

        i. Describe the scatterplot of identical rankings.

        ii. What would the rank correlation be? Explain your response based on the formula for r_s.

    e. Suppose you had ranked the types of music exactly opposite of the way your partner ranked the music.

        i. Describe the scatterplot of opposite rankings.

        ii. What would the rank correlation be? Verify by computing r_s.

5 Examine the plots below, showing paired rankings of favorite movies. Match each rank correlation below with the appropriate scatterplot. The scales on each scatterplot are the same.

    a. $r_s = 0.1$ **b.** $r_s = 0.8$ **c.** $r_s = -0.9$ **d.** $r_s = 0.5$

I

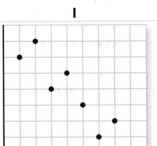

II

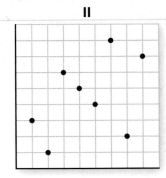

III

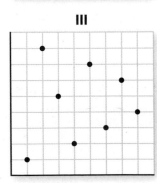

IV

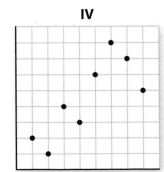

6 The formula for Spearman's rank correlation,

$$r_s = 1 - \frac{6\Sigma d^2}{n(n^2 - 1)},$$

includes a sum of squared differences, Σd^2. Such a sum is used in statistics to measure how much *variability* exists. In this problem, you will investigate one reason why the differences d are squared before they are summed.

a. What formulas have you seen before that involve a sum of squared differences?

b. Compute Σd using your table from Problem 3. Compare your results with those of other students.

c. Compute Σd again for a different pair of rankings. Again, compare your results with those of other students.

d. What does squaring the values of d in Spearman's formula accomplish?

Summarize
the Mathematics

In this investigation, you learned to describe association between two rankings by giving direction and strength and by computing Spearman's rank correlation.

a Sketch a scatterplot showing a strong positive association. A weak negative association. No association. A perfect negative association.

b Compare Spearman's rank correlation formula with the one invented and agreed on by your class in Problem 2. What are the advantages and disadvantages of each?

c How can positive rank correlation be seen in a list of paired ranks? In a scatterplot? In the value of r_s?

d How can negative rank correlation be seen in a list of paired ranks? In a scatterplot? In the value of r_s?

e Why are differences sometimes squared in statistics formulas that include a sum?

Be prepared to explain your ideas and examples to the class.

✓Check Your Understanding

A couple decides to measure their compatibility by ranking their favorite leisure activities. The rankings are given below in the table.

	Mallisa	Matt
Watch TV	4	7
Read	5	2
Exercise	2	4
Talk to friends	7	3
Go to a movie	6	5
Go to dinner	1	6
Go to the mall	3	1

a. Make a scatterplot for the two rankings with appropriate scales and labels on the two axes.

b. Predict whether the value of r_s will be closer to -1, 0, or 1. Use the scatterplot to help explain your answer.

c. Calculate the value of r_s. Show your work.

d. What would you conclude about this couple's compatibility?

Investigation 2   Shapes of Clouds of Points

In Investigation 1, you learned to describe the association between two rankings as shown in a scatterplot by giving its *direction* (positive, negative, or none) and its *strength* (strong, moderate, weak, none) or by reporting the correlation. In this investigation, you will examine more closely patterns in a scatterplot. As you complete the following problems, make notes of answers to this question:

What are the common shapes of bivariate data displayed on a scatterplot?

1 Bivariate data are **linear** if they form an *oval* or *elliptical cloud*. Two examples of linear relationships are shown below.

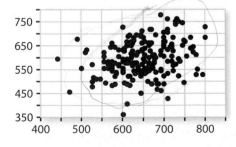 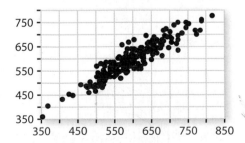

a. Place a sheet of paper over each plot. Sketch the axes and an oval that contains the points.

b. Describe the direction and strength of the relationship in the first plot. In the second plot.

c. When the relationship is linear, you can summarize the relationship with a straight line. However, a line would not be an appropriate summary of the relationships shown below. How could you describe the shape, trend (center), and strength (spread) of these distributions?

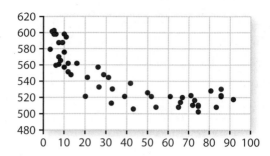 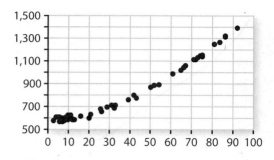

d. If the points tend to fan out at one end, the relationship is said to **vary in strength**. Both of the plots below vary in strength. Sketch a plot where the pattern varies in strength but the points cluster about a line. That is, the shape is not curved.

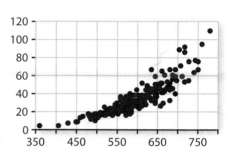 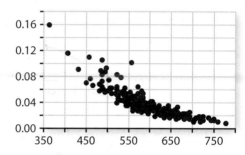

2 The scatterplot below shows the time between two consecutive eruptions of the Old Faithful geyser in Yellowstone National Park plotted against the duration of the first eruption.

Old Faithful Eruption Times

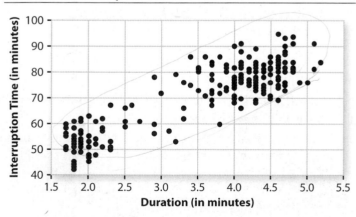

Source: Samprit Chatterjee, et al. *A Casebook for a First Course in Statistics and Data Analysis.* Wiley, 1995.

a. What is the shape of this distribution? Is it appropriate to summarize the trend with a line?

b. Is the relationship positive or negative? Is it strong, moderate, or weak? Does the strength of the relationship vary?

c. Can you give a reason why the duration of the first eruption might have an effect on the time until the next eruption?

d. When examining a scatterplot, you should also look for *clusters* of points and for *outliers* that lie away from the main cloud of points. Do you see clusters or outliers in the scatterplot of the geyser data?

3 The state of Alaska has the largest population of black bears in the U.S.—approximately 100,000. The scatterplot below gives the weight in pounds and length in inches of a large sample of black bears.

a. Describe the shape of this plot.

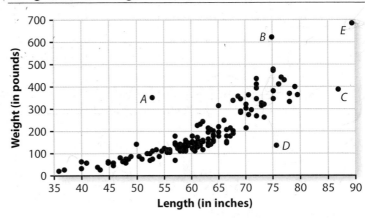

Lengths and Weights of Black Bears

b. This scatterplot illustrates the types of outliers that can occur:

- an outlier for length only

- an outlier for weight only

- both an outlier for length and an outlier for weight

- not an outlier for length and not an outlier for weight, but an outlier when length and weight are jointly considered

For each labeled point on the scatterplot, tell which type of outlier it is. Then describe the bear.

4 The plot at the top of the next page is called a **scatterplot matrix**, a matrix whose entries are scatterplots. In each scatterplot, one dot represents one of the 50 states, the District of Columbia, or Puerto Rico. The five variables are as follows:

- Dropout% percentage of 16 to 19 year olds who are not enrolled in school and have not graduated from high school

- Med Age median age (in years)

- PerCapIn per capita (per person) income

- %Poverty percentage of the population below the poverty level

- %ColGrad percentage of people at least 25 years old who have earned bachelor's degrees or higher

Population Characteristics

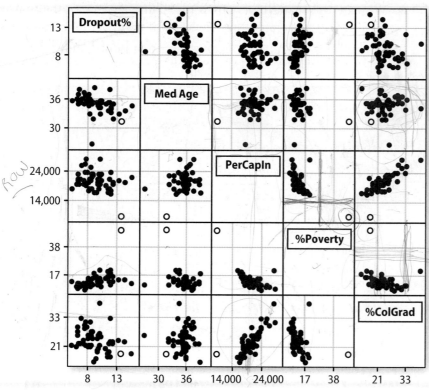

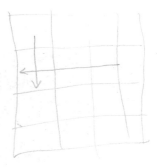

Source: 2000 U.S. Census

a. The points in the plot in the second row and fifth column have *percentage college graduate* plotted on the *x*-axis and *median age* plotted on the *y*-axis. Is there a strong positive, a strong negative, or almost no association between these two variables?

b. Describe the location(s) of the scatterplots within the matrix for which *percentage below poverty level* is the variable graphed on the *x*-axis. On the *y*-axis.

c. The state with the lowest median age is Utah. Estimate the median age in Utah. Estimate the per capita income.

d. The open circle on each plot represents Puerto Rico. The scatterplot in the first row and second column shows that Puerto Rico has a relatively high dropout percentage and a relatively low median age, but it is not an outlier. What can you tell about Puerto Rico from each scatterplot identified below? If it is an outlier, give the type of outlier.

   i. the scatterplot in the third row and fourth column

   ii. the scatterplot in the fourth row and fifth column

   iii. the scatterplot in the first row and fifth column

e. If you ignore Puerto Rico, which pair of variables has the strongest positive association?

f. If you ignore Puerto Rico, which pairs of variables have negative association?

g. Which scatterplot has a curved shape? Does it show varying strength?

⑤ Look back at the scatterplot matrix in Problem 4.

    a. Why are the scatterplots down the main diagonal of the matrix not included? What would they look like if they were included?

    b. Which pairs of scatterplots give the same information?

Summarize
the Mathematics

In this investigation, you learned how to describe the pattern in a scatterplot.

(a) When describing the cloud of points on a scatterplot, what information should you give?

(b) What are the types of outliers that you may see on a scatterplot?

(c) When would a scatterplot matrix be useful?

Be prepared to share your ideas and reasoning with the class.

✔Check Your Understanding

Each point on the scatterplot below represents a state or the District of Columbia. The variables are the percentage of ninth graders who graduate from high school four years later and the percentage of people who are unemployed. (Two states are missing because their graduation rate was not available.)

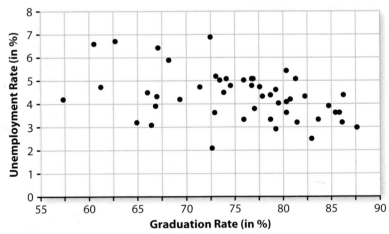

Source: U.S. Department of Education and U.S. Bureau of Labor Statistics, 2006

    a. Describe the shape of this distribution. Include the direction of the relationship, the strength, and whether the strength varies.

    b. Do you see any unusual features?

    c. Nebraska has the highest graduation rate. Estimate this rate from the plot. Is Nebraska's unemployment rate about what you would expect, given its graduation rate?

    d. Michigan has the highest unemployment rate. Estimate this rate from the plot. Is Michigan clearly an outlier?

Applications

① Which are the best steel roller coasters in the United States? The rankings below are from an *Amusement Today* annual survey of roller-coaster riders.

The table and scatterplot show the top ten roller coasters from the 1999 survey and the order that those same coasters appeared in the 2006 survey.

Roller Coaster Rankings

Roller Coaster	1999 Rank	2006 Relative Rank
Magnum XL-200, Cedar Point, OH	1	1
Montu, Busch Gardens, FL	2	2
Steel Force, Dorney Park, PA	3	5
Alpengeist, Busch Gardens, VA	4	6
Kumba, Busch Gardens, FL	5	8
Raptor, Cedar Point, OH	6	4
Desperado, Buffalo Bill's Resort, NV	7	10
Mind Bender, Six Flags Over Georgia, GA	8	7
Mamba, Worlds of Fun, MO	9	9
Superman, Ride of Steel, Six Flags Darien Lake, NY	10	3

Source: *Amusement Today,* August 2000; www.amusementtoday.com/2006gtasteel.html

Steel Coasters

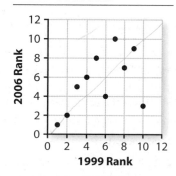

a. Why might the ranks change from year to year?

b. Is the relationship positive or negative? Strong or weak? Estimate the rank correlation by examining the scatterplot.

c. Calculate the rank correlation. Compare it to your estimate.

2 The following are the 10 consumer products that emergency room patients in the United States most often say are related to the cause of their injuries.

Bathtubs and showers
Beds
Bicycles
Cabinets, racks, and shelves
Chairs
Containers and packaging
Knives
Ladders
Sofas
Tables

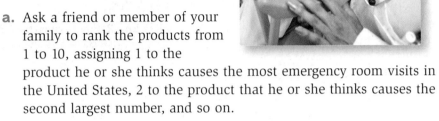

a. Ask a friend or member of your family to rank the products from 1 to 10, assigning 1 to the product he or she thinks causes the most emergency room visits in the United States, 2 to the product that he or she thinks causes the second largest number, and so on.

b. The actual ranking is given below. Make a scatterplot comparing these rankings to those collected in Part a.

Product	Rank
Bathtubs and showers	8
Beds	2
Bicycles	1
Cabinets, racks, and shelves	6
Chairs	5
Containers and packaging	7
Knives	3
Ladders	9
Sofas	10
Tables	4

c. Compute the rank correlation between your friend's or family member's ranking and the actual ranking. Was your friend or family member relatively successful or relatively unsuccessful in matching the actual ranks?

③ The population ranks for the 10 largest countries in the world for the year 2000 are given in the table below. Also given is the projected rank for each country, relative to the other ten countries, for the years 2025 and 2050.

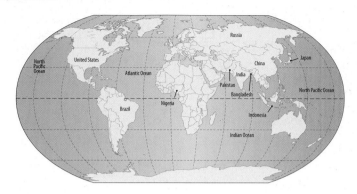

Population Rankings, Largest Countries in 2000

Country	2000 Population (in millions)	2000 Population Rank	2025 Projected Relative Rank	2050 Projected Relative Rank
Bangladesh	129.2	8	8	8
Brazil	170.1	5	6	7
China	1,277.6	1	1	2
India	1,013.7	2	2	1
Indonesia	212.1	4	4	6
Japan	126.7	9	10	10
Nigeria	111.5	10	7	5
Pakistan	156.5	6	5	3
Russia	146.9	7	9	9
United States	273.8	3	3	4

Source: *The New York Times 2001 Almanac.* New York, NY: The New York Times, 2000.

a. Examine this scatterplot for the (*2000, 2050*) rankings. Write two observations that you can make from looking at the scatterplot.

b. What is your estimate of the rank correlation for the (*2000, 2050*) rankings? Check your estimate by computing the rank correlation.

c. Would you expect the correlation between the 2000 ranking and the projected 2025 ranking to be larger or smaller than the one you computed in Part b? Explain. Compute this correlation to see if you were correct.

Population Rankings

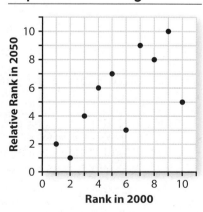

4 The table and scatterplot below show data for 25 countries. The variables are a measure of the carbon dioxide emissions (in metric tons) per person and the number of years a newborn can expect to live.

Country	Carbon Dioxide Emissions (in metric tons) per Person	Life Expectancy at Birth
Australia	5.2	80.3
Brazil	0.5	71.4
Canada	5.0	80.0
China	0.7	72.0
France	1.9	79.4
Germany	2.8	78.5
India	0.3	64
Indonesia	0.4	69.3
Iran	1.5	69.7
Italy	2.2	79.5
Japan	2.6	81
Korea, South	2.6	76.7
Mexico	1.0	74.9
Netherlands	4.4	78.7
Poland	2.0	74.7
Russia	3.0	66.8
Saudi Arabia	3.4	75.2
South Africa	2.5	44.1
Spain	2.3	79.4
Taiwan	3.3	77.1
Thailand	0.8	71.7
Turkey	0.8	72.1
Ukraine	2.0	68.8
United Kingdom	2.6	78.3
United States	5.4	77.4

Source: *2006 and 2007 Statistical Abstract of the U.S.* Tables 1318 and 1325.

Life Expectancy and Carbon Dioxide Emissions by Country

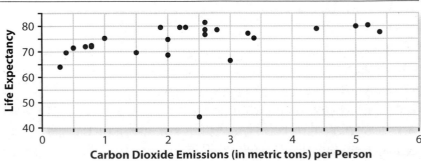

a. What is the shape of the distribution? Describe the association between the two variables. Can you use Spearman's r_s to quantify the strength? Why or why not?

b. Do you think the value of one of the two variables causes or otherwise influences the value of the other? Explain your reasoning.

c. Are there any outliers? If so, which type?

5 The *Places Rated Almanac* ranks metropolitan areas according to a variety of categories including:

- crime—violent crime and property crime rates
- health care—the supply of health care services (such as number of specialists or breadth of hospital services)
- education—the number of available educational opportunities beyond high school

Some characteristics of the 15 largest metropolitan areas in the United States are ranked in the table and following scatterplot matrix. For crime, health care, and education, a rank of 1 is best.

Philadelphia

Los Angeles

Boston

Rankings of Metropolitan Areas

Metro Area	Population	Crime	Health Care	Education
Los Angeles, CA	1	14	9	12
New York, NY	2	15	4	5
Chicago, IL	3	13	7	2
Philadelphia, PA	4	4	5	7
Washington, DC	5	5	1	3
Detroit, MI	6	9	13	14
Houston, TX	7	7	10	13
Atlanta, GA	8	11	11	9
Boston, MA	9	3	2	1
Dallas, TX	10	12	14	6
Riverside, CA	11	10	15	15
Phoenix, AZ	12	8	12	8
Minneapolis, MN	13	1	3	4
San Diego, CA	14	6	8	10
Orange County, CA	15	2	6	11

Source: Savageau, David and Ralph D'Agostino. *Places Rated Almanac,* Millennium Edition. New York: Macmillan, 2000.

Rankings of Metropolitan Areas

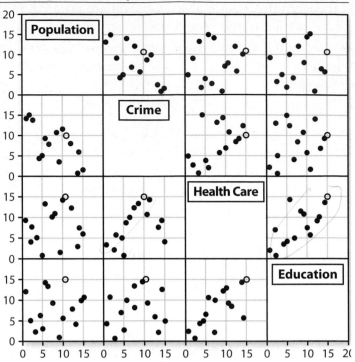

a. Examine the scatterplot matrix shown above. Describe the location(s) of the scatterplots for which *health care* is the variable graphed on the *x*-axis. On the *y*-axis.

b. The open circle on each plot represents the same city. Which city is this? For which variables does this city tend to be ranked toward the best? Toward the worst?

c. Which pair of variables appears to have the strongest positive correlation? Suggest some reasons why this correlation might be so strong.

d. Find a pair of variables with an obvious negative correlation. Write a sentence that describes this relationship.

e. Find the missing values of the rank correlation r_s in the rank correlation matrix below.

	Population	Crime	Health Care	Education
Population	1.000	___	0.168	0.161
Crime	___	1.000	0.475	0.154
Health Care	0.168	0.475	1.000	0.700
Education	0.161	0.154	0.700	1.000

f. Why are the entries along the diagonal of the rank correlation matrix in Part e all 1s?

g. How is this rank correlation matrix related to the scatterplot matrix?

Connections

6 If $r_s = 1$ or $r_s = -1$, all points fall on a line. Write equations for these two lines.

7 Bianca and Pearl each throw a dart at a larger version of the grid shown here. Bianca's dart lands at the point with coordinates (3, 4). Pearl's dart lands at the point (−5, 1).

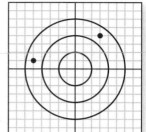

   a. Use the distance formula to determine how far Bianca's dart is from Pearl's dart.

   b. Whose dart is farther from the center?

   c. How is the distance formula like the formula for Spearman's rank correlation?

8 Are the two expressions Σd^2 and $(\Sigma d)^2$ equivalent? Give an example to support your answer.

9 How many scatterplots are there on a scatterplot matrix with 3 variables? With 4 variables? With 5 variables? With k variables?

Reflections

10 Alicia made a scatterplot of two sets of rankings (*set A, set B*) and calculated the rank correlation. Mario made a scatterplot of (*set B, set A*) rankings and calculated the rank correlation.

   a. Are there any cases when the scatterplots will be the same? Explain.

   b. Will Alicia and Mario find the same correlation? Explain why or why not.

11 Write a conclusion that can be drawn from the following situations.

   a. Two reporters ranked the 7 candidates for mayor according to the number of votes they thought the candidates would get in the primary election. The correlation between the rankings of the two reporters was 0.8.

   b. Two judges ranked 10 skaters according to their performance. The correlation between the rankings of the two judges was −0.2.

   c. Two managers ranked a set of employees on job effectiveness. The correlation between their rankings was 0.5.

   d. The two managers in Part c ranked the same employees on efficiency. The correlation between their rankings was −0.7.

(12) Sketch a scatterplot that shows:

 a. a strong negative linear relationship.

 b. a moderate positive curved relationship.

 c. a weak negative linear relationship.

(13) How is the word "linear" used differently in algebra and in statistics?

Extensions

(14) The more things there are to rank, the larger the number of possible ways to rank them.

 a. Suppose that four soccer teams are to be ranked on goals scored for a season. How many possible ways can such a ranking be done?

 b. Make a chart that shows how many different rankings are possible if there are 1, 2, 3, 4, and 5 soccer teams to be ranked.

 c. Describe any patterns in your chart.

 d. If there are k soccer teams, how many different rankings are possible?

(15) Give an example to illustrate that, for a given number n of items ranked by two people, the largest possible value of Σd^2 is $\dfrac{n(n^2 - 1)}{3}$. What does this imply about the rank correlation?

(16) Suppose you are given two rankings of five items as shown below.

Item	First Ranking	Second Ranking
1	p	v
2	q	w
3	r	x
4	s	y
5	t	z

 a. What is the sum $p + q + r + s + t$ equal to? What is the sum $v + w + x + y + z$ equal to?

 b. Use your answer from Part a to show that $\Sigma d = 0$, no matter what the rankings are.

 c. Do you think a similar argument could be given to show that for two sets of rankings of 10 items, $\Sigma d = 0$? Why or why not?

17 Maurice Kendall, a British statistician (1907–1983), developed an alternative method to measure the strength of association between two rankings. **Kendall's rank correlation** r_k is given by the formula

$$r_k = 1 - \frac{2c}{\frac{n}{2}(n-1)}.$$

Here, n is the number of items being ranked. To find c, write the ranks for each item side-by-side and connect the ranks as shown below for $n = 4$.

Item	First Ranking	Second Ranking
A	1	2
B	2	4
C	3	3
D	4	1

The number of crossings of the lines is c. Here, $c = 4$.

a. Does a large number of crossings indicate general agreement or general disagreement in the ranks?

b. Find Kendall's rank correlation for the roller coaster data in Applications Task 1 (page 269).

c. Compute Kendall's correlation when there is perfect agreement between the ranks 1 to 5. Compute Kendall's correlation when there is completely opposite ranking of five items.

d. Are Spearman's and Kendall's rank correlations equivalent? That is, do they always give the same value? Explain your answer.

e. Investigate whether r_k always lies between -1 and 1.

Review

18 The table below gives some information about various fast-food hamburgers.

Company	Name	Fat (in grams)	Protein (in grams)
Hardee's	Hamburger	12	14
	Thickburger	57	30
Wendy's	Jr. Hamburger	9	15
	Classic Single	20	25
Burger King	Hamburger	12	15
	Whopper	39	28
McDonald's	Hamburger	9	13
	Quarter Pounder	18	24
	Big Mac	30	25
Carl's Jr.	Kid's Hamburger	18	25
	Famous Star	32	24

Source: www.wendys.com; www.mcdonalds.com; www.burgerking.com; www.hardees.com; www.carlsjr.com (December 2006).

a. Make plots to compare the distribution of fat with the distribution of protein.

b. Describe the shape of the distribution of the number of grams of fat.

c. Using the *more than 1.5 • IQR from the nearest quartile* rule, are there any outliers in the distribution of the number of grams of fat?

d. Is the mean number of grams larger for fat or for protein? Can you tell from the plots or do you need to compute? Is the mean a suitable measure of center for either distribution?

19 Suppose you have a piece of cloth that has an area of 2 square yards.

a. Give two different possible dimensions for this piece of cloth.

b. How many square feet of material do you have?

c. How many square inches of material do you have?

20 Match each equation with the correct graph. All graphs are drawn using the same scales on the axes. Be prepared to explain how you can do this without using your calculator.

a. $y = -x^2 + 4$ b. $y = -x^2 + 4x$ c. $y = -x^2 + 4x - 4$

d. $y = x^2 + 4$ e. $y = -x^2 - 4x$

I

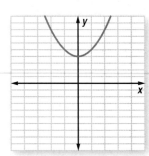

II

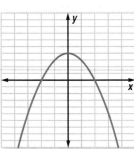

III

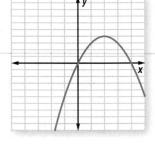

IV

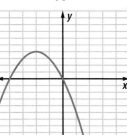

V

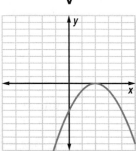

21 Rewrite each expression without parentheses.

a. $(a + 5)(a - 5)$

b. $a(18 - 4a) + 3a^2$

c. $6a^2 + 4a - (2a + 4)(a + 6)$

d. $5(2a + 10) - 6(12 - a)$

e. $(a + 9)^2$

22 Refer back to the fast-food hamburger data in Review Task 18.

  a. Which hamburger has the largest deviation (in absolute value) from the mean for fat? Find and interpret that deviation.

  b. Which hamburger has the largest deviation (in absolute value) from the mean for protein? Find and interpret that deviation.

  c. Is the standard deviation of the number of grams larger for fat or for protein? Answer this question in two ways: by looking at the plots you made in Review Task 18 and also by computing the standard deviations.

23 Three vertices of square *ABCD* are $A(-2, 10)$, $B(10, 6)$, and $C(6, -6)$.

  a. Find the coordinates of the center of the square.

  b. Find the coordinates of vertex *D*.

  c. Find the reflection image of square *ABCD* across the *y*-axis. How are square *ABCD* and its image related by shape and size?

  d. Find the image of square *ABCD* under a size transformation of magnitude $\frac{1}{2}$ with center at the origin. How are square *ABCD* and its image related by shape and size?

24 In 2006, the number of students enrolled at Mission Hills High School was 1,500. By 2008, the number of students enrolled had increased to 1,815.

  a. By what percentage did the student enrollment increase between 2006 and 2008?

  b. If the percentage growth was the same in each of the two years, by what percent did the student enrollment increase each year?

25 Rewrite each expression in an equivalent form by using the distributive property and combining like terms.

  a. $\frac{1}{4}(3x + 5) + \frac{2}{3}(6x - 10)$

  b. $-15(x^2 + 4) - 11(2x + 3x^2)$

  c. $(6 - 4x)^2$

  d. $x(2x + 7) + 3(8x - 9) - 15$

26 Using the fast-food hamburger data in Review Task 18, make a scatterplot and graph the regression line for predicting protein from fat.

  a. How many grams of protein are predicted for a hamburger with 20 grams of fat?

  b. Interpret the slope of the regression line.

  c. Does a line appear to be a good summary for these data?

Least Squares Regression and Correlation

Leonardo da Vinci (1452–1519) was both an artist and a scientist. During the period 1484–1493, he combined these skills in the analysis of human proportion. As part of this work, he wrote instructions for other artists on how to draw the human body. One of his rules was that the kneeling height should be three-fourths of the span of the outstretched arms. Explore this idea by collecting measurements from the students in your class. Save the data as you will need it later in this lesson.

Complete a table like that below. Measure to the nearest inch.

Student	Arm Span	Kneeling Height

Think About This Situation

Think about how you would use your calculator or data analysis software to explore da Vinci's rule.

a Examine a scatterplot of your (*arm span, kneeling height*) data. Describe any patterns that you see.

b Describe the direction and strength of the relationship between arm span and kneeling height. How might you compute a measure of the strength of the association?

c If da Vinci is correct, what should the equation of the regression line be?

d Find the equation of the regression line for your class data. If a student has a 60-inch arm span, what would this equation predict for his or her kneeling height? What does the slope mean in the context of your data? Is there any reason to conclude that da Vinci is not correct?

e Find the point on the scatterplot that the regression line fits least well. How do you think the line would change if you temporarily remove this point from the data set? Try it.

f How do you think your calculator or computer software finds the equation of the regression line?

In previous units, you used your calculator or computer software to find a regression line to summarize the linear relationship between two variables. In this lesson, you will investigate some properties of the regression line and learn how its equation is calculated. You will also learn to describe the *strength* of an elliptical cloud of points by computing a new type of correlation that can be used even if the variables are not ranks.

Investigation 1 How Good Is the Fit?

There are several possible criteria you could use to determine which line through an elliptical cloud of points is the "best-fitting" line. You might choose the line passing through the most points, or you might choose the one with the smallest average distance from the points. In this investigation, you will further explore the method used by most calculators and data analysis software—the method of least squares. As you work on the following problems, look for answers to this question:

> *How is the least squares regression line determined?*

1 With the high price of gasoline in the U.S., motorists are concerned about the gas mileage of their cars. The table below gives the curb weights and highway mileage for a sample of 2007 four-door compact sedans, all with automatic transmissions.

Compact Cars

Car	Curb Weight (in lbs)	Highway mpg
Audi A4	3,450	32
Chevrolet Cobalt	3,216	32
Ford Focus	2,636	34
Honda Civic	2,690	40
Honda Civic Hybrid	2,875	51
Hyundai Accent	2,403	36
Kia Spectra	2,972	35
MAZDA3	2,811	34
Mercedes-Benz C280	3,460	28
Nissan Sentra	2,897	36
Saturn ION	2,805	32
Subaru Impreza	3,067	28
Suzuki Aerio	2,716	31
Toyota Corolla	2,595	38
Toyota Yaris	2,326	39
Volkswagen Rabbit	2,911	30

Source: www.edmunds.com

a. What is the weight of the Kia Spectra in *hundreds of pounds*? Of the Toyota Yaris?

b. Use data analysis software or your calculator to make a scatterplot of the points (*curb weight, highway mpg*). Enter each weight in 100s of pounds, so the slope will not be so close to 0. Does a line appear to be an appropriate summary of the relationship?

c. Find the regression equation and graph the line on the scatterplot.

d. Select the best interpretation of the slope of the regression line from the choices below. Explain your choice.

  • If the weight of a car is increased by 100 pounds, then we predict that the car's highway gas mileage will decrease by about 0.75 mpg.

  • If one model of car is 100 pounds heavier than another model, then we predict that its highway gas mileage will be 0.75 mpg less.

e. A compact car that is not in the table, the Acura TSX, has a weight of 3,345 lbs. Use each of the following to predict the highway mpg for the Acura TSX.

  i. the regression line on the scatterplot

  ii. the equation of the regression line

2 For a car, like the Acura TSX, that *was not used* in calculating the regression equation, the difference between the actual (*observed*) value and the value predicted by the regression equation is called the **error in prediction**:

  error in prediction = observed value − predicted value

a. The Acura TSX has highway mpg of 31. What is the error in prediction for the Acura TSX?

b. The Volkswagen Jetta has a curb weight of 3,303 lbs.

  i. Use the regression equation to predict the highway mpg for the Jetta.

  ii. The Jetta actually has highway mpg of 32. What is the error in prediction for the Jetta?

3 For a car that *was used* in calculating the regression equation, the difference between the observed value and predicted value is called the **residual**:

  residual = observed value − predicted value

a. Estimate the residual from the plot and then compute the residual for the Honda Civic Hybrid. For the Subaru Impreza.

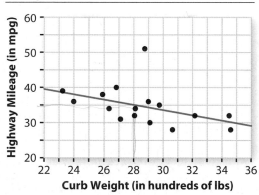

Compact Cars

b. Are points with negative residuals located above or below the regression line?

c. Find a car with a negative residual that is close to zero. What does that residual tell you about the predicted highway mileage?

(4) The equation of the regression line for the three points in the table below is $y = 2x - \frac{4}{3}$.

x	1	2	3
y	1	2	5

a. Graph this line on a scatterplot showing the three points. Draw line segments that show the size of the residuals.

b. Complete a copy of the table below using the equation of the regression line. What do you notice about the sum of the residuals? (This is always true about the sum of the residuals from the least squares regression line.)

x	y	Predicted y	Residual	Squared Residual
1	1			
2	2			
3	5			
	Total			

c. The line that goes through the points (1, 1) and (2, 2) also fits these three points reasonably well.

  i. Write the equation of this line.

  ii. Graph this line on a scatterplot showing the three points. Draw in the residuals for this line.

  iii. Complete a copy of the table above using this new equation to predict values of y.

d. Find the equation of a third line that fits these three points reasonably well. Graph this line on a scatterplot showing the three points. Draw in the residuals. Complete another copy of the table using your new equation to predict values of y.

e. Which of the three equations gave the smallest sum of squared residuals?

f. Compare your answer with that of others who may have used a different equation for the third line. What appears to be true about the regression line?

Your work in Problem 4 illustrates a general rule: The **regression line** or **least squares regression line** is the line that has a smaller **sum of squared errors** (residuals), or **SSE**, than any other line.

Summarize the Mathematics

In this investigation, you explored the least squares method for fitting a line to points on a scatterplot.

a How can you find a residual from the scatterplot? From the equation of the fitted line? What is the sum of the residuals?

b What is the difference between an error of prediction and a residual?

c How is the idea of a sum of squared differences important to least squares regression?

d What is the meaning of the term "least squares"?

Be prepared to explain your ideas to the class.

✔ Check Your Understanding

The data below come from a study of nine Oregon communities in the 1960s, when nuclear power was relatively new. The study compared exposure to radioactive waste from a nuclear reactor in Hanford, Washington, and the death rate due to cancer in these communities.

Community	Index of Exposure	Cancer Deaths (per 100,000 residents)
Umatilla	2.5	147
Morrow	2.6	130
Gilliam	3.4	130
Sherman	1.3	114
Wasco	1.6	138
Hood River	3.8	162
Portland	11.6	208
Columbia	6.4	178
Clatsop	8.3	210

Source: *Journal of Environmental Health*, May–June 1965.

Radioactive Waste Exposure

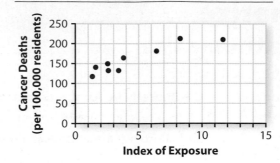

a. Describe the direction and strength of the relationship.

b. Find the equation of the regression line. What does the slope mean in the context of these data?

c. Use the equation to predict the cancer death rate for Hood River. Find and interpret the residual for Hood River.

d. What is the sum of the squared residuals? Can you find a different line that gives a smaller sum? Explain.

Investigation 2   Behavior of the Regression Line

You have seen that the least squares regression line makes the sum of the squares of the residuals as small as possible. The regression line has other important properties. As you work on the following problems, look for answers to this question:

What are some of the properties of the least squares regression line?

1 The point with coordinates $(\bar{x}, \bar{y})$ can be thought of as the balance point for bivariate data. This point is called the **centroid**. The regression line always goes through the centroid.

a. Shown below are the three points in Problem 4 of Investigation 1. Calculate $\bar{x}$ and $\bar{y}$. Then verify that the point with coordinates $(\bar{x}, \bar{y})$ is on the regression line.

x	1	2	3
y	1	2	5

b. Refer back to the table of data on compact car highway mileage in Problem 1 of Investigation 1 (page 282).

  i. Calculate the mean curb weight in 100s of pounds and the mean highway mpg of that group of cars.

  ii. Verify that the point (*mean curb weight, mean highway mpg*) lies on the regression line.

② Now, use data analysis software to examine the idea of finding the line on a scatterplot that minimizes the sum of squared errors (residuals), or SSE.

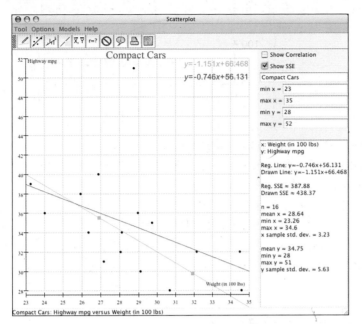

Compact Cars: Highway mpg versus Weight (in 100 lbs)

a. Using the moveable line capability of the software, visually find a line that you think best fits the compact car (*curb weight in hundreds of pounds, highway mpg*) data.

b. Compare the line you found visually and its equation with the regression line and its equation.

c. The Honda Civic Hybrid is an outlier and so may have a large effect on the location of the regression line. To investigate the effect of this point, first remove from the plot the lines you found in Part a.

  i. Delete the point for the Honda Civic Hybrid from the data set. How do the regression line and equation change?

  ii. Replace the point for the Honda Civic Hybrid and then delete the point for the Mercedes-Benz C280, which is the second heaviest car. How do the regression line and equation change in this case?

  iii. Does the Honda Civic Hybrid or the Mercedes-Benz C280 have more influence on the regression line and equation?

An **influential point** is a special type of outlier. It strongly influences the equation of the regression line or the correlation. When such a point is removed from the data set, the slope or y-intercept of the regression line changes quite a bit. The interpretation of "quite a bit" depends on the real-life situation. You will further examine the idea of an influential point in the next two problems.

3 The data in the following table and scatterplot are from a larger set of data collected by a student for a science fair project. The student measured characteristics of a sample of horses because she wanted to see how various measurements were related to the length of the horse's stride.

Horse	Height (in hands)	Hip Angle While Running (in degrees)
Charm	16.1	47.7
Hugs	16.2	48.8
Otis	16.2	51.1
Cosmo	16.2	51.0
Gaspe	16.3	39.4
Sam	16.1	47.1
Pi	16.1	50.6
Binky	16.0	43.7
Bella	16.1	48.8
Prima	16.1	48.7
Bandit	16.0	44.8
Blackie	16.1	48.9

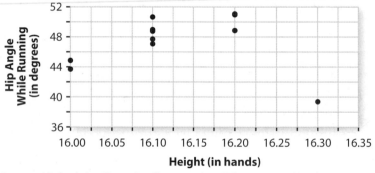

Source: AP Statistics discussion list, posted on February 03, 2006, www.mathforum.org/kb/forum.jspa?forumID=67

a. Which horse is the outlier?

b. What do you think will happen to the slope and intercept of the regression line if this horse is removed from the data set?

c. Do the computations needed to check your conjecture.

d. Is this horse influential?

4 The most famous scandal in baseball history occurred at the 1919 World Series. Eight players for the Chicago White Sox were accused of throwing the series to the Cincinnati Reds. The players were acquitted of criminal charges but banned from professional baseball for life.

The 1919 season and World Series batting averages for the nine White Sox players who had 10 or more at bats in the World Series are given in the following table and scatterplot. These include five of the accused players. The equation of the regression line for predicting the World Series batting average from the season batting average is $y = 1.99x - 0.36$.

Chicago White Sox

Player	Season Batting Average	World Series Batting Average	Accused? N=no/Y=yes
Eddie Collins	.319	.226	N
Shano Collins	.279	.250	N
Happy Felsch	.275	.192	Y
Chick Gandil	.290	.233	Y
Shoeless Joe Jackson	.351	.375	Y
Nemo Leibold	.302	.056	N
Swede Risberg	.256	.080	Y
Ray Schalk	.282	.304	N
Buck Weaver	.296	.324	Y

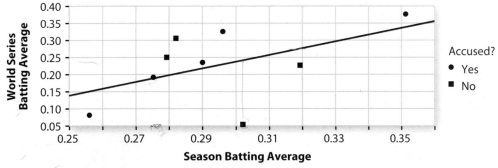

Source: www.baseball-reference.com/postseason/1919_WS.shtml

a. Which player did the worst in the World Series, compared to what would be predicted? Was he one of the accused players?

b. Do the accused players appear to have a different pattern than the players who were not accused?

c. Select the three players who you think might have influential points on the scatterplot. For each player, predict how the regression line would change if he were removed from the data set.

d. Remove one of the three players you identified in Part c, and recompute the equation. Put that player back, remove the second player, and recompute the equation. Put that player back, remove the third, and recompute the equation. Which player was most influential on the slope of the regression line?

Shoeless Joe Jackson

e. How is the effect of removing the point for Nemo Leibold different from the effect of removing either Shoeless Joe Jackson or Swede Risberg?

f. The scatterplot below shows all Cincinnati Reds players who had 10 or more at bats in the Series (none were accused) along with the four White Sox players who had 10 or more at bats and were not accused. It also includes Shoeless Joe Jackson who was accused (circular dot).

i. Describe the relationship between *season batting average* and *World Series batting average* for these players. Is this about what you would expect?

ii. How would the slope of the regression line change if the point for Shoeless Joe were removed from this data set?

Batting Averages

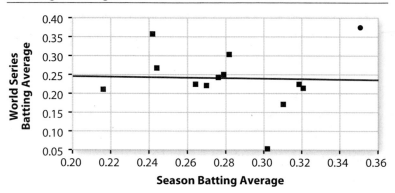

g. The Shoeless Joe Jackson Society is "devoted to the goal of seeing Joe claim his rightful place in the Baseball Hall of Fame." From the evidence that you have seen here, make a case for Shoeless Joe that the Society could use toward their goal.

Summarize
the Mathematics

In this investigation, you explored some properties of the regression line.

a) What point is always on the regression line?

b) What is the difference between an outlier and an influential point?

Be prepared to explain your ideas to the class.

✓ Check Your Understanding

Refer to the Check Your Understanding on page 285 of the previous investigation.

a. Find the mean index of exposure and the mean cancer death rate. Then verify that the centroid is on the regression line.

b. Which community is a potential influential point? Remove this point, find the new regression line, and describe how the regression line changes.

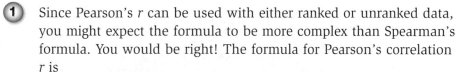

Investigation 3   How Strong Is the Association?

In previous investigations, you have learned how to describe the *shape* of a scatterplot and how to use the regression line as the *center* of an elliptical cloud of points. In this investigation, you will explore a way of describing the *strength* of an elliptical cloud of points by computing a correlation.

A formula for calculating a measure of linear association between pairs of values (x, y) that are ranked or unranked was developed by British statistician Karl Pearson (1857–1936). The resulting correlation, called Pearson's r, can be interpreted in the same way as Spearman's rank correlation, r_s. The correlation indicates the direction of the association between the two variables by whether it is positive or negative. It indicates the strength by whether it is near 1 or −1 versus near 0. For ranked data with no ties, Pearson's and Spearman's formulas give the same value.

As you work on the problems in this investigation, make notes of answers to this question:

How can you compute and interpret Pearson's correlation?

Karl Pearson

1 Since Pearson's r can be used with either ranked or unranked data, you might expect the formula to be more complex than Spearman's formula. You would be right! The formula for Pearson's correlation r is

$$r = \frac{\Sigma(x - \bar{x})(y - \bar{y})}{(n - 1)s_x s_y}.$$

Here, $\bar{x}$ is the mean of the x values, $\bar{y}$ is the mean of the y values, s_x is the standard deviation of the x values, s_y is the standard deviation of the y values, and n is the number of data pairs.

a. Examine the formula. To use it, how would you proceed?

b. Now, examine the plot shown at the right. Which has the larger mean, the values of x or the values of y? Which has the larger standard deviation? What value of r would you expect for these points?

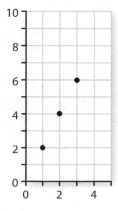

c. When using complex formulas, it is often helpful to organize intermediate calculations in a table.

  i. Compute $\bar{x}$ and $\bar{y}$. Then complete a copy of the table below.

x	y	$x - \bar{x}$	$y - \bar{y}$	$(x - \bar{x})(y - \bar{y})$
1	2			
2	4			
3	6			
Sum (Σ)				

  ii. Compute $n - 1$, s_x and s_y.

  iii. Calculate r by substituting the appropriate sums in the formula. Compare the value you calculated with your prediction in Part b.

d. What value of r would you predict for the points on the scatterplot at the right? Check your prediction by computing the value of r and comparing it to your prediction.

e. What value do you get for r if you reverse the coordinates of each point in Part d? Why does this make sense?

f. Create a set of six points that have correlation -1. Create a set of six points that have correlation close to 0.

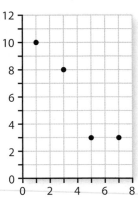

2 In this problem, you will explore how Pearson's formula works. Examine the formula for the correlation r.

$$r = \frac{\Sigma(x - \bar{x})(y - \bar{y})}{(n - 1)s_x s_y}$$

a. Recall that $(x - \bar{x})$ is called a *deviation from the mean*. What is $\Sigma(x - \bar{x})$? What is $\Sigma(y - \bar{y})$? Is this necessarily true of $\Sigma(x - \bar{x})(y - \bar{y})$? Explain.

b. As long as the values of x are not all the same and the values of y are not all the same, the denominator of the formula for r gives a positive number. Explain why this is true.

c. On the scatterplot at the right, horizontal and vertical lines are drawn through $(\bar{x}, \bar{y})$. For the points in region A:

  • Is $(x - \bar{x})$ positive or negative?

  • Is $(y - \bar{y})$ positive or negative?

  • Is $(x - \bar{x})(y - \bar{y})$ positive or negative?

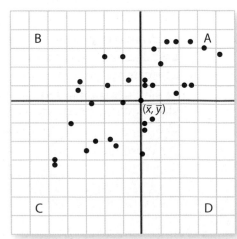

d. Fill in each space on a copy of the table below with "positive" or "negative."

Region	Value of $x - \bar{x}$	Value of $y - \bar{y}$	Value of $(x - \bar{x})(y - \bar{y})$
A			
B			
C			
D			

e. Explain why Pearson's formula will give a positive value of r for the points on the scatterplot in Part c.

f. Explain why Pearson's formula will give a negative value of r for the points on the scatterplot below.

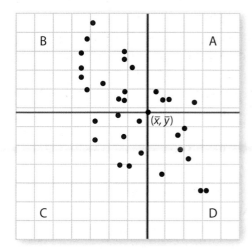

3 For most sets of real data, calculation by hand of Pearson's r is tedious and prone to error. Thus, it is important to know how to calculate r using your graphing calculator or data analysis software.

a. Refer to the scatterplot of arm span and kneeling height for the members of your class that you collected in the Think About This Situation for Lesson 2. Estimate the correlation.

b. Learn to use your calculator or data analysis software to find the value of r for these data and check your estimate in Part a.

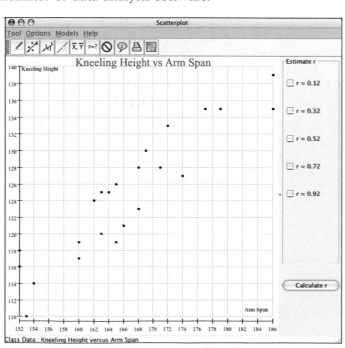

Match each correlation with the appropriate plot. Then write a sentence that describes the association between the two variables in the plot.

a. $r = -0.4$

b. $r = 0.5$

c. $r = -0.8$

d. $r = 0.94$

I

Vehicles

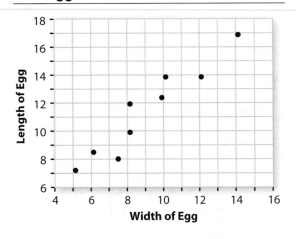

II

Office Workers

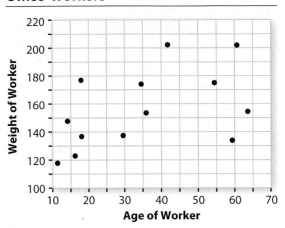

III

Bird Eggs

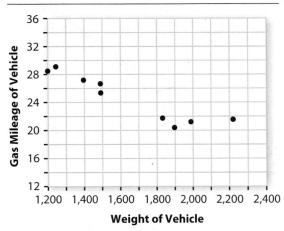

IV

High School Seniors

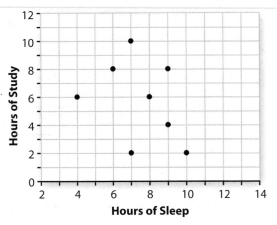

5 In this problem, you will explore the effect of a change of scale on the correlation, r. Shown below is nutritional information on fast-food hamburgers.

How Hamburgers Compare

Company	Burger	Calories	Fat (in grams)	Protein (in grams)	Sodium (in mg)
Hardee's	Hamburger	310	12	14	560
	Thickburger	850	57	30	1,470
Wendy's	Jr. Hamburger	280	9	15	590
	Classic Single	420	20	25	880
Burger King	Hamburger	290	12	15	560
	Whopper	670	39	28	1,020
McDonald's	Hamburger	260	9	13	530
	Quarter Pounder	420	18	24	730
	Big Mac	560	30	25	1,010
Carl's Jr.	Kid's Hamburger	520	18	25	1,040
	Famous Star	590	32	24	910

Source: www.wendys.com; www.mcdonalds.com; www.burgerking.com; www.hardees.com; www.carlsjr.com (December 2006).

A scatterplot matrix of these data is shown below.

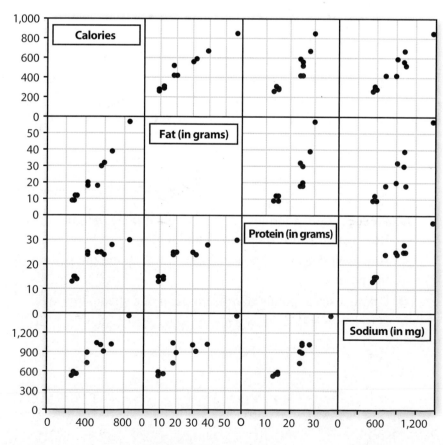

a. Find and interpret the correlation between sodium and calories. Why might this correlation be so strong?

b. Transform the amounts of sodium by converting them to grams. (Recall there are 1,000 milligrams in a gram.) Find r for calories and the transformed values of sodium. What do you notice? Explain why your observation makes sense.

c. Now, transform the numbers of calories by subtracting 200 from each value. Find r for sodium and the transformed values of calories. What do you notice? Explain why your observation makes sense.

(6) As with the regression equation, you can tell if an outlier is influential on the correlation by temporarily removing it from the data set and seeing how much the correlation changes.

a. Which hamburger is an outlier in the (*sodium, calories*) data set? Is this outlier an influential point with respect to the correlation? With respect to the slope of the regression line?

b. For each of the plots below, identify the outlier. Indicate whether removing the point will make the correlation stronger, weaker, or unchanged.

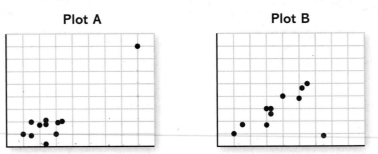

Plot A   **Plot B**

(7) Belinda conjectured that "a high correlation between two variables means that the variables are linearly related."

a. Use your calculator to find the correlation for the following points. From your calculated value, would you expect the scatterplot to have a linear pattern?

x	0	8	1	7	3	6	5	4	2	−2	−1
y	0	65	2	50	8	35	25	15	4	3	1

b. Produce the scatterplot on your calculator. Are the points linear? How would you respond to Belinda?

c. How well does $y = 0.97x^2 + 0.2x + 0.5$ model the pattern in the points?

d. Create a set of points that has a parabolic shape, but the correlation is 0.

8 The scatterplot below shows the heights of 1,078 fathers and their sons. Karl Pearson collected the data around the year 1900.

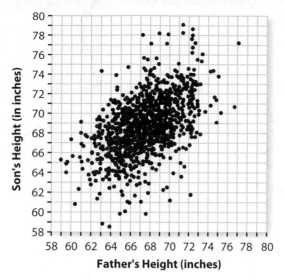

a. Make an estimate of the correlation. Would you say this is a strong correlation?

b. Does this correlation mean that a linear model is not appropriate for these data? Explain your reasoning.

9 Write a summary of what you can conclude from Problems 7 and 8. Compare your conclusions with that of others and resolve any differences.

Summarize
the Mathematics

In this investigation, you learned how to compute and interpret Pearson's correlation for paired (x, y) data.

a Explain why computation of $\Sigma(x - \bar{x})(y - \bar{y})$ determines whether r is positive or negative. How is the sign of r related to the regression equation?

b Explain why it is important to examine a scatterplot of a set of data even though you know the correlation.

c What is the effect on r if you:

   i. reverse the coordinates of the points?

   ii. multiply or divide each value by the same positive number?

   iii. add the same number to each value?

   iv. add an outlier to the data set?

Be prepared to share your ideas and reasoning with the class.

✔Check Your Understanding

The table and scatterplot below give the marriage rates and divorce rates for the countries listed in the *Statistical Abstract of the United States*. Marriage and divorce rates are the number per 1,000 people aged 15–64.

Country	Marriage Rate	Divorce Rate
United States	11.7	6
Canada	6.8	3.3
Japan	8.8	3.4
Denmark	10.4	4.3
France	7.2	3.3
Germany	7.1	3.7
Ireland	7.6	1
Italy	6.9	1.1
Netherlands	7.7	3
Spain	7.4	1.5
Sweden	6.6	3.7
United Kingdom	7.3	4.1

Source: *Statistical Abstract of the United States*, 2006, Table 1320.

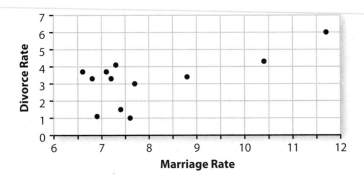

a. Estimate the correlation, and then compute the regression equation and correlation for (*marriage rate, divorce rate*).

b. Interpret the slope of the regression line in the context of this situation.

c. Which country appears to be an influential point? How will the regression equation and correlation change if this country is removed from this data set?

d. Remove this country, and recompute the correlation and regression equation. How influential is this country?

e. Convert the marriage and divorce rates for the United States to the rates per 1,000,000 people. If you do this for all countries and then recompute the regression equation and correlation, will they change?

Association and Causation

Reports in the media often suggest that research has found a cause-and-effect relationship between two variables. For example, a newspaper article listed several "weird" things that are associated with whether or not a student graduates from college. The following excerpt is about one of them.

Five Weird Ways to College Success

Don't smoke.

[Alexander] Astin and [Leticia] Oseguera [of UCLA] examined the graduation rates of 56,818 students at 262 colleges, a huge sample, and reported that smoking had one of the largest negative associations with degree completion.

Source: Jay Mathews, *The Washington Post*, June 13, 2006, www.washingtonpost.com/wp-dyn/content/article/2006/06/13/AR2006061300628.html

As you work on the problems in this investigation, look for answers to the following question:

> *When you have an association between two variables,*
> *how can you determine if the association is a result*
> *of a cause-and-effect relationship?*

1 The 12 countries listed below have the highest per person ice cream consumption of any countries in the world. As shown in the following table and scatterplot, there is an association between the number of recorded crimes and ice cream consumption.

Country	Ice Cream Consumption per Person (in liters) per Year	Recorded Crimes per 100,000 Inhabitants per Year
New Zealand	26.3	12,591
United States	22.5	9,622
Canada	17.8	8,705
Australia	17.8	6,161
Switzerland	14.4	4,769
Sweden	14.2	13,516
Finland	13.9	7,273
Denmark	9.2	1,051
Italy	8.2	4,243
France	5.4	6,765
Germany	3.8	8,025
China	1.8	131

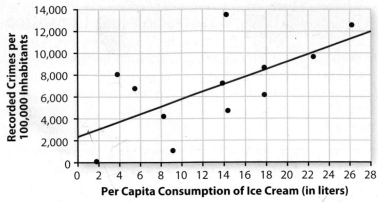

Source: United Nations Office on Drugs and Crime www.unodc.org/unodc/
crime_cicp_surveys.html and www.foodsci.uoguelph.ca/dairyedu/icdata.html.
Their source: *The Latest Scoop*, 2000 Edition, Int. Dairy Foods Assn.

a. For the data above, the regression line is $y = 343x + 2,500$, and
the correlation is 0.637. Interpret the slope of the regression line in
the context of this situation.

b. Do these data imply that if a country wants to decrease the
crime rate, it should ask people to eat less ice cream? Explain
your reasoning.

c. The following scatterplot shows the variables reversed on the axes.
The regression equation is now $y = 0.00118x + 4.77$. Interpret the
slope of this regression line. Can you now say that if the crime rate
increases, then people will eat more ice cream?

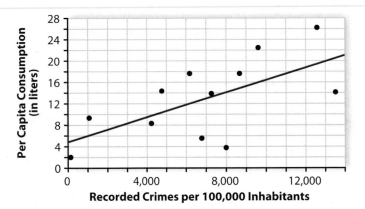

d. A **lurking variable** is a variable that lurks in the background and
affects *both* of the original variables. What are some possible
lurking variables that might explain this association between crime
rate and consumption of ice cream?

There are several reasons why two variables may be correlated:

- The two variables, *A* and *B*, have a **cause-and-effect relationship**.
  That is, an increase in the value of *A*, called the **explanatory** (or
  independent) **variable**, tends to cause an increase (or decrease) in
  the value of *B* called the **response** (or **dependent**) **variable**.

A B

- The two variables, *A* and *B*, have nothing directly to do with each other. However, an increase in the value of a lurking variable *C* tends to cause the values of each of the two variables to increase together, to decrease together, or one to increase and the other to decrease.

A •   • *B*
    ↘ ↙
     C

- Even though the correlation between the two variables is actually zero or close to zero in the entire population, you get a nonzero correlation just by chance when you have a small number of observations.

A •   • *B*

2 The association for each pair of variables below is strong. Decide which of the three reasons above best explains the association. Then, where appropriate, draw a directed graph to indicate the relationship between the two variables.

  a. Number of hours of studying each week and GPA

  b. Reading ability of a child and his or her shoe size

  c. Value of a car and its age

  d. Degree of baldness of a man and probability of a heart attack in the next year

  e. The median household income in the U.S. and skin cancer rate over the years

  f. Number of people attending a movie and income from ticket sales

  g. Number of letters in first name and age for a group of three adult women

3 Suppose you ask everyone in your community who has a phone for the number of letters in his or her last name and for the sum of the last four digits of his or her phone number.

  a. Should there be positive association, negative association, or no association? Why?

  b. Collect the information above from five members of your class or from a local telephone book, and compute the correlation *r* between *number of letters in last name* and *sum of last four digits of phone number.*

  c. Did you get the correlation that you predicted in Part a? Explain why this should or should not be the case.

4 Look back at the article excerpt from the *Washington Post* on page 299.

  a. What variables are said to be associated in the article? Which is considered the explanatory variable and which is the response variable?

  b. What are some possible reasons for the association that are not mentioned in the article?

c. Answer the questions in Parts a and b for the following *Los Angeles Times* article.

Tall Men Display Greater Risk of Skin Disease

A poll conducted by University of Washington researchers in Seattle found that men taller than 6 feet, 1 inch had almost $2\frac{1}{2}$ times the risk of developing melanoma, an often fatal form of skin cancer, as those who were shorter than 5-foot-8.

Source: *Los Angeles Times*, January 14, 2002, page S2.

5 In the following study, the researchers tried to "control" for lurking variables by taking them into account.

Mind Games May Keep the Brain Sharp

An absorbing book or a challenging crossword puzzle may keep your mind more than busy. It may keep it healthy, too, according to a 21-year study of mental breakdown in old age. ...

In the Einstein College study of 469 elderly people, those in the top third in mental activity had a 63 percent lower risk of dementia than the bottom third. Taking part in a single activity one day a week reduced the risk by 7 percent.

The use-it-or-lose-it notion is not a new idea. Other researchers have discovered evidence that mental activity may guard against dementia. But it is hard to prove since early dementia without obvious symptoms may cause people to slack off their hobbies. If this is so, dementia affects hobbies—and maybe not the reverse.

The researchers tried to minimize that possibility by considering only those who were dementia-free for seven years after joining the study. They also tried to eliminate the potential role of education and intelligence in guarding against dementia.

The study also took physical exercise into account. Nearly all physical activities, including stair climbing and group exercises, appeared to offer no protection against dementia. The only exception was frequent dancing, perhaps because dance music engages the dancer's mind, suggested lead researcher Joe Verghese, a neurologist at Einstein College.

Source: www.cnn.com/2003/HEALTH/conditions/06/19/avoiding.dementia.ap/index.html

a. Identify the explanatory and response variables.

b. Name the lurking variables that the researchers considered.

c. Describe how the lurking variables might be taken into consideration.

d. Give at least one other possible lurking variable that is not mentioned.

6 By now, you may be wondering how anyone could ever know whether an association means that one variable causes the other or whether there is a lurking variable that causes both. The only way to find out for sure is to conduct an **experiment**. In an experiment, volunteer subjects are **randomly assigned** to two or more different **treatments**.

For example, suppose you want to decide if a cup of tea causes reduction in pain from a tension headache. You cannot just give a cup of tea to people with a tension headache and see if it goes away because some headaches go away over time without any treatment at all. So, you randomly divide your group of volunteers into those who get a cup of tea and those who get a cup of hot water. By randomizing, you hope to balance the people whose headaches go away quickly without any treatment at all between the two treatment groups. So if the headaches of those with tea tend to go away more quickly, you will know that it is the tea that caused it, not just sitting awhile to have a drink or not just the extra hot liquid.

a. For the study "Mind Games May Keep the Brain Sharp" from Problem 5, describe how you could conduct an experiment to decide whether one variable actually causes the other.

b. For each of these studies, explain why it is impossible to do an experiment to determine cause-and-effect.

  i. "Five Weird Ways to College Success" from page 299

  ii. "Tall Men Display Greater Risk of Skin Disease" from Problem 4 Part c

Summarize
the Mathematics

In this investigation, you learned how to distinguish between correlation and cause-and-effect in situations that involved an association between two variables.

a Describe a situation involving two variables for which the correlation is strong, but there is no cause-and-effect relationship.

b Describe a situation involving two variables for which the correlation is strong and where a change in one variable causes a change in the other variable.

c Explain what is meant by the often-repeated statement, "Correlation does not imply causation."

d When you make a scatterplot, on which axis should you put the explanatory variable?

e How can you be certain whether an association means that there is a cause-and-effect relationship between two variables?

Be prepared to share your ideas and examples with the class.

✓ Check Your Understanding

The following study compared state voter turnout rate to 12 social, economic, and government policy indicators.

Voter Turnout Correlates to Quality of Life

A new study suggests that your vote may count after all, even if every candidate you favor goes down to defeat on Election Day.

A study by the Durham-based Institute for Southern Studies reveals that states with the highest rates of voter turnout also have higher rates of employment and a smaller gap in incomes between the rich and poor.

"Very clearly, it pays to vote," said study author Bob Hall. "There's more reasons to vote than you may think. It may actually influence the quality of life in a broad way."

Source: *The Charlotte Observer*, October 29, 1996.

a. What are the explanatory and response variables implied by this article?

b. Do you think it could just as well be the other way around? Explain.

c. The article goes on to say that some people believe this study may not have considered enough of the variables that determine whether voters go to the polls. What are some possible lurking variables in this situation?

d. Describe an experiment that would determine if higher voter turnout improves the quality of life.

Applications

1 The following table and *plot over time* (years in this case) give the hippopotamus population on the Luangwa River in Zambia for various years between 1970 and 1984.

Year	Number of Hippos
1970	2,815
1972	2,919
1975	2,342
1976	4,501
1977	5,147
1978	4,765
1979	5,151
1981	4,884
1982	6,293
1983	6,544

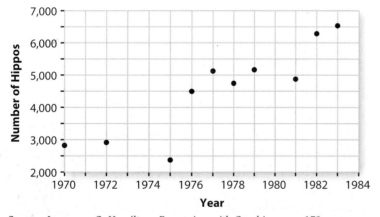

Source: Lawrence C. Hamilton, *Regression with Graphics*, page 179.

a. Find the equation of the regression line and graph it on a copy of the scatterplot.

b. What is the slope of the regression line? Interpret this slope in the context of these data.

c. For which year is the residual largest in absolute value? Estimate this residual using the scatterplot. Then find the value of this residual using the regression equation. Finally, interpret this residual.

d. Use the regression equation to predict the hippopotamus population for the current year. How much faith do you have in this prediction?

2 The age of a tree can often be determined by counting its rings. However, in tropical forests, annual tree rings do not always exist. Researchers measured the diameter of 20 large trees from a central Amazon rain forest and found their ages using carbon-14 dating. The results appear in the following table and scatterplot. The regression equation for predicting the age of a tree from its diameter is $y = 4.39x - 19$.

Diameter (in cm)	Age (in years)	Diameter (in cm)	Age (in years)
180	1,372	115	512
120	1,167	140	512
100	895	180	455
225	842	112	352
140	722	100	352
142	657	118	249
139	582	82	249
150	562	130	227
110	562	97	227
150	552	110	172

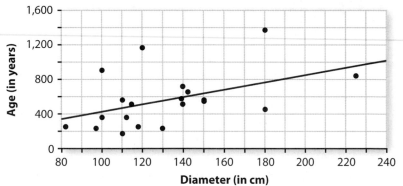

Source: *Statistics for the Life Sciences, 3rd Ed.*, Myra L. Samuels and Jeffrey A. Witmer, pages 575–576, 2003. Their source: Jeffrey Q. Chambers, Niro Higuchi & Joshua P. Schimel. Ancient trees in Amazonia, *Nature, 391* (1998) 135–136.

a. Interpret the slope of the regression line in the context of these data.

b. Use the regression equation to predict the age of a tree that is 125 cm in diameter.

c. For which tree diameter is the residual largest? Estimate the value of this residual from the scatterplot. Then find the value of this residual using the regression equation. Finally, interpret this residual.

d. Does it appear that the age of a tree can reasonably be predicted from measuring its diameter?

3 The following table and scatterplot show the average gestation periods (length of pregnancy) and average life spans of various mammals. The regression equation for predicting *average longevity* from *gestation* is $y = 0.0425x + 6.2$.

Gestation and Life Span of Some Mammals

Mammal	Gestation (in days)	Average Longevity (in years)	Mammal	Gestation (in days)	Average Longevity (in years)
Baboon	187	20	Goat	151	8
Black Bear	219	18	Gorilla	258	20
Beaver	105	5	Horse	330	20
Bison	285	15	Leopard	98	12
Cat	63	12	Lion	100	15
Chimpanzee	230	20	Moose	240	12
Cow	284	15	Rabbit	31	5
Dog	61	12	Sheep	154	12
African Elephant	660	35	Squirrel	44	10
Fox (red)	52	7	Wolf	63	5

Source: *World Almanac and Book of Facts 2001.* Mahwah, NJ: World Almanac, 2001.

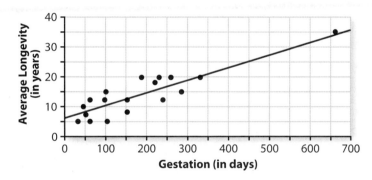

a. Does a line appear to be an appropriate model of this situation?

b. What is the slope of the regression line? What does the slope indicate in the context of these data?

c. Use the regression line to predict the average life span of elk that have a gestation time of 250 days. How much faith would you have in the prediction?

d. Domestic pigs have a 112-day gestation period and live for an average of 10 years. Find and interpret the error of prediction for the domestic pig.

e. Verify that the regression line contains the centroid $(\bar{x}, \bar{y})$.

f. Identify a potential influential point in these data and determine how influential it is with respect to the regression equation.

4 The following table and plot over time give the federal minimum wage in dollars in the United States for the years when Congress passed an increase in the minimum wage. The regression equation for predicting the minimum wage given the year is $y = 0.1027x - 200.26$.

Year	Federal Minimum Wage (in dollars)
1955	0.75
1956	1.00
1961	1.15
1963	1.25
1967	1.40
1968	1.60
1974	2.00
1975	2.10
1978	2.65
1979	2.90
1980	3.10
1981	3.35
1990	3.80
1991	4.25
1996	4.75
1997	5.15
2007	5.85

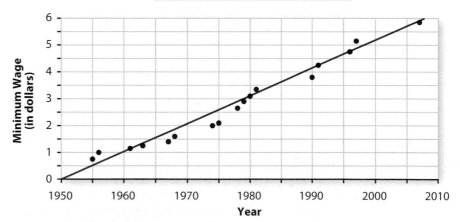

a. Is a line a reasonable model for these data?

b. Verify that the regression line contains the centroid $(\bar{x}, \bar{y}) \approx (1977.53, 2.77)$.

c. What is the slope of the regression line? What does it mean in the context of these data?

d. Check that the sum of the residuals is 0. Then find the sum of the squared residuals.

e. Use the regression line to predict the minimum wage for the current year. What was your error in prediction?

5 A table and scatterplot showing the amount of fiber and the number
of calories in one cup of various kinds of cereal are shown below.

Cereal Nutrition Information

Cereal	Calories	Fiber (in gm)	Cereal	Calories	Fiber (in gm)
Alpha-Bits	133.5	1.5	Honey Graham Oh's	149	1
Apple Jacks	115.5	0.5	Honey Nut Cheerios	114.5	1.5
Cap'n Crunch	143	1	Kix	85.5	0.5
Cheerios	109.5	2.5	Lucky Charms	116	1
Cocoa Puffs	119	0	Product 19	110	1.5
Corn Chex	113.5	0.5	Puffed Rice	53.5	0
Corn Flakes	102	0.5	Raisin Bran (Kelloggs)	196.5	8
Froot Loops	117.5	0.5	Rice Krispies	99.5	0.5
Frosted Mini-Wheats	186.5	6	Special K	114.5	6.5
Golden Grahams	154	1	Total	140.5	3.5
Grape Nuts	389	11	Trix	122.5	0.5
Grape Nuts Flakes	144.5	4	Wheaties	110	2

Source: www.cereal.com/nutrition/compare-cereals.html

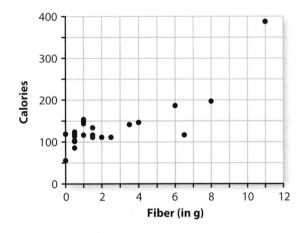

a. Describe the relationship between the grams of fiber and the
calories in a serving of cereal.

b. Which of the following do you estimate is closest to the
correlation?

$r = -0.8$ $r = -0.3$ $r = 0.5$ $r = 0.8$

c. Which cereal is a potential influential point? What will happen to
the slope of the regression line if it is removed from the data set?

6 The average length and weight of five different kinds of seals are given below.

Seal Sizes

Seals	Length (in ft)	Weight (in lbs)
Ribbon Seal	4.8	176
Bearded Seal	7.0	660
Hooded Seal	8.0	880
Common Seal	5.2	220
Baikal Seal	4.2	187

Source: *Grzimek's Encyclopedia, Mammals* V4. New York: McGraw-Hill, 1990.

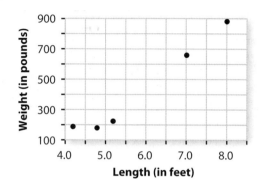

a. Estimate the correlation between the average length and weight of the seals.

b. Calculate the correlation. How close is *r* to your estimate?

c. If you include the Northern Elephant seal at 14.4 feet long and 5,500 pounds, how do you think the correlation will be affected? Check your conjecture.

d. Do you think a line is a good model of the data? Why or why not?

e. Suppose in the table above, you converted each length to meters and each weight to kilograms. (A foot is 0.3048 meters, and a pound is about 0.454 kg.) What would be the correlation? Explain.

7 Consider the following two situations involving possible lurking variables.

a. Examine the following plots of mean earnings and years of schooling for men and women who are year-round, full-time workers, 25 years and older.

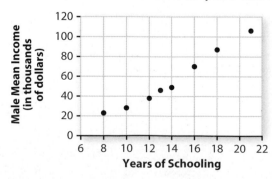

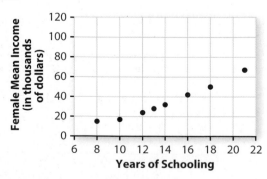

Source: U.S. Census Bureau, Current Population Survey, Annual Social and Economic Supplement. www.census.gov/hhes/www/income/histinc/p22.html

i. As you can see, for people in the United States, there is a high correlation between number of years of schooling S and yearly income I. One theory is that the correlation is high because jobs that pay well tend to require many years of schooling. Model this theory by a directed graph.

ii. Some people have suggested that there is a lurking variable P, which is the economic status of the person's parents. That is, a person whose parents have more money tends to have the opportunities to earn more money. He or she also tends to be able to stay in school longer. Model this theory by a directed graph.

b. Examine the following report of research in which some possible lurking variables have been controlled.

Schooling Pays Off on Payday

Workers earn more from their investment in education than had been thought, a new study says. Students can increase their future income by an average of 16% for each year they stay in school, the study reports. Researchers Alan Krueger and Orley Ashenfelter, both of Princeton, based their estimate on interviews with 250 sets of twins. They correlated differences in wages and years of schooling within sets of twins.

Source: Todd Wallack, *USA Today*, September 1993.

i. How have the researchers controlled for some lurking variables? Which lurking variables have been controlled by this method?

ii. What lurking variable(s) has not been controlled?

8 The following article appears to claim that unemployment allows people to live longer.

Study Links Job Loss, Longer Life

As the economy enters another year of expansion and low unemployment, new research suggests that loss of a job may actually contribute to a healthier, longer life for at least some Americans. Christopher Ruhm, a professor of economics at the University of North Carolina at Greensboro, has concluded in a study that higher unemployment may lead to lower overall mortality rates and reduce fatalities from several major causes of death. The new study, which looks at state-level data compiled between 1972 and 1992, suggests that a 1 percentage point rise in the unemployment rate lowers the total death rate by 0.5 percent.

Source: *San Diego Union-Tribune*, January 27, 1997. Reprinted by permission of Reuters.

a. What variable is said to be the explanatory variable? The response variable?

b. Suppose you were to graph these data on a scatterplot.

    i. What would each point represent?

    ii. What variable would go on the *x*-axis? On the *y*-axis?

    iii. What would be the slope of the regression line?

c. Name a lurking variable that might explain the relationship between higher unemployment rates and lower death rates.

9 In each of the following news clips, a study is reported that revealed an association between two variables. Comment on the validity of the conclusion and whether or not you think there is a cause-and-effect relationship between the two variables.

a. *USA Today* (June 14, 2001) reported a study by researcher Lilia Cortina of the University of Michigan-Ann Arbor that rudeness in the workplace is damaging mental health and lowering productivity. "As encounters with uncivil behavior rose, so did symptoms of anxiety and depression. … Incidents of rude behavior were tied to less job satisfaction for the employee and lower productivity.

b.

> ### Study Links Parental Bond to Teenage Well-Being
> #### by Judy Foreman
>
> A study published in the Journal of the American Medical Association finds that strong emotional connection to a parent is the factor most strongly associated with teenagers' "well-being", as measured by health, school performance, and avoidance of risky behavior. The correlations were found to hold regardless of family income, education, race, and the specific amount of time a parent spends with a child or family structure.
>
> From an initial 1995 survey of 90,000 students in grades 7 through 12, the study focused on 12,000 teenagers, who were interviewed individually at home in 1995 and again in 1996. The study was praised for its breadth and depth, and the data are expected to be a continuing source of material for investigation.
>
> Among the findings already reported here are the following. High parental expectation for school performance were associated with lower incidence of risky behavior. Feeling that at least one adult at school treats them fairly was associated with lower risk in every health category studied except for pregnancy. Students with easy access to guns, alcohol, tobacco at home were more likely to use them or to engage in violence.

Source: *The Boston Globe*, 10 September 1997, A1.

c.

Why Your Credit History Affects Your Insurance Rates
by Carrie Teegardin and Ann Hardie

By shuffling a customer's debt and bill-paying records through a complicated computer program, insurers believed they could predict with amazing accuracy which customers were most likely to get into an auto accident and file a claim.

The computer program boiled each customer's history down to a new version of a credit score and called it an "insurance score." Customers with bad scores were bigger risks than customers with good scores, insurers said, so it was only fair that their policies cost more.

Like most people, Golick couldn't then—and can't now—explain the connection. Why would information about credit card bills and mortgage payments predict someone's driving habits?

"I work in this business. It is not obvious to me," said Golick, who in addition to his legislative job is an attorney for Allstate Insurance Co. "I do know that the data is conclusive that there is absolutely a correlation."

The mysterious correlation was so strong that it prompted Golick, who handles regulatory matters for Allstate's Southeast region, to take action. In 2003, he sponsored legislation that allows insurers to use credit information when pricing auto and homeowners insurance—but keeps the formulas they use secret from consumers.

Source: *Atlanta Journal-Constitution*, December 12, 2006, www.ajc.com/business/content/business/stories/2006/12/09/1210bizcreditmain.html

Connections

10 Make a scatterplot of the points (1, 1), (2, 2), and (3, 5). Plot the regression line $y = 2x - \frac{4}{3}$. Draw line segments on your graph to show the residuals for each point. Illustrate the geometry of the term *squared residuals* by drawing on the graph an appropriate square for each residual.

11 Consider the set of points (1, 3), (2, 2), (3, 5), and (6, 5).

a. Using the equation $y = x + 0.75$, find the sum of the residuals and the sum of the squared residuals.

b. Using the equation $y = 0.5x + 2.25$, find the sum of the residuals and the sum of the squared residuals.

c. One of the two equations is the regression equation. Tell which one it is and how you know.

12 In this task, you will discover one reason why the sum of the *squared* residuals and not the sum of the *absolute values* of the residuals is used in defining the regression line.

    a. Find the centroid of the points below.

x	y
0	0
0	1
1	0
1	1

    b. Find the equations of three different lines that go through the centroid.

    c. For each of your three lines in Part b:

        i. find the sum of the residuals.

        ii. find the sum of the absolute values of the residuals.

        iii. find the sum of the squared residuals.

    d. Which of your three lines has:

        i. the smallest sum of residuals?

        ii. the smallest sum of absolute residuals?

        iii. the smallest sum of squared residuals?

    e. What is one helpful result of squaring the residuals when finding the regression line?

13 Imagine a scatterplot of points (x, y) and a second scatterplot of the transformed points $(-x, y)$. Make conjectures about answers to the questions in Parts a–c.

    a. How do the plots of (x, y) and $(-x, y)$ differ?

    b. How are the correlations related?

    c. How are the regression lines related?

    d. Test your conjectures with a set of ordered pairs (x, y) and a transformed set $(-x, y)$.

14 Create a set of five ordered pairs (x, y) for which the values of y are all even, positive integers and the points are not all collinear.

    a. Plot your points.

    b. Find the correlation, the regression line, and the sum of the squared errors.

    c. Transform the values using the rule $(x, y) \rightarrow (x, 0.5y)$. Make a scatterplot of the transformed values. Then find the regression line and recalculate the correlation and the sum of the squared errors.

    d. Compare and explain the results of Parts b and c.

15 For a project, Diana is examining the question of whether she can use linear regression to predict the height of a daughter from the height of the mother. She did all of her measurements in inches and has computed the mean height of the mothers, the mean height of the daughters, the value of r, and the equation of the regression line. Her science teacher suggested that she report her results in centimeters rather than in inches. (Recall there are approximately 2.54 centimeters in an inch.)

   a. How can Diana most easily find the mean height in centimeters of the mothers and the daughters?

   b. If Diana reports her results in centimeters rather than in inches, how is the value of r affected?

   c. If Diana reports her results in centimeters rather than in inches, how does the equation of the regression line change?

   d. If the heights of the mothers are left in inches but the heights of their daughters are reported in centimeters, how is the value of r affected?

16 In this task, you will compare Pearson's correlation and Spearman's rank correlation for two sets of data.

   a. Compute Pearson's correlation for the roller coaster rankings in Applications Task 1 on page 269 in Lesson 1. Spearman's rank correlation was 0.515. Compare the two correlations.

   b. Refer to the sizes of seals in Applications Task 6, page 310. Rank the seals according to length. Rank the seals according to weight. Using the ranks, compute and compare Pearson's correlation and Spearman's rank correlation. Then, compare these with the correlation computed using the actual lengths and weights in Applications Task 6 Part b.

   c. When might you want to rank data before computing a correlation? When would you not want to rank data before computing a correlation?

Reflections

17 What can you say about the shape of a cloud of points:

   a. if the points tend to be above the regression line on the left and right and below it in the middle?

   b. if the points tend to be below the regression line on the left and right and above it in the middle?

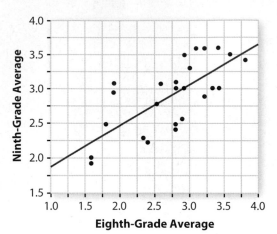

Ninth-Grade Average (vertical axis: 1.5, 2.0, 2.5, 3.0, 3.5, 4.0)

Eighth-Grade Average (horizontal axis: 1.0, 1.5, 2.0, 2.5, 3.0, 3.5, 4.0)

18 The eighth and ninth grade point averages (GPAs) for a sample of 25 students are given in the scatterplot. The line on the scatterplot is the regression line. Its equation is $y = 0.58x + 1.33$.

a. Which of the following is the best interpretation of the slope of the regression line? Explain your reasoning.

• If a student raises his or her eighth-grade GPA by 1 point, then we expect the student to raise his or her ninth-grade GPA by 0.58.

• If a student has an eighth-grade GPA of 1, then we expect the ninth-grade GPA to be 1.58.

• If one student has an eighth-grade GPA that is 1 point higher than another student's, then his or her ninth-grade GPA tends to be only 0.58 higher.

b. Roughly how large would you expect the error of prediction to be in the case of a student with an eighth-grade GPA of 2.2?

19 Many formulas in statistics are based on just a few fundamental ideas. Think about how the formulas in the following pairs are similar.

a. In what way is the formula for the standard deviation similar to the formula for the sum of squared residuals (errors)?

b. The formula for Spearman's rank correlation r_s involved a sum of squared differences. Does the formula for Pearson's correlation r include any sums of squared differences? Explain.

c. The formula below gives the slope of the regression line (see also Extensions Task 25, page 318). How is this formula similar to the formula for Pearson's correlation?

$$b = \frac{\Sigma(x - \bar{x})(y - \bar{y})}{\Sigma(x - \bar{x})^2}$$

20 The following data were collected in an experiment in which students threw a ball straight up in the air and measured the height of the ball over a series of time intervals.

Time (in seconds)	0.0	0.2	0.4	0.6	0.8	1.0	1.2	1.4
Height (in meters)	1.1	2.3	3.2	3.6	3.7	3.5	2.7	1.6

a. Calculate the correlation for these data.

b. As you saw in the Course 1 *Quadratic Functions* unit, you can predict the height h of an object at time t if you know its initial height h_0 at time 0 and the velocity v_0 at which it was initially thrown straight up using the rule:

$$h = h_0 + v_0 t - 4.9t^2$$

In this rule, height is in meters, time is in seconds, and velocity is in meters per second.

i. What is h_0 for the data collected by these students?

ii. Use this value and one other point in the table above to estimate v_0.

iii. Write a rule that describes the pattern in these data using your estimated coefficients.

iv. How well does your rule model the students' data?

c. What might explain why there is a low correlation in Part a but fairly good agreement with your rule in Part b?

21 Each year, hundreds of thousands of people come to see the blooming of the cherry trees around the Jefferson Memorial in Washington, D.C. The *Peak Bloom Date* is defined as the day in which 70 percent of the blossoms of the Yoshino Cherry trees are open. The date when the Yoshino cherry blossoms reach peak bloom varies from year to year, depending on weather conditions.

The plot over time below shows the days after March 1 when the cherry trees hit peak bloom for the years beginning in 1980.

Peak Cherry Tree Blooming

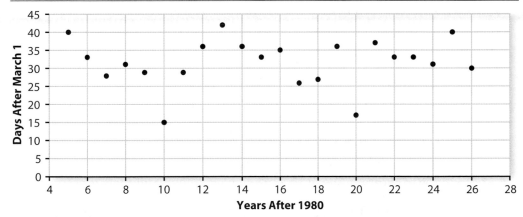

Source: National Park Service. www.nps.gov/cherry/updated.htm

a. Estimate the correlation between day of peak bloom and year.

b. Should a line be fit to these data?

c. As is typical of plots over time, there is a pattern in these data, apart from any linear trend. How would you describe it?

Extensions

22 Experiment with data analysis software and sets of ordered pairs (x, y) to create the following examples.

a. An example where a line is a good model for the scatterplot even though there are relatively large residuals

b. An example where a line is not a good model for the scatterplot even though there are relatively small residuals

23 In this task, you will explore the relationship between the slope of a line through the centroid and the sum of squared residuals.

a. Find the centroid of the four points (1, 1), (2, 3), (3, 4), and (6, 8).

b. Find the equation of the line that goes through the centroid and has a slope of 0. Compute the sum of squared residuals (SSE) for that line.

c. Repeat Part b using the slopes in the table below. Fill in the values of the SSE.

Slope	0	0.5	1	1.5	2	2.5
SSE						

d. Plot the pairs (*slope, SSE*). What do you observe?

e. Estimate the slope that will give the smallest SSE.

f. Check your answer to Part e by finding the equation of the regression line.

24 In this task, you will explore why the regression line is sometimes called the "line of averages." Shown below is a plot of the data from the Check Your Understanding on page 285 of the exposure to radioactive waste from a nuclear reactor in Hanford, Washington, and the rate of deaths due to cancer in these communities. The vertical lines divide the set of data points into thirds.

Radioactive Waste Exposure

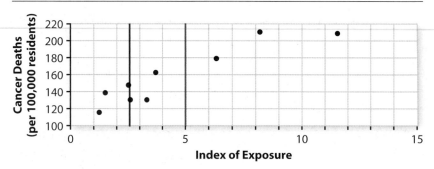

a. Compute the centroid of the points in each of the three vertical strips using the data given on page 285 and plot them on a copy of the plot.

b. Draw in the regression line.

c. How close are the three centroids to the regression line? Could you have done a good job of approximating the regression line just by drawing in a line that goes almost through the three centroids?

25 Formulas that can be used to find the slope b and y-intercept a of the regression line $y = a + bx$ are

$$b = \frac{\Sigma(x - \bar{x})(y - \bar{y})}{\Sigma(x - \bar{x})^2} \qquad \text{and} \qquad a = \bar{y} - b\bar{x}.$$

a. Use these formulas to find the equation of the regression line for the points (1, 1), (2, 2), and (3, 5). Check your computations by finding the regression equation using your calculator or data analysis software.

b. What fact is reflected in the formula for a?

26 Examine these four sets of points and their scatterplots.

Set 1		Set 2		Set 3		Set 4	
x	**y**	**x**	**y**	**x**	**y**	**x**	**y**
10	8.04	10	9.14	10	7.46	8	6.58
8	6.95	8	8.14	8	6.77	8	5.76
13	7.58	13	8.74	13	12.74	8	7.71
9	8.81	9	8.77	9	7.11	8	8.84
11	8.33	11	9.26	11	7.81	8	8.47
14	9.96	14	8.1	14	8.84	8	7.04
6	7.24	6	6.13	6	6.08	8	5.25
4	4.26	4	3.1	4	5.39	19	12.5
12	10.84	12	9.13	12	8.15	8	5.56
7	4.82	7	7.26	7	6.42	8	7.91
5	5.68	5	4.74	5	5.73	8	6.89

Source: Anscombe, F.J., Graphs in Statistical Analysis, *American Statistician, 27,* 17–21.

Set 1

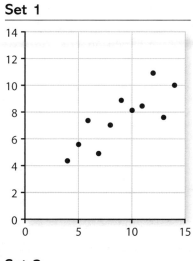

Set 2

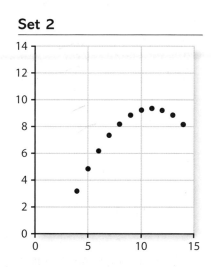

Set 3

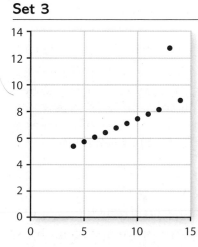

Set 4

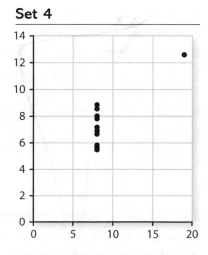

a. Which scatterplot do you predict would have the largest correlation? Which would have the regression line with the largest slope?

b. Check your predictions in Part a by computing the correlations and slopes.

c. What does this analysis illustrate?

27 Not surprisingly, the correlation is related to the slope of the regression line and to the SSE. Verify that the formulas in Parts a and b hold for the points (1, 1), (2, 2), and (3, 5) and the regression line $y = 2x - \frac{4}{3}$. (Refer to your work in Problem 4 of Investigation 1.)

a. The following formula can be used to find r if you know the SSE, the values of y, and whether the trend is positive or negative.

$$r^2 = 1 - \frac{SSE}{\Sigma(y - \bar{y})^2}$$

b. A relationship among the slope b, the correlation r, the standard deviation of the values of x, and the standard deviation of the values of y is expressed by the formula:

$$b = r\frac{s_y}{s_x}$$

c. Interpret the formula in Part b for the case when the values of x and the values of y have the same standard deviation.

Review

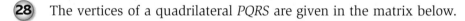

28 The vertices of a quadrilateral *PQRS* are given in the matrix below.

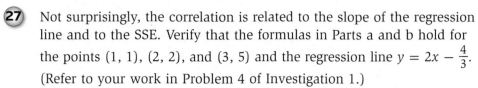

$$\begin{array}{cccc} P & Q & R & S \\ \begin{bmatrix} 0 & 2 & 8 & 1 \\ 4 & 6 & 3 & -1 \end{bmatrix} \end{array}$$

a. Find the coordinates of the image if quadrilateral *PQRS* is reflected across the line $y = x$. Draw a sketch of quadrilateral *PQRS* and this image.

b. Find the coordinates of the image if quadrilateral *PQRS* is rotated 180° about the origin. Draw a sketch of quadrilateral *PQRS* and this image.

c. Find the coordinates of the image if quadrilateral *PQRS* is first reflected across the *y*-axis and then size transformed with center at the origin and scale factor of 3. Draw a sketch of quadrilateral *PQRS* and this image.

29 Consider the graphs of $y = 4x - 40$ and $2x + y = 18$.

a. Describe the shape of these graphs, and explain why the graphs intersect.

b. Find the coordinates of the point of intersection of these two graphs. Show your work.

30 Factor the following binomial expressions.

   a. $x^2 + 4x$

   b. $5x^2 - 25x$

   c. $-6x^2 + 4x$

   d. $18x - 12x^2$

31 Answer the following questions about percent change.

   a. Mr. and Mrs. Reyes bought a house for $150,000. Three years later, it was sold for $165,000. By what percent did the house increase in value?

   b. If all recording companies agree to lower the average price of compact discs from $15 to $12, by what percent will the average price be marked down?

   c. The membership fee at the Sierra Vista Community Recreation Center is expected to increase by 4% per year for the next several years. If the membership fee is $253 this year, what will it be in three years?

32 Without using your calculator, determine if the numbers in each pair are equivalent.

   a. $5\sqrt{5}$ and $\sqrt{125}$

   b. $\sqrt{288}$ and $12\sqrt{2}$

   c. $6\sqrt{5}$ and $2\sqrt{45}$

   d. $\sqrt{8}\,\sqrt{10}$ and $\sqrt{16}\,\sqrt{5}$

33 Draw a graph of each quadratic function, and identify the coordinates of the vertex of each parabola.

   a. $y = -x^2$

   b. $y = x^2 + 3$

   c. $y = -x^2 + 4x$

   d. $y = (x - 2)(x - 8)$

Looking Back

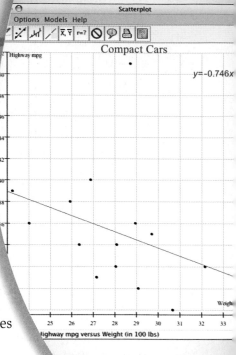

In this unit, you learned how to interpret scatterplots and how to describe their shape, their center using the regression line, and their strength (spread from the regression line) using Pearson's correlation coefficient. You learned that the regression line is the line that minimizes the sum of the squared residuals. You discovered that influential points, which can be detected visually on a scatterplot, can make a marked difference in the correlation coefficient and the regression line. You also examined why a high correlation between two variables does not imply that a change in one of the variables tends to cause a change in the other. The tasks that follow give you an opportunity to pull together the important ideas and methods of this unit.

① A study was conducted to determine if babies bundled in warm clothing learn to crawl later than babies dressed more lightly. The parents of 414 babies were asked the month their child was born and the age that the child learned to crawl. The table below and the scatterplot on the next page give the average daily outside temperature when the babies were six months old and the average age in weeks at which those babies began to crawl.

Crawling Age

Birth Month	Average Outside Temperature at Age 6 Months (in °F)	Age Began to Crawl (in weeks)
January	66	29.84
February	73	30.52
March	72	29.70
April	63	31.84
May	52	28.58
June	39	31.44
July	33	33.64
August	30	32.82
September	33	33.83
October	37	33.35
November	48	33.38
December	57	32.32

Source: Benson, Janette. *Infant Behavior and Development,* 1993.

Crawling Age

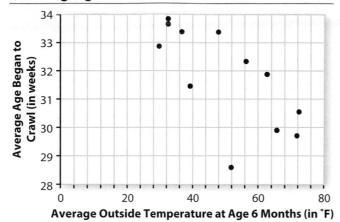

a. Does it appear from the scatterplot that babies who are six months old during cold months of the year learn to crawl at a later age on average than babies who are six months old during warmer months?

b. Approximately how many babies are represented by each point on the scatterplot?

c. What is the shape of the cloud of points?

d. Find the least squares regression line for predicting age from temperature, and graph it on a copy of the scatterplot.

e. Interpret the slope of the regression line in the context of these data.

f. Find the point that has the largest residual (in absolute value).

   i. In what month were these babies born?

   ii. Estimate the residual for that point from the scatterplot.

   iii. Compute this residual using the regression equation and the data in the table.

   iv. Is this point an outlier in terms of x, in terms of y, in terms of both x and y, or only in terms of x and y jointly? Explain.

g. Is the point you identified in Part f an influential point? Explain your reasoning.

2 Respond to the following questions in the context of the baby-crawling study.

a. What variable is suggested as responsible for the association between temperature at six months and the age the babies began to crawl?

   i. Is this a lurking variable? If so, in what sense?

   ii. Make a directed graph to illustrate how this variable operates.

b. What other explanations might be given for this association? Make a directed graph to illustrate each of your possibilities.

3 Refer again to the baby-crawling study in Task 1.

    a. Estimate the correlation.

    b. Compute the correlation using the list capabilities of your calculator and the formula

$$r = \frac{\Sigma(x - \bar{x})(y - \bar{y})}{(n - 1)s_x s_y}.$$

    c. Compute the sum of the squared residuals (SSE).

    d. Explain the relationship of the SSE to the regression equation.

    e. Is the month of May influential with respect to the correlation? Explain.

4 Suppose you transform the data in the baby-crawling study by converting the temperature in Fahrenheit to temperature in Celsius and the age the babies began to crawl to days rather than weeks. This formula may be used to convert the temperatures to degrees Celsius:

$$C = \frac{5}{9}(F - 32)$$

    a. How, if at all, will this transformation change the correlation?

    b. How, if at all, will this transformation change the slope and intercept of the regression line?

    c. Make these transformations to check your answers to Parts a and b.

Summarize
the Mathematics

In this unit, you studied regression and correlation—ways of summarizing the center and strength of an elliptical pattern of paired data.

a What does "linear" mean in statistics? Does a weak correlation necessarily mean there is no linear relationship?

b Describe how the idea of a sum of squared differences is used in statistics.

c What is an influential point?

d What transformations can you make on a set of data and not change the value of the correlation coefficient?

e Give an example to illustrate that a strong correlation does not imply a cause-and-effect relationship.

Be prepared to explain your responses to the class.

✔Check Your Understanding

Write, in outline form, a summary of the important mathematical concepts and methods developed in this unit. Organize your summary so that it can be used as a quick reference in future units and courses.

NONLINEAR FUNCTIONS AND EQUATIONS

In earlier *Core-Plus Mathematics* units, you studied nonlinear functions that are useful in solving problems related to projectile motion, exponential growth and decay, and profit of business ventures. In this unit, you will extend your ability to use quadratic functions to solve scientific, technical, and business problems that involve algebraic systems comprised of linear and nonlinear functions. You will also develop understanding of logarithms—another important tool for modeling nonlinear patterns and for solving problems related to exponential functions.

Key ideas will be developed through your work on problems in three lessons.

Lessons

1 Quadratic Functions, Expressions, and Equations

Use function notation to express relationships between variables. Use the connection between quadratic functions and graphs to solve design problems. Extend understanding and skill in expanding and factoring quadratic expressions and solving quadratic equations.

2 Nonlinear Systems of Equations

Develop numeric, graphic, and symbolic reasoning strategies for solving systems of equations that locate intersections of linear and inverse variation function graphs and intersections of linear and quadratic function graphs.

3 Common Logarithms and Exponential Functions

Use common logarithms to represent patterns of change in quantities, like sound intensity and pH of liquids, and to solve equations that involve exponential functions.

Quadratic Functions, Expressions, and Equations

Luge is a winter sports event in which competitors slide down an ice-covered course at speeds of up to 70 mph, lying on their backs on a small sled. Luge races are timed to the thousandth of a second, reflecting the narrow margin between victory and defeat.

Along with the skill of the athletes, gravity is the major factor affecting the time of each run. Theory about the effects of gravity on falling objects (ignoring friction) predicts the following run times for a 1,000-meter run, straight downhill.

Vertical Drop (in meters)	Run Time (in seconds)
10	143
20	101
30	82
40	71
50	64
60	58
70	54
80	51
90	48
100	45

Inspecting the (*vertical drop, run time*) data makes it clear that *run time* depends on *vertical drop* of the luge course.

a What kind of function would you expect to provide a mathematical model for the relationship of those variables?

b How would you go about finding a rule for such a function?

c How would you expect run times to change if, like most luge courses, the path downhill had a number of sharp, banked curves?

In work on investigations of this lesson, you will learn about mathematical notation for expressing functions. Then you will apply that new knowledge to extend your understanding of quadratic functions and your skill in solving problems that involve quadratics.

Investigation 1 — Functions and Function Notation

The graph and table below show the dependence of run time y on vertical drop x in a luge run. For each x value, there is exactly one corresponding value of y. In such cases, we say that y **is a function of** x or that the relationship between run time y and vertical drop x is a **function**.

Theoretical 1,000-m Luge Run Times

Vertical Drop (in meters)	Run Time (in seconds)
10	143
20	101
30	82
40	71
50	64
60	58
70	54
80	51
90	48
100	45

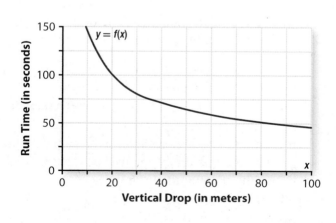

To show that y is a function of x, mathematicians and other professional users of mathematics commonly write "$y = f(x)$." Then facts and questions about the function can also be written in symbolic shorthand form. For example, to express the fact that a vertical drop of 30 meters will lead to a run time of 82 seconds, you could write "$f(30) = 82$" and say "f of 30 equals 82." (Written in this form, "$f(30)$" does *not* mean "f times 30.")

As you work on this investigation, keep in mind these basic questions:

What types of relationships between variables are called functions?

How is function notation used to express facts and questions about functions and the situations they describe?

1 Use information in the table and graph relating downhill run times to vertical drop of the luge course to answer the following questions. In each case, explain what the answer to the question tells about the luge run variables.

  a. How is the fact $f(50) = 64$ shown on the graph?

  b. What value of y satisfies the equation $y = f(40)$?

  c. What value of x satisfies the equation $51 = f(x)$?

  d. What value of y satisfies the equation $y = f(10)$?

  e. What value of x satisfies the equation $45 = f(x)$?

Bouncing Balls Gravity has the same effect on all objects, whether it is an athlete on an ice-covered luge course or a ball dropped from a tall building. However, gravity brings luge athletes down to stay, while it has to bring the bouncing balls down again and again. The following graph shows how the height of one type of bouncing ball changes over time after it is dropped onto a hard surface.

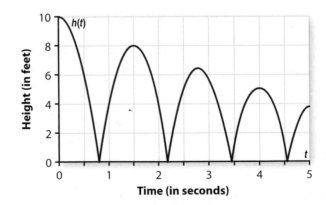

It is common to choose letter names for functions and variables that indicate the quantities involved in the relationship. In this case, it is helpful to use the notation $h(t)$ to represent the height of the ball in feet t seconds after it is dropped.

2 Use the graph above to estimate answers for the following questions about the function relating height of the ball to time. In each case, explain what the answer tells about the bouncing ball.

  a. What information is expressed by $h(0) = 10$?

  b. What is the value of $h(4)$?

c. What values of t satisfy the equation $h(t) = 8$?

d. What is the value of $h(1)$?

e. What values of t satisfy the equation $h(t) = 0$?

3 Recall that in a relationship between two variables x and y, y is a function of x when there is exactly one y value corresponding to each given x value.

a. Explain why the height of a bouncing ball is a function of time since it was dropped.

b. Explain why light intensity is a function of distance from light source to receiving surface.

c. Explain why length of the shadow cast by a tall building is a function of the time of day and time of the year.

d. Explain why the sales tax on a purchase is a function of the selling price of the item.

Sound Intensity The intensity of sound, like intensity of light, is inversely proportional to the square of the distance from the source.

4 The rule $I(d) = \dfrac{8}{d^2}$ gives intensity of sound from a stereo speaker (in watts per square meter) as a function of distance (in meters) from the speaker.

a. Evaluate and explain the meaning of each of the following:

   i. $I(1)$ **ii.** $I(2)$

  iii. $I(0.5)$ **iv.** $I(10)$

b. Consider the equation $I(d) = 0.5$.

   i. What values of d will satisfy the equation?

  ii. What do those values tell you about the sound?

Nonfunctions At this point, it may seem that all relations between variables are functions. That is not the case. You can recognize "nonfunctions" by studying patterns in tables and graphs of (x, y) values and rules relating variables.

5 The relationships between variables in Parts a–d are *not* functions. For each, explain why y is not a function of x.

a. The relationship between height (in cm) y and age (in years) x for a group of 20 young people that is shown in the scatterplot at the right

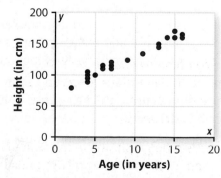

b. The relation between fuel efficiency y (in mpg) and weight x of a car or truck (in lbs)

c.

x	−1	0	1	1	2
y	3	5	7	9	11

d.

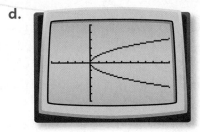 

e. Explain why time is not a function of height in the bouncing ball experiment.

Domain and Range For any function $f(x)$, it is customary to refer to the variable x as the *input* and the function value $f(x)$ as the *output*. Many functions are only defined for some input values. Those numbers make up what is called the **domain** of the function. Similarly, only some numbers will occur as outputs of the function. Those numbers make up what is called the **range** of the function.

There are two ways to think about domain and range for a function that models a real-world situation. On one hand, you might ask what numbers are realistic or *practical* as inputs and outputs. On the other hand, you might ask what numbers *theoretically* can be used as inputs for a given function rule and what numbers will result as outputs (regardless of whether they make sense in the specific situation).

6 Consider again the function rule $I(d) = \dfrac{8}{d^2}$ relating sound intensity and distance from a speaker.

a. Which variable represents inputs? Which variable represents outputs?

b. Give an example of a number that theoretically cannot be used as an input to the rule. Give an example of a number that cannot possibly be an output from the rule.

c. Evaluate each of the following, if possible.

  i. $I(-1)$ **ii.** $I(-2)$ **iii.** $I(0)$ **iv.** $I(-10)$

d. Find the values of d that will satisfy the following equations, if possible.

  i. $I(d) = 8$ **ii.** $I(d) = -0.5$ **iii.** $I(d) = 0$ **iv.** $I(d) = 32$

e. When $I(d)$ is used as a model of the relationship between sound intensity and distance from a source, what might seem to be practical limits on the domain and range of the function?

f. What are the theoretical domain and range of the function $I(d)$? In other words, what values of d can be used as inputs to the rule for $I(d)$ and what range of output values will occur over that domain?

g. Look at a table and a graph of the function $I(d)$ and explain how such tables and graphs can be used to help determine the domain and range of a function.

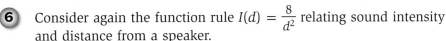

Summarize
the Mathematics

In this investigation, you gained experience in identifying functions and using function notation.

a What does it mean for y to be a function of x?

b What examples would you use to illustrate the difference between relationships of variables that are functions and those that are not?

c How can you tell from a graph whether the relationship of variables displayed is a function or not?

d How can you determine the theoretical domain and range of a function? The practical domain and range?

Be prepared to share your thinking with the whole class.

✓ Check Your Understanding

The graph below shows the height of a basketball from a player's free throw attempt on its path to the basket that is 15 feet from the foul line.

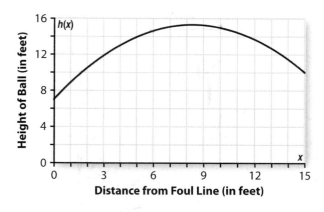

a. Explain why the height of the ball is a function of distance from the foul line.

b. Explain why the distance from the foul line is not a function of height of the ball.

c. If $h(x)$ gives the height in feet of the ball when it is above a spot x feet from the foul line:

   i. What information is expressed by $h(5) = 14$?

   ii. What value of y satisfies the equation $y = h(10)$? What does that fact tell about the flight of the player's shot?

   iii. What value(s) of x satisfy the equation $10 = h(x)$? What do the value(s) tell about the flight of the player's shot?

iv. What are the practical domain and range of the function as shown on the graph?

v. What kind of function rule would you expect to provide a good algebraic model of the relationship shown in the graph?

vi. What would be the theoretical domain and range of such a function?

 **Investigation 2** **Designing Parabolas**

In the *Quadratic Functions* unit of Course 1, you explored quadratic patterns of change and developed skill in reasoning with the various representations of those patterns—tables, graphs, and symbolic rules. In this investigation, you will extend your understanding and skill in use of quadratic functions. As an example, consider the type of problem often faced by architects.

Developers of a new Magic Moments restaurant were intrigued by the design of a restaurant at the Los Angeles International Airport. In a meeting with their architect, they showed her a picture of the airport restaurant and asked if she could design something similar for them.

Like the airport structure, the Magic Moments restaurant was to be suspended above the ground by two giant parabolic arches—each 120 feet high and meeting the ground at points 200 feet apart. To prepare plans for the restaurant building, the designers had to develop and use functions whose graphs would match the planned arches.

One way to tackle this design problem is to imagine a parabola drawn on a coordinate grid as shown below. Any parabola can be described as the graph of some quadratic function.

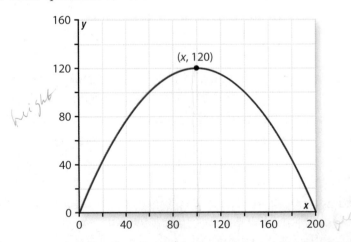

As you work on the problems of this investigation and the next, look for answers to this question:

> *What strategies can be used to find functions that*
> *model specific parabolic shapes?*

1 Using ideas from your earlier study of quadratic functions and their graphs, write the rule for a function with parabolic graph that contains points (0, 0), (200, 0), and a maximum point whose y-coordinate is 120. Use the hints in Parts a–d as needed.

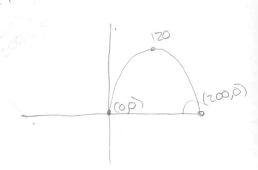

a. The graph of the desired function has x-intercepts (0, 0) and (200, 0). How do you know that the graph of the function $f(x) = x(x - 200)$ has those same x-intercepts?

b. What is the x-coordinate of the maximum point on this graph?

c. Suppose that $g(x)$ has a rule in the form $g(x) = k[x(x - 200)]$, for some particular value of k. What value of k will guarantee that $g(x) = 120$ at the maximum point of the graph?

d. Write the rule for $g(x)$ in equivalent expanded form using the k value you found in Part c.

2 The logo chosen for Magic Moments continued the parabola theme with a large letter M drawn using two intersecting parabolas. The idea is shown in the next graph.

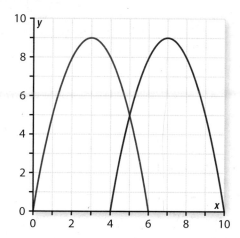

a. Modify the strategy outlined in Problem 1 to find a quadratic function that will produce the leftmost parabola in the M.

- Start with a function $f(x)$ that has x-intercepts at (0, 0) and (6, 0).
- Find coordinates of the maximum point on the graph.
- Find the rule for a related function $g(x)$ that has the same x-intercepts as $f(x)$ but passes through the desired maximum point.
- Write the function rule for $f(x)$ using an equivalent expanded form of the quadratic expression involved.

b. Use a similar strategy to find a function $g(x)$ that will produce the rightmost parabola.

- Start with a function that has a graph with x-intercepts (4, 0) and (10, 0).
- Then adjust the rule so that the function graph also passes through the required maximum point.
- Finally, write the function rule using an equivalent expanded form of the quadratic expression involved.

Your solutions to Problems 1 and 2 illustrate important and useful connections between rules and graphs for quadratic functions. Work on Problems 3, 4, and 5 will develop your understanding and skill in using those ideas.

3 Explain why the next diagram does or does not show the graph of $f(x) = (x - 3)(x + 1)$.

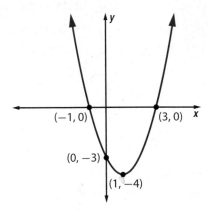

4 Use reasoning alone to sketch graphs of the following functions. Label these key points with their coordinates on the graphs:

- x-intercept(s),

- y-intercept, and

- maximum or minimum point.

Then check results of your reasoning using a graphing tool.

a. $f(x) = (x + 3)(x - 1)$

b. $f(x) = (x - 1)(x - 3)$

c. $f(x) = (x + 1)(x + 5)$

d. $f(x) = -2(x + 2)(x - 3)$

e. $f(x) = 0.5(x - 6)^2$

5 Write rules for quadratic functions whose graphs have the following properties. If possible, write more than one function rule that meets the given conditions.

a. x-intercepts at $(4, 0)$ and $(-1, 0)$

b. x-intercepts at $(7, 0)$ and $(1, 0)$ and graph opening upward

c. x-intercepts at $(7, 0)$ and $(1, 0)$ and minimum point at $(4, -10)$

d. x-intercepts at $(-5, 0)$ and $(0, 0)$ and graph opening downward

e. x-intercepts at $(3, 0)$ and $(-5, 0)$ and maximum point at $(-1, 8)$

f. x-intercepts at $(3.5, 0)$ and $(0, 0)$ and graph opening upward

g. x-intercepts at $(4.5, 0)$ and $(1, 0)$ and y-intercept at $(0, 9)$

h. x-intercepts at $(m, 0)$ and $(n, 0)$

i. only one x-intercept at $(0, 0)$

j. only one x-intercept at $(2, 0)$ and y-intercept at $(0, 6)$

Summarize
the Mathematics

In this investigation, you explored the ways in which factored forms, graphs, and intercepts are related for quadratic functions.

a How can the factored expression of a quadratic function be used to locate the x-intercept(s) and y-intercept of its graph?

b How can the factored expression of a quadratic function be used to locate the maximum or minimum point of its graph?

c How can the x-intercept(s), y-intercept, and maximum or minimum point of a parabola be used to write a rule for the corresponding quadratic function?

Be prepared to share your ideas, strategies, and reasoning with the class.

✓ Check Your Understanding

Use your understanding of connections between rules and graphs for quadratic functions to complete the following tasks.

a. Use reasoning alone to sketch graphs of these functions. Label x-intercepts, y-intercepts, and maximum or minimum points with their coordinates. Then check your sketches with a graphing tool.

   i. $f(x) = x(x + 6)$

   ii. $f(x) = -(x + 3)(x - 5)$

   iii. $f(x) = 3(x - 2)(x - 6)$

b. Find rules for quadratic functions with graphs meeting these conditions. In any case where it is possible, write more than one rule that meets the given conditions.

   i. x-intercepts at $(4, 0)$ and $(-1, 0)$ and opening upward

   ii. x-intercepts at $(2, 0)$ and $(6, 0)$ and maximum point at $(4, 12)$

   iii. only one x-intercept at $(-3, 0)$ and y-intercept at $(0, 18)$

The standard form of rules for quadratic functions is $f(x) = ax^2 + bx + c$. But, as you have seen in problems of Investigation 2, rules for quadratics often occur naturally as products of linear expressions. Those factored quadratic expressions reveal useful connections between the functions and their graphs.

For example, the next diagram shows graphs of the functions $f(x) = x^2 + 2x - 8$ and $g(x) = -x^2 + 5x$. The rules for those functions can also be expressed as $f(x) = (x + 4)(x - 2)$ and $g(x) = -x(x - 5)$, forms that reveal the x-intercepts of each graph.

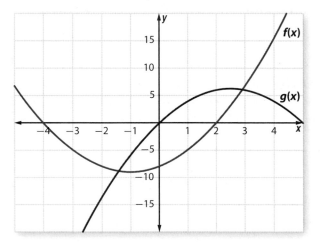

As you work on the problems of this investigation, look for answers to these questions:

What reasoning can be used to expand products of linear factors into equivalent standard form?

How can standard-form quadratic expressions be written as products of linear factors?

1 The next diagram illustrates a visual strategy for finding products of linear expressions.

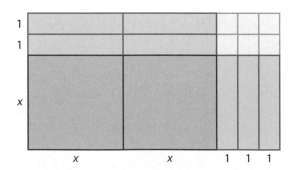

a. How does the diagram show that $(x + 2)(2x + 3) = 2x^2 + 7x + 6$?

b. What similar diagram would help to find the expanded form of $(x + 3)(x + 1)$, and what is that expanded form?

2 In earlier work with quadratic expressions like $-3x(4x - 5)$, you have seen how the distributive property can be applied to write the equivalent form $-12x^2 + 15x$. A group of students at Spring Valley High School claimed that by using the distributive property twice, they could expand other factored quadratic expressions. Check the steps in their example below and then apply similar reasoning to expand the expressions in Parts a–f.

$$(x + 5)(x - 7) = (x + 5)x - (x + 5)7$$
$$= (x^2 + 5x) - (7x + 35)$$
$$= x^2 - 2x - 35$$

a. $(x + 5)(x + 6)$ b. $(x - 3)(x + 9)$

c. $(x + 10)(x - 10)$ d. $(x - 5)(x + 1)$

e. $(x + a)(x + b)$ f. $(x + 7)(2x + 3)$

3 Look back at your work in Problem 2. Compare the standard-form results to their equivalent factored forms in search of a pattern that you can use as a shortcut in expanding such products. Describe in words the pattern that can be used to produce the expanded forms.

4 The next six expressions have a special form $(x + a)^2$ in which both linear factors are the same. They are called *perfect squares*. Find an equivalent expanded form for each expression. Remember: $(x + a)^2 = (x + a)(x + a)$.

a. $(x + 5)^2$ b. $(x - 3)^2$

c. $(x + 7)^2$ d. $(x - 4)^2$

e. $(x + a)^2$ f. $(3x + 2)^2$

5 Compare the standard-form results to their equivalent factored forms in Problem 4. Find a pattern that you can use as a shortcut in expanding such perfect squares. Describe in words the pattern that can be used to produce the expanded form.

6 Write each of these quadratic expressions in equivalent expanded form.

a. $(x + 6)(x - 6)$ b. $(x + 6)(x - 3)$

c. $(2x + 5)(2x - 5)$ d. $(x - 2.5)(x + 2.5)$

e. $(8 - x)(8 + x)$ f. $(x - a)(x + a)$

7 Look back at your work in Problem 6. Compare the standard-form results to their equivalent factored forms to find a pattern that you can use as a shortcut in expanding products like those in Part a and Parts c–f. Describe the pattern in words.

Factoring Quadratic Expressions In many problems that involve quadratic functions, the function rule occurs naturally in standard form $f(x) = ax^2 + bx + c$. In those cases, it is often helpful to rewrite the rule in equivalent factored form to find the x-intercepts of the graph and then the maximum or minimum point.

For example, the height of a gymnast's bounce above a trampoline is a function of time after the takeoff bounce. The function might have rule:

$$h(t) = -16t^2 + 24t$$
$$= -8t(2t - 3)$$

This information makes it easy to find the time when the gymnast hits the trampoline surface again and when she reaches her maximum height.

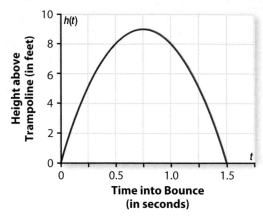

$h(t) = 0$ when $-8t(2t - 3) = 0$
when $-8t = 0$ or when $2t - 3 = 0$
when $t = 0$ or when $t = 1.5$

Maximum value of $h(t)$ is $h(0.75)$ or 9.

In general, it is more difficult to write a quadratic expression like $ax^2 + bx + c$ in equivalent factored form than to expand a product of linear factors into equivalent standard form. In fact, it is not always possible to write a factored form for given standard-form quadratic expressions (using only integers as coefficients and constant terms in the factors).

8 To find a factored form for a quadratic expression like $x^2 + 5x + 6$, you have to think backward through the reasoning used to expand products of linear factors.

Suppose that $(x + m)(x + n) = x^2 + 5x + 6$.

a. How is the number 6 related to the integers m and n in the factored form?

b. How is the number 5 related to the integers m and n in the factored form?

c. What is a factored form for $x^2 + 5x + 6$?

9 Find equivalent factored forms for each of these standard-form quadratic expressions.

a. $x^2 + 7x + 6$ **b.** $x^2 + 7x + 12$

c. $x^2 + 8x + 12$ **d.** $x^2 + 13x + 12$

e. $x^2 + 10x + 24$ **f.** $x^2 + 11x + 24$

g. $x^2 + 9x + 8$ **h.** $x^2 + 6x$

i. $x^2 + 9x + 18$ **j.** $3x^2 + 18x + 24$

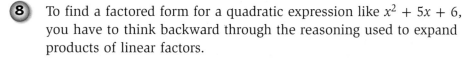

What general guidelines do you see for factoring expressions like these?

10 When a quadratic expression involves differences, finding possible factors requires thinking about products and sums involving negative numbers. Write these expressions in equivalent forms as products of linear expressions.

a. $x^2 - 7x + 12$

b. $x^2 + 5x - 6$

c. $x^2 - 8x + 12$

d. $x^2 - x - 12$

e. $x^2 - 10x + 24$

f. $x^2 + 10x - 24$

g. $x^2 - 9x + 8$

h. $x^2 - 4x$

i. $x^2 - 7x - 18$

j. $2x^2 + 13x - 7$

What general guidelines do you see for factoring expressions like these?

11 The examples here involve some of the special cases you studied in the practice of expanding products. Where possible, write each given expression as the product of linear expressions.

a. $x^2 - 9$

b. $x^2 - 81$

c. $x^2 + 16$

d. $x^2 + 10x + 25$

e. $x^2 - 6x + 9$

f. $x^2 + 16x + 64$

g. $4x^2 - 49$

h. $9x^2 + 6x + 1$

What general guidelines do you see for factoring expressions like these? The two special forms are called *difference of squares* and *perfect square* quadratic expressions.

Using Computer Algebra Tools When the factoring task involves an expression like $-10x^2 + 240x - 950$, things get more challenging. Fortunately, what *is* known about factoring quadratic expressions has been converted into routines for computer algebra systems. Thus, some simple commands will produce the desired factored forms and the insight that comes with them.

For example, the next screen display shows how a CAS would produce a factored form of $-10x^2 + 240x - 950$.

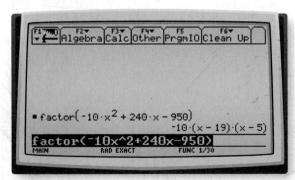

12 Use a computer algebra system to write each expression in equivalent form as products of linear factors. Then use your own reasoning to check the accuracy of the CAS results.

a. $x^2 + 2x - 24$

b. $x^2 - 6x + 5$

c. $-x^2 + 8x - 15$

d. $2x^2 - 7x - 4$

e. $2x^2 + 15x + 18$

f. $3x^2 - 7x - 6$

g. $-3x^2 + 8x$

h. $5x + 3x^2$

Summarize
the Mathematics

In this investigation, you discovered strategies for expanding and factoring expressions that represent quadratic functions.

a How do you go about expanding a product of two linear expressions like $(x + a)(x - b)$?

b What shortcut can be applied to expand products in the form $(x + a)^2$? In the form $(x + a)(x - a)$?

c How would your answers to Parts a and b change if the coefficients of x were numbers other than 1?

d How do you go about finding a factored form for quadratic expressions like $x^2 + bx + c$?

e How would you modify your strategy in Part d for quadratic expressions of the form $ax^2 + bx + c$?

Be prepared to explain your ideas and strategies to others in your class.

✔ Check Your Understanding

Write the following expressions in equivalent expanded or factored forms.

a. $(x + 7)(x - 4)$
b. $(x - 5)(x + 5)$
c. $(x - 3)^2$
d. $x^2 - 100$
e. $x^2 + 9x + 20$
f. $x^2 + 3x - 10$
g. $x^2 + 7x$
h. $5x(x - 3)$

Investigation 4 — Solving Quadratic Equations

In situations that involve quadratic functions, the interesting questions often require solving equations. For example,

When a pumpkin is dropped from a point 50 feet above the ground, it will hit the ground at the time t that satisfies the equation $50 - 16t^2 = 0$.

To find points where the main cable of a suspension bridge is 20 feet above the bridge surface, you might need to solve an equation like
$$0.02x^2 - x + 110 = 20.$$

You can always estimate solutions for these equations by scanning tables of (x, y) values or by tracing coordinates of points on function graphs. In some cases, you can get exact solutions by reasoning algebraically—without use of calculator tables or graphs. As you work on the problems of this investigation, look for answers to this question:

What strategies can be used to solve quadratic equations by factoring and the quadratic formula?

1 Find, if possible, exact solutions for each of the following equations algebraically. Record steps in your reasoning so that someone else could retrace your thinking.

a. $5x^2 + 12 = 57$

b. $-5x^2 + 12 = -33$

c. $3x^2 - 15 = 70$

d. $8 - 2x^2 = x^2 + 5$

e. $7x^2 - 24 = 18 - 2x^2$

f. $x^2 + 4 = 2x^2 + 9$

What general guidelines would you suggest for solving equations like these?

2 Solve the following equations algebraically. Record steps in your reasoning so that someone else could retrace your thinking.

a. $x(5 - x) = 0$

b. $5x^2 + 15x = 0$

c. $16x - 4x^2 = 0$

d. $3x + 5x^2 = -7x$

What general guidelines would you suggest for solving equations like these?

3 Each of the following equations involves a quadratic expression that can be factored into a product of two linear expressions. Solve each equation and record steps in your reasoning so that someone else could retrace your thinking.

a. $x^2 + 2x - 24 = 0$

b. $x^2 - 6x + 5 = 0$

c. $-x^2 + 8x - 15 = 0$

d. $x^2 + 10x + 21 = 0$

e. $2x^2 - 12x + 18 = 0$

f. $x^2 + 6x - 16 = 0$

When you know several strategies for solving different kinds of equations, the key step in working on any particular problem is matching your strategy to the equation form. Use what you have learned from work on Problems 1–3 to apply effective strategies for solving the quadratic equations in Problem 4.

4 Solve each of these equations by algebraic reasoning. Record steps in your reasoning so that someone else could retrace your thinking. Be prepared to explain how you analyzed each given problem to decide on a solution strategy.

a. $x^2 + 6x + 5 = 0$

b. $6x + x^2 = 0$

c. $x^2 + 12x + 20 = 0$

d. $7x + x^2 + 12 = 0$

e. $9 = -7 + 4x^2$

f. $x^2 + 3x + 4 = 0$

g. $2x^2 + 3x + 1 = 0$

h. $2x^2 - 5x = 12$

5 In addition to expanding and factoring expressions, you can also use a computer algebra system to solve quadratic equations directly. The screen display below shows a CAS solution to the equation $2x^2 - 9x - 5 = 0$.

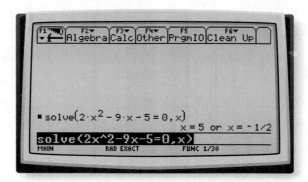

Use a calculator or computer algebra system to solve these equations.

a. $x^2 + 5x + 3 = 0$ **b.** $2x^2 - 5x - 12 = 0$

c. $-10x^2 + 240x - 950 = 0$ **d.** $x^2 - 6x + 10 = 0$

The Quadratic Formula Revisited In solving the equations of Problem 5, you saw how some equations that look fairly complex actually have simple whole number solutions, while equations that look quite simple end up with irrational number solutions or even no real number solutions at all.

When quadratic expressions involve fractions or decimals, mental factoring methods are not often easy to use. You will recall from *Core-Plus Mathematics* Course 1 that for those cases and for situations where you do not have access to a computer algebra system, there is a **quadratic formula** for finding solutions.

For any equation in the form $ax^2 + bx + c = 0$, the solutions are

$$x = \frac{-b}{2a} + \frac{\sqrt{b^2 - 4ac}}{2a} \text{ and } x = \frac{-b}{2a} - \frac{\sqrt{b^2 - 4ac}}{2a}.$$

To use the quadratic formula in any particular case, all you have to do is

Be sure that the equation is in the form prescribed by the formula.

Identify the values of a, b, and c.

Enter those values where they occur in the formula.

6 Test your understanding and skill with the quadratic formula by using it to solve the following equations. In each case:

- Give the values of a, b, and c that must be used to solve the equations.

- Evaluate $\frac{-b}{2a}$ and $\frac{\sqrt{b^2 - 4ac}}{2a}$.

- Evaluate $x = \frac{-b}{2a} + \frac{\sqrt{b^2 - 4ac}}{2a}$ and $x = \frac{-b}{2a} - \frac{\sqrt{b^2 - 4ac}}{2a}$.

- Check that the solutions produced by the formula actually satisfy the equation.

- Graph the related quadratic function to see how the solutions appear as x-intercepts.

a. $5x^2 + 3x - 2 = 0$

b. $x^2 - 5x - 6 = 0$

c. $x^2 - 7x + 10 = 0$

d. $x^2 - x - 12 = 0$

e. $10 - x^2 - 3x = 0$

f. $2x^2 - 12x + 18 = 0$

g. $13 - 6x + x^2 = 0$

h. $-x^2 - 4x - 2 = 2$

(7) Now look back at your work on Problem 6 in search of connections between the quadratic formula calculations and the graphs of the corresponding function rules.

For a quadratic function with rule in the form $f(x) = ax^2 + bx + c$:

a. What information about the graph is provided by $\frac{-b}{2a} + \frac{\sqrt{b^2 - 4ac}}{2a}$ and $\frac{-b}{2a} - \frac{\sqrt{b^2 - 4ac}}{2a}$?

b. What information about the graph is provided by the expression $\frac{-b}{2a}$?

c. What information about the graph is provided by the expression $\frac{\sqrt{b^2 - 4ac}}{2a}$?

Solution Possibilities In solving equations like $ax^2 + bx + c = 0$ with the quadratic formula, the key steps are evaluating $\frac{-b}{2a}$ and $\frac{\sqrt{b^2 - 4ac}}{2a}$. Even if the coefficients a and b and the constant term c are integers, the formula can produce solutions that are not integers.

(8) Use the quadratic formula to check each of the following claims about equations and solutions.

a. The solutions of $x^2 + 5x - 6 = 0$ are *integers* 6 and -1.

b. The solutions of $6x^2 + x - 2 = 0$ are *rational numbers* $\frac{1}{2}$ and $\frac{-2}{3}$.

c. The solutions of $x^2 - 6x + 4 = 0$ are *irrational numbers* $3 + \sqrt{5}$ and $3 - \sqrt{5}$.

d. The equation $x^2 + 5x + 7 = 0$ has no *real number* solutions.

(9) Study the results of your work in Problem 8 to find answers to the questions below.

a. What part of the quadratic formula calculations shows whether there will be 2, 1, or 0 real number solutions? How is that information revealed by the calculations?

b. What part of the quadratic formula calculations shows whether the solutions will be rational or irrational numbers? How is that information revealed by the calculations?

Summarize
the Mathematics

In this investigation, you developed strategies for solving quadratic equations by algebraic reasoning without the aid of calculator or computer tables, graphs, or symbol manipulation programs.

a How can you solve quadratic equations like $ax^2 + b = c$ using algebraic reasoning?

b How can you solve equations like $ax^2 + bx = 0$ using algebraic reasoning?

c How can you solve equations like $ax^2 + bx + c = 0$ when the expression $ax^2 + bx + c$ can be written in equivalent form as the product of two linear expressions?

d In what situations does it make sense to use the quadratic formula to solve an equation?

e How does the quadratic formula show whether a given quadratic equation will have 2, 1, or 0 real number solutions? How will this information appear in a graph and in the calculations leading to the solutions?

Be prepared to share your strategies and thinking with the class.

✔ Check Your Understanding

Solve these equations algebraically—without the quadratic formula, if possible. Show your reasoning in each case.

a. $4x^2 + 7 = 31$

b. $8x^2 + 24x = 0$

c. $x^2 + 15x - 16 = 0$

d. $2x^2 + x = -1$

e. $3x^2 - 10x + 12 = 9$

Applications

1 Consider $I(p) = p(50 - p)$, which gives the expected income in dollars from a bungee jump attraction when the ticket price is p dollars.

   a. Is I a function of p? Explain why or why not.

   b. What does $I(10) = 400$ tell about the bungee attraction?

   c. What are the theoretical domain and range of the function?

   d. What is a practical domain and range of the function in this context?

   e. How does a graph of the function $I(p)$ help determine both the theoretical and practical domains and ranges?

2 Examine the following tables to determine which show y as a function of x. For each table that is not a function, explain why not.

   a.

x	1	2	3	4	5	6	7	8	9
y	3	5	7	9	11	13	15	17	19

   b.

x	1	2	3	4	5	6	7	8	9
y	3	5	7	9	11	9	7	5	3

   c.

x	9	4	1	0	1	4	9	16	25
y	−3	−2	−1	0	1	2	3	4	3

   d.

x	1	2	3	4	5	4	3	2	1
y	3	5	7	9	11	9	7	5	3

3 Examine the following graphs to determine which show y as a function of x. For each graph that is not a function, explain why not.

   a.

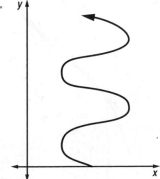

   b.

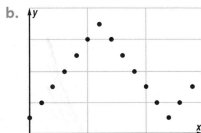

c. 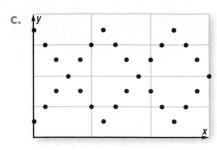   **d.**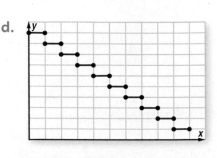

4 If $h(t)$ represents height (in feet above the ground) of a skydiver t seconds into a drop, what do each of the following statements tell about the diver's flight?

a. $h(0) = 10,000$

b. $h(25) = 7,500$

c. $h(10) - h(5) = -2,500$

5 The Zelden Athletic Shoe Company estimates that its production costs (per pair of shoes made) for a new model endorsed by a star athlete will be a function of the number of pairs of shoes that it makes. The rule relating cost to number of pairs of shoes is $c(x) = 29 + \dfrac{25,000,000}{x}$. Calculate and explain the meaning of:

a. $c(1)$ **b.** $c(1,000,000)$

c. $c(2,500,000)$ **d.** The value of x for which $c(x) = 40$.

6 Describe the theoretical domains and ranges for the following functions. That is, explain the values of x that can be used as inputs and the values of y that can occur as outputs.

a. $f(x) = 5 + 3x$ **b.** $g(x) = x^2 + 4$ **c.** $h(x) = 1.5^x$

d. $j(x) = 3\sqrt{x}$ **e.** $k(x) = -x^2 + 2$ **f.** $m(x) = \dfrac{3}{x}$

7 The reflectors in flashlights, like the headlamp model below, are often parabolic in shape so that the bulb sends maximum light forward. In one such headlamp, the reflectors have a diameter of 10 centimeters and are 5 centimeters deep.

Find rules for three different quadratic functions that will each give a graph in the parabolic shape required for the headlamp reflector. Design the three function rules so that their graphs look like those in the following diagrams.

a. Open up with vertex below *x*-axis

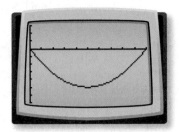

b. Open up with vertex on *x*-axis

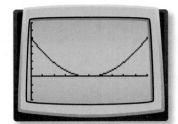

c. Open down with vertex above *x*-axis

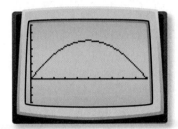

8 The dish antennas used to receive satellite television signals vary in size but are all in the shape of parabolas. Suppose that you were asked to design such a parabolic dish that had a diameter of 4 feet and a depth of 1 foot.

a. Write the rule for a quadratic function with graph that has *x*-intercepts at (0, 0) and (4, 0) and minimum point at (2, −1).

b. Write the rule for a quadratic function with graph that has *x*-intercepts at (0, 0) and (4, 0) and maximum point at (2, 1).

9 Find coordinates of x-intercepts, y-intercept, and maximum or minimum points on the graphs of these quadratic functions. Show how the answers can be obtained by reasoning with the symbolic forms. Think strategically about when and how your answers for one part can help you with the next part.

a. $f(x) = x(7 - x)$

b. $f(x) = -x(7 - x)$

c. $f(x) = (x - 3)(x + 5)$

d. $f(x) = -(x - 3)(x + 5)$

e. $f(x) = (x - 3)(x - 8)$

f. $f(x) = -(x - 3)(x - 8)$

g. $f(x) = 2(x - 3)(x - 8)$

h. $f(x) = -2(x - 3)(x - 8)$

10 Write rules for quadratic functions with graphs that meet these conditions:

a. x-intercepts at $(0, 0)$ and $(6, 0)$ with graph opening upward

b. x-intercepts at $(-2, 0)$ and $(6, 0)$ with graph opening upward

c. x-intercepts at $(-4, 0)$ and $(-6, 0)$ with graph opening downward

d. x-intercepts at $(2, 0)$ and $(6, 0)$ with minimum point at $(4, -8)$

e. x-intercepts at $(2, 0)$ and $(6, 0)$ with maximum point at $(4, 2)$

f. x-intercepts at $(-2, 0)$ and $(6, 0)$ with y-intercept at $(0, -60)$

11 Write these products in equivalent $ax^2 + bx + c$ form.

a. $(x + 7)(x - 3)$

b. $(x - 7)(x + 3)$

c. $(x - 7)(x - 3)$

d. $(x + 3)(x - 3)$

e. $(3 + x)(x - 3)$

f. $(x + 7)(x + 7)$

g. $(2x + 7)(2x - 7)$

h. $(2x + 7)^2$

i. $(5x - 3)(4 + 2x)$

12 Write these quadratic expressions in equivalent form as products of linear factors, where possible.

a. $x^2 + 7x + 10$

b. $x^2 - 7x + 10$

c. $x^2 + 4x - 12$

d. $x^2 - 64$

e. $x^2 + 6x + 9$

f. $64 - x^2$

g. $x^2 - 9x + 20$

h. $2x^2 - 8$

13 Solve these quadratic equations by factoring, where possible.

a. $x^2 + 7x + 10 = 0$

b. $x^2 - 7x + 10 = 0$

c. $x^2 + 4x - 10 = 2$

d. $x^2 - 64 = 0$

e. $x^2 + 6x - 9 = 0$

f. $x^2 - 9x + 20 = 20$

g. $2x^2 - 10x = 0$

h. $x^2 - 15x + 50 = 0$

14 Solve these quadratic equations by use of the quadratic formula.

a. $2x^2 - 10x - 48 = 0$

b. $2x^2 - x + 8 = 0$

c. $6x^2 + 7x - 10 = -5$

d. $3x^2 - 10x + 7 = 0$

e. $4x^2 + 12x + 9 = 0$

f. $-2x^2 + 8x - 3 = 2$

15 For each part, write a quadratic equation that has the indicated solutions.

   a. $x = 5$ and $x = -2$ **b.** $x = -5$ and $x = -2$

   c. $x = 0.5$ and $x = \frac{2}{3}$ **d.** $x = 1$ and $x = \frac{1}{2}$

Connections

16 The graphs below illustrate three different relationships between variables. Match each graph with the description in Parts a–c it seems most likely to represent. In each case, decide whether the graph shows y as a function of x.

   a. Age (in years) and height (in centimeters) for a group of 20 young people of various ages

   b. Age (in years) and IQ for the same group of 20 young people

   c. Age (in years) and average height (in centimeters) for young people of various ages

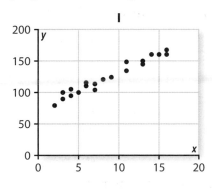

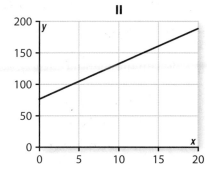

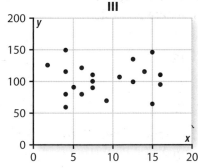

17 The transformations you studied in Unit 3, *Coordinate Methods*, are geometric functions. Under a transformation, each preimage point is paired with exactly one image point. You can use function notation when describing these transformations.

   a. Consider $T(x, y) = (x + 5, y - 12)$.

      i. Find $T(1, 1)$.

     ii. Find $T(-20, 15)$.

      iii. Find $T(0, 0)$.

      iv. Find (x, y) so that $T(x, y) = (-32, 18)$.

      v. What type of transformation is T?

  b. Consider $M(x, y) = (y, x)$.

      i. Find $M(2, 2)$.

      ii. Find $M(16, -11)$.

      iii. Find $M(0, 0)$.

      iv. Find (x, y) so that $M(x, y) = (23, -30)$.

      v. What type of transformation is M?

  c. Consider $S(x, y) = (8x, 8y)$.

      i. Find $S(5, 5)$.

      ii. Find $S(-12, 10)$.

      iii. Find $S(0, 0)$.

      iv. Find (x, y) so that $S(x, y) = (2, -6)$.

      v. What type of transformation is S?

18 When elementary school students first learn the standard algorithm for multiplication, they are often encouraged to record their work for a calculation like 65×42 in a form like this:

$$
\begin{array}{r}
65 \\
\times\ 42 \\
\hline
10 \\
120 \\
200 \\
2{,}400 \\
\hline
2{,}730
\end{array}
$$

  a. Expand the product $(60 + 5)(40 + 2)$ to show why that "beginner's" multiplication algorithm works.

  b. Show how to calculate 73×57 using the "beginner's" multiplication algorithm and how expansion of $(70 + 3)(50 + 7)$ explains why the steps work.

19 Use the algebraic principle that $(m - n)(m + n) = m^2 - n^2$ for any numbers m and n to explain these shortcuts for what seem to be complex arithmetic calculations.

  a. $95 \times 105 = 10{,}000 - 25$ or $9{,}975$

  b. $93 \times 107 = 10{,}000 - 49$ or $9{,}951$

  c. $991 \times 1{,}009 = 1{,}000{,}000 - 81$ or $999{,}919$

20 How can areas of regions in the diagram below be used to give a visual proof that $(x + q)(x + p) = x^2 + (q + p)x + qp$ for any positive numbers p and q?

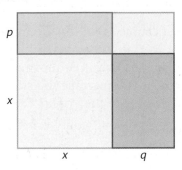

21 Consider matrices $M = \begin{bmatrix} 2 & 3 \\ 4 & 5 \end{bmatrix}$ and $N = \begin{bmatrix} 3 & 4 \\ 2 & 5 \end{bmatrix}$.

a. Compare $7M + 7N$ and $7(M + N)$.

   i. Are the results equal?

   ii. Which expression requires fewer calculations to evaluate?

b. Compare $M(M + N)$ and $M^2 + MN$.

   i. Are the results equal?

   ii. Which expression requires fewer calculations to evaluate?

22 In the *Coordinate Methods* unit, you learned that points on a circle with radius r and center (p, q) have coordinates that satisfy the equation $(x - p)^2 + (y - q)^2 = r^2$. In a precalculus textbook, the following equation was given as an equation of a circle:

$$x^2 + y^2 + 6x - 4y = 23$$

Rebecca believed she could reason as follows to determine the center and radius of the circle and then easily sketch it.

> $x^2 + y^2 + 6x - 4y = 23$ is equivalent to $(x^2 + 6x) + (y^2 - 4y) = 23$, which is equivalent to $(x^2 + 6x + 9) + (y^2 - 4y + 4) = 23 + 9 + 4$, which is equivalent to $(x + 3)^2 + (y - 2)^2 = 36$. So, the circle has center at $(-3, 2)$ and radius 6.

a. Explain Rebecca's strategy and how it is related to your earlier work with recognizing the form of *perfect square* quadratic expressions that are the expanded form of expressions like $(x + a)^2$.

b. Sketch the circle in a coordinate plane.

c. Use similar reasoning to identify the center and radius of each circle below, and then sketch the circle in a coordinate plane.

   i. $x^2 + y^2 + 12x - 2y = -21$

  ii. $x^2 + y^2 + 8y = 9$

23 The area of a square with sides of length x is given by the formula $A = x^2$. Use what you have learned about expanding perfect square expressions to answer the following questions about the way that square areas change as the sides are lengthened.

a. Kei and Matsu claimed that if you increase the side lengths in a square from x to $x + 2$, the area will increase by $2^2 = 4$; if you increase side lengths to $x + 3$, the area will increase by $3^2 = 9$; and, in general, if you increase the side lengths to $x + k$, the area will increase by k^2. Are they right? Explain.

b. What algebraic argument would prove or disprove the conjecture by Kei and Matsu?

c. What visual proof would prove or disprove their conjecture?

24 Use the meaning of square root to help solve each equation.

a. $\sqrt{x - 4} = 10$

b. $\sqrt{3x - 5} = 4$

c. $\sqrt{x - 1} + 3 = x$

Reflections

25 When mathematics students first meet the function notation $f(x)$, they often assume that it means "f times x." Why do you suppose that is such a common difficulty, and what could you do to keep the special meaning of the notation clear?

26 What seems to be the difference between claiming that a relationship between variables x and y *is a function* and suggesting that *y depends on x*?

27 Explain the difference in meaning of these equations: $f(a) = 0$ and $f(0) = a$.

28 Consider the graph of $f(x) = (x - 3)(x + 4)$.

a. What are the x-intercepts of the graph?

b. Why does the graph of the function $g(x) = (x - 3)(x + 4) + 2$ have different x-intercepts than $f(x)$?

c. Why does the graph of the function $h(x) = 2(x - 3)(x + 4)$ have the same x-intercepts as $f(x)$?

29 Based on your earlier studies, you know that two points determine a line. Given the coordinates of the points, you can write an equation $y = ax + b$ for the line.

a. How many points do you think are needed to determine the equation of a parabola?

b. Does it make any difference which points you are given? Explain your thinking.

30 When students are asked to expand quadratic expressions, there are some common errors.

   a. What do you think is the most common error in expanding $(x + a)^2$, and how would you help someone see the error and understand the correct expansion?

   b. What do you think is the most common error in expanding $(x + a)(x + b)$, and how would you help someone see the error and understand the correct expansion?

31 When you need to solve a quadratic equation, how do you decide whether to try factoring, to use the quadratic formula, to use a CAS, or to use a table or a graph of the related quadratic function?

32 When attempting to solve a quadratic equation of the form $x^2 + bx + c = 0$ by factoring, how do you approach the task of finding linear factors whose product is the given quadratic expression?

33 Two students, Brody and Lydia, were arguing about how to solve $(x + 3)(x + 1) = 24$. Brody figured that since $6 \cdot 4 = 24$, he could use the equations $x + 3 = 6$ and $x + 1 = 4$ to find the solution. Lydia insisted that his method will not always give him the right answer. What do you think about Brody's approach?

Extensions

34 In testing the effect of platform height on roll time for the *On a Roll* experiments from Unit 1, *Functions, Equations, and Systems*, it makes sense to use a single ramp length in all rolls. Suppose that a 10-foot ramp length was the choice.

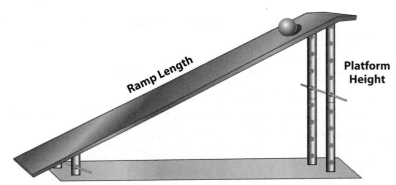

   a. What rule can be used to calculate the distance from the base of the platform to the end of the ramp for any platform height?

     i. Why does this rule describe a function relating distance from platform base to ramp end and platform height?

     ii. What is the practical domain of this function? What is the theoretical domain?

     iii. What is the practical range of this function? What is the theoretical range?

b. Use the function rule from Part a to produce a table showing how the distance from platform base to ramp end changes as the platform height increases from 0 to 10 feet in steps of 1 foot. Plot these (*height, distance*) values. Then add the function graph to your plot.

c. Use the function rule from Part a to write a rule giving ramp slope as a function of platform height.

> **i.** Use that rule to produce a table showing the slope of the ramp for platform heights from 0 to 10 feet in steps of 1 foot.
>
> **ii.** Plot these (*height, slope*) values. Add to your plot the function giving the slope for any platform height.

35 Consider the following functions.

$$f(x) = 3x^2 \qquad g(x) = -2x^3 \qquad h(x) = \frac{5}{x} \qquad k(x) = -\frac{4}{x^2}$$

a. Write sentences describing, in terms of direct and inverse variation, how each of the above functions vary with *x*.

b. What is the value of each function when $x = 1$? When $x = -1$? Express your answers using function notation.

c. When a function correspondence holds in both directions (when each value of *x* corresponds to exactly one value of the function, *and* each value of the function corresponds to exactly one value of *x*), we say that the function is **one-to-one**.

> **i.** Determine whether or not each of the above functions is one-to-one.
>
> **ii.** How could you tell by looking at a function rule that describes a direct or inverse variation whether the function is one-to-one?
>
> **iii.** How could you tell by looking at a graph of a function whether the function is one-to-one?

36 In the Course 1 *Patterns in Shape* unit, you used congruent triangles to prove that if the diagonals of a parallelogram are congruent, then the parallelogram is a rectangle. In this task, you will establish that result by algebraic reasoning with coordinates.

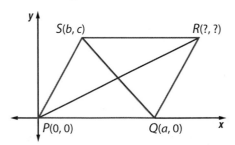

a. What must the coordinates of point *R* be if the quadrilateral is a parallelogram?

b. Write expressions for the length of diagonal $\overline{PR}$ and for the length of diagonal $\overline{QS}$.

c. Use the given fact $\overline{PR} \cong \overline{QS}$ and your expressions in Part b to write a convincing argument that $\square PQRS$ is a rectangle.

37 Solve $x^2 - x - 42 > 0$ by first factoring the quadratic expression.

38 Write expanded forms for each of these functions with rules that are given as products of linear factors. Then graph the results and compare the pattern of those graphs to other functions you have worked with in this lesson.

a. $y = x(x - 1)(x - 2)$

b. $y = x(x + 3)(x - 3)$

c. $y = (x + 2)(x - 1)(x - 4)$

39 Consider line segments of lengths $2m$, $m^2 - 1$, and $m^2 + 1$ where m is any positive integer greater than 1.

a. If $m = 3$, can a triangle be formed with the given lengths as sides? What about if $m = 5$? If $m = 10$?

b. For each triangle that could be formed in Part a, is the triangle a right triangle? Explain your reasoning.

c. Write a general argument that shows that for any positive integer $m > 1$, segments of lengths $2m$, $m^2 - 1$, and $m^2 + 1$ will form a triangle.

d. Show that for any positive integer $m > 1$, $(2m)^2 + (m^2 - 1)^2 = (m^2 + 1)^2$. What can you conclude about the triangles in Part c?

e. Develop a spreadsheet that can be used to make a list of *Pythagorean triples* (a, b, c)—three numbers that are lengths of the legs and hypotenuse of a right triangle. Then, in the same row that each triple appears in the spreadsheet, calculate $a^2 + b^2$ and c^2 to check your work.

40 Look back at your work for Extensions Task 39. If m and n are any two positive integers with $m > n$, will the three numbers $m^2 - n^2$, $2mn$, and $m^2 + n^2$ be a Pythagorean triple? Write a convincing argument justifying your answer.

41 Provide justifications for each step in the following proof that the quadratic formula will find any solutions for equations in the form $ax^2 + bx + c = 0$, $a \neq 0$.

If $ax^2 + bx + c = 0$:

then $a\left(x^2 + \frac{b}{a}x + \frac{c}{a}\right) = 0$; (1)

then $x^2 + \frac{b}{a}x + \frac{c}{a} = 0$; (2)

then $x^2 + \frac{b}{a}x = \frac{-c}{a}$; (3)

then $x^2 + \frac{b}{a}x + \frac{b^2}{4a^2} = \frac{-c}{a} + \frac{b^2}{4a^2}$; (4)

then $\left(x + \frac{b}{2a}\right)^2 = \frac{b^2}{4a^2} + \frac{-c}{a}$; (5)

then $\left(x + \frac{b}{2a}\right)^2 = \frac{b^2}{4a^2} + \frac{-4ac}{4a^2}$; (6)

then $\left(x + \frac{b}{2a}\right)^2 = \frac{b^2 - 4ac}{4a^2}$; (7)

then $x + \frac{b}{2a} = \frac{\sqrt{b^2 - 4ac}}{2a}$ or $x + \frac{b}{2a} = \frac{-\sqrt{b^2 - 4ac}}{2a}$. (8)

So, $x = \frac{-b}{2a} + \frac{\sqrt{b^2 - 4ac}}{2a}$ or $x = \frac{-b}{2a} - \frac{\sqrt{b^2 - 4ac}}{2a}$. (9)

42 The diagram below shows two intersecting circles in a coordinate plane. One has radius 2 and is centered at the origin (0, 0). The other has radius 4 and is centered at the point (4, 0).

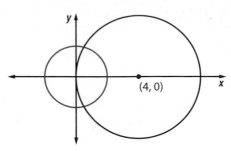

a. What is your best guess for the coordinates of points where the two circles intersect?

b. Write equations for the two circles.

c. To find the intersection points of the circles, you need to solve the system of equations from Part b. Adapt the *method of elimination* from your earlier work with systems of linear equations to solve this system of quadratic equations.

Review

43 You may recall that a number is rational if it can be written as a ratio of two integers. The decimal form of a rational number is either a terminating or a repeating decimal. Some examples of rational numbers are $\frac{2}{3}$, $1.\overline{256}$, -12, and 0. If a real number is not rational, it is irrational. Two examples of irrational numbers that you are familiar with are $\sqrt{2}$ and π. Classify each of the following numbers as rational or irrational.

a. $\sqrt{7}$

b. $\frac{136}{5}$

c. $\sqrt{250}$

d. $\sqrt{\frac{1}{9}}$

e. $(\sqrt{5})^2$

f. $-\frac{2\pi}{3\pi}$

g. $\sqrt{\frac{18}{4}}$

h. $\sqrt{0.81}$

44 Find matrix M such that $3M + \begin{bmatrix} -2 & 4 \\ 5 & -1 \end{bmatrix} = \begin{bmatrix} 1 & 19 \\ -1 & 8 \end{bmatrix}$.

45 Write each of these radical expressions in an equivalent form with the smallest possible positive integer under the radical sign.

a. $\sqrt{250}$

b. $\frac{\sqrt{24}}{2}$

c. $\sqrt{54}\sqrt{12}$

d. $\frac{\sqrt{100 - 4(16)}}{4}$

46 The graph below displays the average tuition costs for four-year and two-year public colleges by state. The equation of the regression line is $y = 0.365x + 400$.

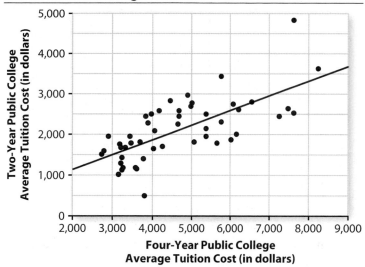

2003–2004 College Costs

Source: nces.ed.gov/programs/digest/d04/tables/dt04_314.asp

a. Explain the meaning of the slope of the regression line.

b. Is the regression coefficient for this set of data positive or negative? Explain your reasoning.

c. In California, the average tuition at four-year public colleges is $3,785 and the average tuition cost at two-year public colleges is $485. Find the residual for this point.

47 Solve each of the following linear inequalities. Graph each solution on a number line.

a. $2x + 5 < 17$

b. $-2x + 5 < 17$

c. $8 - 5x \geq 43$

d. $7x + 3 > 4x - 9$

e. $\frac{x}{2} + 7 < 12$

f. $\frac{3x}{2} + 7 \leq 12$

48 Use rules of exponents to determine the value of x in each equation.

a. $(2^{30})^5 = 2^x$

b. $(30^x)(30^{16}) = 30^{64}$

c. $\frac{5^x}{5^{40}} = 5^{70}$

49 Express each of these relationships among variables in at least two equivalent forms.

a. y is inversely proportional to x with constant of proportionality 5.

b. z is directly proportional to x and inversely proportional to y with constant of proportionality -3.

50 The points $A(0, 0)$ and $B(4, 8)$ are two vertices of a triangle.

a. If the coordinates of vertex C are $(-3, 6)$, is $\triangle ABC$ a right triangle? Explain your reasoning.

b. If the coordinates of vertex D are $(5, -2)$, is $\triangle ABD$ an isosceles triangle? Explain your reasoning.

c. Is the line with equation $y = -\frac{1}{2}x + 8$ the perpendicular bisector of $\overline{AB}$? Explain your reasoning.

51 Solve these proportions.

a. $\frac{17}{x} = 25$

b. $487 = \frac{x}{4.9}$

c. $\frac{6 + x}{3} = \frac{9 + x}{5}$

52 Two octahedral dice with the numbers 1 through 8 on the sides are rolled. The sum of the numbers is found.

a. What is the probability that the sum will be less than 3?

b. What sum has the greatest probability of being rolled?

c. What is the probability that the sum is an even number and less than 5?

LESSON 2

Nonlinear Systems of Equations

When summer approaches, high school and college students start thinking about finding summer jobs. In some cities and towns, students can find work in shops, restaurants, and seasonal service jobs like lawn mowing, farm work, or lifeguarding. But often, there are not enough jobs to ensure that everyone who wants to work will be employed.

Want Ads — Help Wanted

Fast Food—Restaurant seeks summer help; cashiers, cooks, cleanup; 20 hours per week all shifts available. Call 555-5678.

Natural Lawns—Summer Help needed. $7.50 per hour. No prior experience required. Call 24 hours 1-800-555-1589.

Camp Staff—Playground supervisors and camp counselors age 15 or older. Good pay and lots of fresh air. Call 555-6543.

Child Care—Tiny Tots Day Care Center seeks summer help for child care positions. Hours 7-5 four days per week. References to Box Q.

To help students find summer work, both to earn money and to get work experience, many city and county governments have special summer jobs programs. Students are hired to do cleanup and construction jobs in parks or other community facilities. These programs usually have a fixed budget of funds available to pay the student workers.

Suppose that the Kent County government sets aside $200,000 each summer for student salaries in a youth jobs program.

a What factors should planners of the jobs program consider when they set the pay that will be offered to each student worker?

b What factors should students consider when deciding whether or not to apply for the summer work program?

c What salary for 2 months of summer work would attract you to participate in such a program?

Solving the problems of this lesson will help you develop skill in answering questions like these that involve combinations of linear and nonlinear functions.

Investigation 1   Supply and Demand

In planning the Kent County summer jobs program, county officials must consider the relationships between the pay offered for each student, the number of students who could be hired, and the number of students interested in the work opportunity. Analysis of these relationships involves work with systems of functions. As you complete this investigation, look for answers to this question:

> *What strategies are useful in solving problems that involve links between two functions—one a linear function and one an inverse variation function?*

The problem-solving process involves two major steps. First, you have to identify independent and dependent variables and the functions that relate those variables. Then you have to use the functions to answer questions about the variables.

1 Kent County has $200,000 to spend on student salaries.

    a. How many student workers can be hired if the county pays $2,000 per worker for a summer contract covering eight weeks? What if the county pays only $1,500 per worker? What if the county pays only $1,000 per worker?

    b. If the pay per worker is represented by p, what function $h(p)$ shows how the number of students who could be hired depends on the level of pay offered?

    c. Sketch a graph of the function $h(p)$ and write a brief description of the way $h(p)$ changes as p increases.

2 If the jobs program offers very low pay, then few students will be interested in the work opportunity. After doing a survey in one local high school, Kent County officials arrived at the following estimates of the relation between summer pay rate and number of students they can expect to apply for the jobs.

Summer Jobs Program

Pay Offered (in dollars)	500	1,000	1,500	2,000	2,500
Expected Applicants	55	100	155	210	255

a. Does the pattern in the data table seem reasonable? Why or why not?

b. What function $s(p)$ would be a good model of the relationship between the number of students who will apply for the jobs and the level of pay offered?

c. Sketch a graph of $s(p)$ and write a brief description of what it shows about the way the number of job applicants changes as pay increases.

3 The decision to be made by Kent County summer jobs officials is how much pay to offer for the eight-week summer work contracts. Both the number of students who could be hired and the number of students who would be interested in the summer work depend on the pay rate p.

a. Write equations and inequalities that match the following questions about the jobs program, and then estimate or find exact values for solutions.

    i. For what pay rate(s) will the number of students who can be hired equal the number of students who would be interested in the work?

    ii. For what pay rate(s) will the number of students who can be hired be less than the number of students who would be interested in the work?

    iii. For what pay rate(s) will the number of students who can be hired be greater than the number of students who would be interested in the work?

b. When the head of the Kent County summer jobs program had to report to the county council about program plans, he wanted a visual aid to help in explaining the choice of a pay rate to be offered to student workers. Sketch a graph showing how both $h(p)$ and $s(p)$ depend on p and explain how the graph illustrates your answers to the questions in Part a.

Solving Equations in the Form $ax + b = \frac{k}{x}$ An equation like $0.1x = \frac{200,000}{x}$ can be solved by finding point(s) where graphs of the two functions, $y = 0.1x$ and $y = \frac{200,000}{x}$, intersect. The functions involved are representatives of two important function families—one in which the dependent variable is *linearly* related to the independent variable and one in which the dependent variable is *inversely* related to the independent variable.

In Problems 4 and 5, use tables, graphs, and reasoning about equations to explore the solution possibilities for equations that seek intersection points of graphs for one linear function and one inverse variation function.

4 Consider first, equations in the form $ax = \dfrac{k}{x}$ where a and k are not zero.

    a. Use tables or graphs of functions to estimate solutions for the following equations. In each case, sketch a graph showing the two functions involved in the equation and give coordinates of point(s) that correspond to solutions of the equation.

        i. $1.5x = \dfrac{6}{x}$ **ii.** $-1.5x = \dfrac{6}{x}$

        iii. $-2x = \dfrac{-18}{x}$ **iv.** $2x = \dfrac{-18}{x}$

    b. Recall from work with multivariable relations that any statement in the form $a = \dfrac{b}{c}$ is equivalent to $ac = b$. Use this principle to find exact values of solutions to the equations in Part a by algebraic reasoning. Check your work using a CAS command like **solve(1.5x=6/x,x)**.

    c. Look at the results of your work in Parts a and b. In general, how many solutions can there be for an equation in the form $ax = \dfrac{k}{x}$? Illustrate your answer with sketches of graphs showing the different cases.

5 Now consider equations in the form $ax + b = \dfrac{k}{x}$, where a, b, and k are not zero.

    a. Estimate the solutions to the following equations using tables or graphs of values for the related functions. In each case, sketch a graph showing the two functions involved in the equation and give coordinates of point(s) that correspond to solutions of the equation.

        i. $x - 1 = \dfrac{6}{x}$ **ii.** $x - 6 = \dfrac{-9}{x}$

        iii. $-x + 10 = \dfrac{9}{x}$ **iv.** $-x + 1 = \dfrac{4}{x}$

    b. Use the principle that any statement in the form $a = \dfrac{b}{c}$ is equivalent to $ac = b$ to write the equations in equivalent form without fractions. Then find exact values of the solutions by algebraic reasoning. For example, when solving the first equation, you may want to start like this:

$$x - 1 = \dfrac{6}{x}$$
$$x(x - 1) = 6$$
$$x^2 - x = 6$$
$$x^2 - x - 6 = 0$$

    c. Look at the results of your work in Parts a and b. In general, how many solutions can there be for an equation in the form $ax + b = \dfrac{k}{x}$? Illustrate your answer with sketches showing the different cases.

Summarize
the Mathematics

In this investigation, you developed strategies for solving problems that involve both a linear function and an inverse variation function.

a What strategy would you use to solve a system of equations of the form $y = ax + b$ and $y = \frac{k}{x}$?

b What are the possible numbers of solutions for equations in the form $ax = \frac{k}{x}$? How about equations like $ax + b = \frac{k}{x}$? How are these equations related to quadratic equations?

c How can you estimate the solutions to equations like those in Part b by inspecting tables and graphs of functions? What will graphs look like in each solution case?

d How can you calculate exact values of the solutions for such equations by reasoning with the symbolic expressions involved? By using a computer algebra system *solve* command?

Be prepared to explain your ideas to the class.

✓ Check Your Understanding

Each year, the Wheaton Boys and Girls Club sells fresh Christmas trees in December to raise money for sports equipment. They have $2,400 to use to buy trees for their lot; so the number of trees they can buy depends on the purchase price per tree p, according to the function $n(p) = \frac{2,400}{p}$. Experience has shown that (allowing for profit on each tree sold) the number of trees that customers will purchase also depends on p with function $c(p) = 300 - 6p$.

a. Write equations and inequalities that match the following questions about prospects of the tree sale and then estimate solutions.

  i. For what price per tree will the number of trees that can be bought equal the number of trees that will be sold?

  ii. For what price per tree will the number of trees that can be bought be greater than the number of trees that will be sold?

  iii. For what price per tree will the number of trees that can be bought be less than the number of trees that will be sold?

b. Sketch graphs showing how the supply and demand functions $n(p)$ and $c(p)$ depend on price per tree and explain how the graphs illustrate your answers to the questions of Part a.

In your work on the problems of Investigation 1, you developed strategies for solving equations involving linear and inverse variation functions. As you work on the problems in this investigation, look for answers to this question:

> *What strategies are effective in solving equations that*
> *relate linear and quadratic functions?*

In most businesses, one of the most important tasks is setting prices for the goods or services that are being offered for sale. For example, consider the case of producers who have a contract to bring a musical production to a summer theater.

They have to estimate costs of putting on the show, income from ticket sales and concessions, and the profit that can be made. Values of these variables depend on the number of tickets sold and the prices charged for tickets.

1 Data from a market survey suggest the following relationship between ticket price and number of tickets sold.

Relationship between Ticket Price and Ticket Sales

Price (in dollars)	5	10	15	20	30	40
Tickets Sold	2,300	2,000	1,700	1,500	1,050	500

  a. After plotting the data and experimenting to find a function model for the pattern, the business planners proposed the function $s(p) = 2,500 - 50p$ for this demand pattern.

    i. What do p and $s(p)$ represent in that function?

    ii. Is that function reasonable? Can you produce a better model?

  b. What do the numbers 2,500 and -50 tell about the way ticket sales depend on ticket price?

2 Based on the relationship between ticket price and number of tickets sold, the show planners figured that income could be predicted from ticket price, as well. They reasoned that since income is equal to the product of price per ticket and number of tickets sold, $I(p) = p(2,500 - 50p)$.

  a. Test this function rule by calculating the predicted income from ticket sales in two ways.

    i. First, use the data in Problem 1 to estimate income if the ticket price is set at $10, $20, and $40.

    ii. Then use the function to calculate predicted ticket income for the same ticket prices.

b. Sarrem proposed the function $I(p) = 2{,}500p - 50p^2$ for predicting income from ticket price. Is this equivalent to $I(p) = p(2{,}500 - 50p)$? Why or why not?

3 The next step in making business plans for the production was to estimate operating costs. Some costs were fixed (for example, pay for the cast and rent of the theater), but other costs would depend on the number of tickets sold s (for example, number of ushers and ticket takers needed). After estimating all of the possible operating costs, the function $c(s) = 17{,}500 + 2s$ was proposed.

a. According to that rule, what are the fixed operating costs and the costs per customer?

b. To show how operating costs depend on ticket price, Daniel proposed the function $c(p) = 17{,}500 + 2(2{,}500 - 50p)$. Is this rule correct? Why or why not?

c. Minta suggested that the expression $17{,}500 + 2(2{,}500 - 50p)$ in Part b could be simplified to $22{,}500 - 100p$. Is that correct? Why or why not?

4 The crucial step in business planning came next—finding out the way that ticket price would affect profit. The following graph shows how income and operating cost depend on ticket price and how they are related to each other.

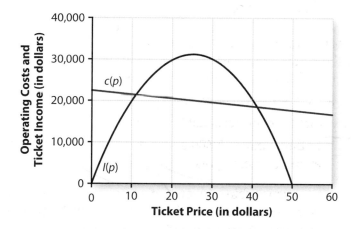

a. Use the graph to estimate answers for the following questions, and explain how you arrive at each estimate.

   i. For what ticket price(s) will operating cost exceed income?

   ii. For what ticket price(s) will income exceed operating cost?

   iii. For what ticket price(s) will income equal operating cost?

b. Use expressions in the income and operating cost functions to write and solve an equation that helps in locating the *break-even* point(s)—the ticket prices for which income exactly equals operating cost.

5 It is likely that the show producers want to do more than break even. They will probably seek maximum profit.

a. Use the income and operating cost functions to write a function showing how profit depends on ticket price. Write the function in two equivalent forms—one that shows the expressions for income and cost and another that is simplest for calculation of profit.

b. Use the profit function to estimate the maximum profit plan—the ticket price that will lead to maximum profit and the dollar profit that will be made at that price.

c. Use the results from Part b to calculate the number of tickets sold and the operating cost in the maximum profit situation.

Solving Equations of the Form $mx + d = ax^2 + bx + c$

The work you did in analyzing business prospects of the summer theater musical production illustrated ways that problems can require solving equations involving linear and quadratic functions. To work effectively in such situations, it helps to know the solution possibilities and how they will be expressed in graphs of the functions involved.

6 Use your table and graph tools and what you know about linear and quadratic functions to explore solution possibilities for equations in the form $mx + d = ax^2 + bx + c$.

a. Sketch function graphs illustrating the possible number of solutions for equations involving linear and quadratic functions. Compare your graphs with those of others and resolve any differences.

b. Solve each of the following equations using factoring or the quadratic formula.

   i. $x^2 - x + 3 = 2x - 1$

   ii. $x^2 - 3x + 2 = x - 2$

   iii. $10x^2 - 28x - 39 = 2x + 1$

c. For each equation in Part b, sketch graphs of the linear and quadratic functions involved and explain how the graphs illustrate the solutions.

Summarize the Mathematics

In this investigation, you developed strategies for solving problems that involve combinations of linear and quadratic functions.

a What strategy would you use to solve a system of equations of the form

$$y = mx + d \text{ and } y = ax^2 + bx + c?$$

b What are the possible numbers of solutions for equations in the form

$$mx + d = ax^2 + bx + c?$$

c How can you estimate the solutions for equations like those in Part b by inspecting tables and graphs of functions? What will graphs look like for the various solution possibilities?

d How can exact values of the solutions be found by reasoning with the symbolic expressions involved? By using a computer algebra system *solve* command?

Be prepared to explain your ideas to the class.

✓ *Check Your Understanding*

Solve each of these equations, sketch graphs showing the functions involved, and label points corresponding to solutions with their coordinates.

a. $x + 2 = x^2 + 3x - 6$

b. $-x + 2 = x^2 + x - 6$

c. $2x + 3 = 4 - x^2$

d. $2x^2 - x = 3x + 16$

On Your Own

Applications

1. The tenth-grade class officers at Columbus High School want to have a special event to welcome the incoming ninth-grade students. For $1,500, they can rent the Big Ten entertainment center for an evening. Their question is what to charge for tickets to the event so that income from ticket sales will be very close to the rental charge.

   a. Complete a table illustrating the pattern relating number of ticket sales n required to meet the "break-even" goal to the price charged p. Then write a rule relating n to p.

Price p (in dollars)	1	3	6	9	12	15
Tickets Sales Needed n	1,500	500				

   b. Study entries in the following table showing the class officers' ideas about how price charged p will affect number of students s who will buy tickets to the event. Then write a rule relating s to p.

Price p (in dollars)	0	3	6	9	12	15
Likely Ticket Sales s	600	540	480	420	360	300

   c. Write and solve an equation that will identify the ticket price(s) that will attract enough students for the event to meet its income goal. Illustrate your solution by a sketch of the graphs of the functions involved with key intersection points labeled by their coordinates.

2. When Coty was working on his Eagle Scout project, he figured he needed 60 hours of help from volunteer workers. He did some thinking to get an idea of how many workers he might need and how many volunteers he might be able to get.

   a. He began by assuming that each volunteer would work the same number of hours. In that case, what function $w(h)$ shows how the number of volunteer workers needed depends on the number of hours per worker h?

   b. Coty estimated that he could get 25 volunteers if each had to work only 3 hours and only 15 volunteers if each had to work 5 hours. What linear function $v(h)$ matches these assumptions about the relationship between the number of volunteers and the number of hours per worker h?

   c. Write and solve an equation that will help in finding the number of hours per worker and number of workers that Coty needs. Illustrate your solution by a sketch of the graphs of the functions involved with coordinate labels on key points.

3 Use symbolic reasoning to find all solutions for these equations. Illustrate each solution by a sketch of the graphs of the functions involved, labeling key points with their coordinates.

a. $x + 5 = \dfrac{6}{x}$

b. $-0.5x = \dfrac{4}{x}$

c. $1.5x = \dfrac{24}{x}$

d. $10 - x = \dfrac{7}{x}$

4 Use symbolic reasoning to find all solutions for the equation $\dfrac{4}{x} + 1 = 2 - x$. Illustrate the solution by a sketch of the graphs of the functions involved, labeling key points with their coordinates.

5 In making business plans for a pizza sale fund-raiser, the Band Boosters at Roosevelt High School figured out how both sales income $I(n)$ and selling expenses $E(n)$ would probably depend on number of pizzas sold n. They predicted that $I(n) = -0.05n^2 + 20n$ and $E(n) = 5n + 250$.

a. Estimate value(s) of n for which $I(n) = E(n)$ and explain what the solution(s) of that equation tell about prospects of the pizza sale fund-raiser. Illustrate your answer with a sketch of the graphs of the two functions involved, labeling key points with their coordinates.

b. Write a rule that gives predicted profit $P(n)$ as a function of number of pizzas sold and use that function to estimate the number of pizza sales necessary for the fund-raiser to break even. Illustrate your answer with a sketch of the graph of the profit function, labeling key points with their coordinates.

c. Use the profit function to estimate the maximum profit possible from this fund-raiser. Then find number of pizzas sold, income, and expenses associated with that maximum profit situation.

6 The stopping distance d in feet for a car traveling at a speed of s miles per hour depends on car and road conditions. Here are two possible stopping distance formulas: $d = 3s$ and $d = 0.05s^2 + s$.

a. Write and solve an equation to answer the question, "For what speed(s) do the two functions predict the same stopping distance?" Illustrate your answer with a sketch of the graphs of the two functions, labeling key point(s) with their coordinates.

b. In what ways are the patterns of change in stopping distance predicted by the two functions as speed increases similar and in what ways are they different? How do the function graphs illustrate the patterns you notice?

7 Use symbolic reasoning to find all solutions for these equations. Illustrate each solution by a sketch of the graphs of the functions involved, labeling key points with their coordinates.

a. $2x = 2x^2 - 4x$

b. $2x^2 - 4x = 4 - 2x$

c. $x^2 - 4x - 5 = 2x + 2$

d. $-3 - x = x^2 + 3x + 1$

8 Give specific examples of an equation involving one linear and one quadratic function that illustrate cases a–c described below. In each case, give a sketch showing how graphs of the two functions involved in the equation are related to each other. Explain how that relationship illustrates the number of solutions to the equation.

a. Two distinct solutions

b. Exactly one solution

c. No solutions with real numbers

9 Find all points of intersection of graphs of the following linear functions with the circle $(x - 4)^2 + (y - 1)^2 = 10$.

a. $y = 2$

b. $y = x + 1$

c. $y = -x - 3$

d. $y = x$

10 Find all points of intersection of the graphs of the following pairs of functions.

a. $y = x^2$ and $y = -4x^2 + 5$

b. $y = x^2 + 6x$ and $y = 0.5x^2$

c. $y = x^2 + 3x - 4$ and $y = -x^2 + x + 6$

Connections

11 In your early study of systems of linear equations, you found the intersection point of graphs for linear functions like $y = mx + n$ and $y = ax + b$. You found that you could solve such systems by setting $mx + n = ax + b$ and solving for x. You used a similar strategy in the investigations of this lesson to solve systems of equations like $y = mx + b$ and $y = \dfrac{k}{x}$ and like $y = mx + b$ and $y = ax^2 + bx + c$. Compare the solution possibilities for these three types of systems by answering Parts a–c.

a. How many solutions can there be for a system of two linear equations with two variables? Draw sketches of graphs showing the different possibilities.

b. How many solutions can there be for a system like $y = mx + b$ and $y = \frac{k}{x}$? Draw sketches of graphs showing the different possibilities.

c. How many solutions can there be for a system like $y = mx + n$ and $y = ax^2 + bx + c$? Draw sketches of graphs illustrating the different possibilities.

12 In Investigation 2, the business planning for the summer theater production involved three different dependent variables: ticket sales income, operating costs, and profit; each related to the independent variable, ticket price. The following graphs show operating costs, ticket sales income, and profit all as functions of ticket price. Explain what each labeled point tells about the business situation and how you would find the coordinates of those points.

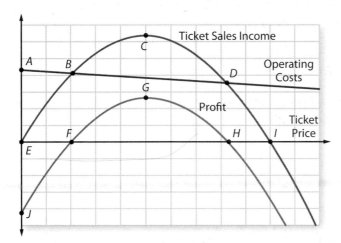

13 Analyze the systems of equations in Parts a–c, giving sketches of graphs for the functions involved to illustrate your answers. Then use the results and other examples you might explore to answer Part d.

a. Estimate solutions for the system $y = 2x + 1$ and $y = 2^x$.

b. Estimate solutions for the system $y = 2x - 1$ and $y = 2^x$.

c. Estimate solutions for the system $y = -2x$ and $y = 2^x$.

d. In general, how many solutions can there be for a system of equations like $y = mx + b$ and $y = a^x$?

14 In previous work with solving equations, you often found it useful to use two basic number properties to write relationships among variables in equivalent but more useful forms. The key ideas were:

$$a + b = c \text{ whenever } a = c - b \quad (1)$$
$$ab = c \text{ whenever } a = c \div b \ (b \neq 0) \quad (2)$$

Use these number properties to answer Parts a–c.

a. What quadratic equation has the same solution(s) as $4x = \frac{36}{x}$?

b. What quadratic equation has the same solution(s) as $5 - x = \frac{6}{x}$?

c. What equation in the form $ax^2 + bx + c = 0$ has the same solution as $x^2 + 9x + 7 = 3x - 1$?

Reflections

15 What seems to be the difference between being asked to solve the system $y = mx + n$ and $y = \frac{k}{x}$ and being asked to solve the equation $mx + n = \frac{k}{x}$?

16 When two different students were asked to solve the equation $\frac{3}{x} = -\frac{2}{x}$, they came up with different answers.

Jim argued that there are no values of x that satisfy the equation. He sketched a graph of the two functions to support his claim.

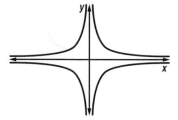

Linda gave the following "proof" that $x = 0$ is the solution.

$$\text{If } \frac{3}{x} = \frac{-2}{x}, \text{ then } \frac{x}{3} = \frac{x}{-2}$$

$$\text{then } \frac{x}{3} + \frac{x}{2} = 0$$

$$\text{then } \frac{5x}{6} = 0$$

$$\text{then } x = 0.$$

a. Which student do you think is right—the student who used the graph or the student who used symbolic reasoning?

b. What is the error in reasoning by the student who got the incorrect answer?

17 When you need to solve an equation in the form $f(x) = g(x)$, what are the values and limitations of using methods that involve:

a. graphs of $f(x)$ and $g(x)$?

b. tables of values for $f(x)$ and $g(x)$?

c. reasoning that uses only the symbolic expressions for each function rule?

18 If you were asked to give advice on symbolic equation solving strategies to another student, what list of steps would you recommend to solve the following equations?

a. $mx = \frac{k}{x}$

b. $mx + n = \frac{k}{x}$

c. $mx + n = ax^2 + bx + c$

Extensions

19 Shown below is a portion of the graph of the system of equations $y = \dfrac{50}{x^2}$ and $y = x^2 - 5x$.

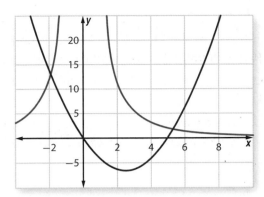

a. Make a copy of these graphs. If possible, color the graph of $y = \dfrac{50}{x^2}$ in yellow.

b. What connections between function rules and graphs allow you to match $y = \dfrac{50}{x^2}$ and $y = x^2 - 5x$ to their graphs?

c. On the x-axis, mark the points corresponding to solutions of the following equation and inequalities. If possible, use the colors suggested.

   i. in blue: $\dfrac{50}{x^2} = x^2 - 5x$

   ii. in red: $\dfrac{50}{x^2} > x^2 - 5x$

   iii. in green: $\dfrac{50}{x^2} < x^2 - 5x$

d. Find estimates of the solutions to the equation and inequalities in Part c.

e. How could you convince someone that your solutions are correct?

20 Determine the possible numbers of solutions (x, y) for each of these kinds of systems of equations. In each case, illustrate your answers by sketching graphs for pairs of functions that illustrate the possibilities.

a. $y = \dfrac{p}{x}$ and $y = \dfrac{q}{x}$

b. $y = \dfrac{k}{x}$ and $y = ax^2 + bx + c$

c. $y = ax^2 + c$ and $y = dx^2$

d. $y = \dfrac{k}{x^2}$ and $y = ax^2 + bx + c$

21 Determine the possible numbers of solutions (x, y) for each of these kinds of systems of equations. In each case, illustrate your answers by sketching graphs for pairs of functions that illustrate the possibilities.

a. $y = ax^2 + bx + c$ and $y = p^x$ ($p > 0$ and $p \neq 1$)

b. $y = \dfrac{k}{x}$ and $y = p^x$ ($p > 0$ and $p \neq 1$)

22 Jackie has found a spreadsheet to be a useful tool in her work in previous units. To explore how a spreadsheet might be used as a tool for estimating solutions to nonlinear systems, she created a spreadsheet to help solve $x - 2 = \dfrac{24}{x}$. The spreadsheet gave results starting like this:

Nonlinear Systems.xls ⬜ 🗗 ❎

◇	A	B	C	D	
1	0	−2	error		
2	1	−1	24		
3	2	⋮	⋮		
4	3				
5	4				
6	⋮				

a. How could you produce such a spreadsheet table to check for solutions from $x = 0$ to $x = 10$ by entering only one number, formulas in 3 cells, and several fill-down or fill-across commands?

b. How could the spreadsheet be modified so that the search began at $x = 0$ and checked as x increased in steps of 0.5 at a time?

c. How could the spreadsheet be modified so that it would help to find solutions to the equation $4x - 3 = x^2 - 2x + 2$?

23 When Tanya and Mike were 15 years old, they had summer jobs. Tanya saved $500 from her earnings, and Mike saved $600 from his earnings.

• Tanya decided to keep her money at home but to add $10 per month from what she earned doing lawn mowing, snow shoveling, and other errands for neighbors.

• Mike decided to invest his savings in a bank account that paid interest at an annual rate of 6% compounded monthly (0.5% monthly interest).

a. What were the values of Tanya's and Mike's savings after 1 month? 2 months?

b. What rules give the values (V_T and V_M) of their savings after m months?

c. At what time (if any), did Tanya's savings grow to become greater than Mike's savings?

d. Illustrate your answer to Part c by sketching graphs of the functions in Part b and by labeling coordinates of key points.

e. How would your answer to Part c change if Tanya decided to save only $5 per month? If Mike was able to invest his savings at a 9% annual interest rate compounded monthly?

24 Throughout this lesson, you have been solving systems of equations. In practical problems, it is common to encounter situations in which one or more of the constraints in a system involve inequalities like $y \le 4x^2$.

Use what you have learned in this lesson as well as what you know about inequalities and graphing to solve the following systems of inequalities. The task is to find all points with coordinates satisfying both inequality conditions. Express your answer with a graph on which solutions are indicated by shading regions of the coordinate plane. (*Hint*: It might help to start with CAS graphs of each system of inequalities.)

a. $y \ge x^2 + 2x - 15$ and $y \le -x^2 - 1$

b. $y \ge x^2 + x - 6$ and $y \le 2x + 4$

c. $y \ge x^2$ and $y \le \dfrac{5}{x}$

d. How could you convince someone that your solutions to the inequality systems are correct?

Review

25 Find rules for the linear functions with graphs meeting the following conditions. Then draw each graph on a coordinate system.

a. Slope of -3.5 and y-intercept at $(0, 2)$

b. Slope of 2.5 and containing the point $(2, 3)$

c. Containing the points $(-1, 2)$ and $(2, -3)$

26 Rewrite each rule in the requested equivalent form.

a. If $y = 4xz$, express x as a function of y and z.

b. If $z = 4x + 2y$, express y as a function of x and z.

c. If $z = \dfrac{x}{y + 2}$, express x as a function of y and z.

27 When owners of A-1 Auto Parts looked for a new delivery truck, they found a small pickup model they liked and got the following offers of lease payment plans from the dealer:

Plan A They could make a down payment of $3,500 and then monthly payments of $250.

Plan B They could make a down payment of only $1,500 and then monthly payments of $330.

They had to make a choice between a higher down payment with lower monthly payments or a lower down payment with higher monthly payments.

a. Write rules giving total lease payment as a function of the number of months in the lease for each plan. Use P_A for total amount paid under lease Plan A, P_B for total amount paid under lease Plan B, and m for the number of months in the lease.

b. Write and solve equations or inequalities that answer these questions about the two lease plans.

    i. For what lease lengths will Plan A be cheaper than Plan B?

    ii. For what lease lengths will Plan A be more expensive than Plan B?

    iii. For what lease length(s) will Plans A and B have the same cost to A-1 Auto Parts?

c. Explain how the answers to the questions in Part b could be found using graphs of P_A and P_B.

28 A fuel storage tank is in the shape of a cylinder. The cylinder has a height of 32 feet and the circumference of the base is 135 feet.

a. What is the radius of the base of the tank?

b. How much fuel will the tank hold?

c. A freighter that delivers fuel can pump the fuel into the tank at 250 gallons per minute. If 1 gallon is equal to 0.134 cubic feet, how long will it take to fill the tank if it is currently empty?

29 Writing numbers in scientific notation is a compact way to represent very large or very small numbers. It can also make multiplying and dividing very large or very small numbers easier.

a. Place the following numbers in order from smallest to largest.

$$321.56 \times 10^4 \qquad 0.00329 \qquad 123{,}537{,}821$$
$$6.2 \times 10^{-5} \qquad 2.1 \times 10^5 \qquad 3.1 \times 10^{-2}$$

b. Rewrite each number in scientific notation and evaluate each product without using your calculator. Then represent the product using scientific notation.

    i. $(300{,}000{,}000)(5{,}000{,}000)$

    ii. $(0.00000006)(3{,}000)$

    iii. $(0.00000012)(0.000005)$

c. Rewrite each number in scientific notation and evaluate each product without using your calculator. Then represent the quotient using scientific notation.

    i. $\dfrac{12{,}000{,}000{,}000}{6{,}000{,}000}$

    ii. $\dfrac{0.000000012}{0.00024}$

    iii. $\dfrac{0.00072}{8{,}000{,}000}$

Common Logarithms and Exponential Equations

Have you ever had someone tell you that you were speaking too loudly (or too softly) or that the volume on a television was turned up too high (or down too low)? While sensitivity to noise can vary from person to person, in general, people can hear sounds over an incredible range of loudness.

Sound intensity is measured in physical units of watts per square centimeter. But the loudness is typically reported in units called decibels. The next table shows intensity values for a variety of familiar sounds and the related number of decibels—as measured at common distances from the sources.

Sound Intensity (in watts/cm^2)	Noise Source	Relative Intensity (in decibels)
10^3	Military rifle	150
10^2	Jet plane (30 meters away)	140
10^1	[Level at which sound is painful]	130
10^0	Amplified rock music	120
10^{-1}	Power tools	110
10^{-2}	Noisy kitchen	100
10^{-3}	Heavy traffic	90
10^{-4}	Traffic noise in a small car	80
10^{-5}	Vacuum cleaner	70
10^{-6}	Normal conversation	60
10^{-7}	Average home	50
10^{-8}	Quiet conversation	40
10^{-9}	Soft whisper	30
10^{-10}	Quiet living room	20
10^{-11}	Quiet recording studio	10
10^{-12}	[Barely audible]	0

Source: *Real-Life Math: Everyday Use of Mathematical Concepts*, Evan Glazer and John McConnell, 2002.

Think About This Situation

Study the sound intensity values, sources, and decibel ratings given in the table.

a Are the intensity and decibel numbers in the order of loudness that you would expect for the different familiar sources?

b What pattern do you see relating the sound intensity values (watts/cm^2) and the decibel numbers?

In this lesson, you will learn about *logarithms*—the mathematical idea used to express sound intensity in decibels and to solve a variety of problems related to exponential functions and equations.

Investigation 1 — How Loud is Too Loud?

Your analysis of the sound intensity data might have suggested several different algorithms for converting watts per square centimeter into decibels. For example,

if the intensity of a sound is $10^x \frac{\text{watts}}{\text{cm}^2}$,

its loudness in decibels is $10x + 120$.

The key to discovery of this conversion rule is the fact that all sound intensities were written as powers of 10. What would you have done if the sound intensity readings had been written as numbers like $3.45 \frac{\text{watts}}{\text{cm}^2}$ or $0.0023 \frac{\text{watts}}{\text{cm}^2}$?

As you work on problems of this investigation, look for an answer to this question:

How can any positive number be expressed as a power of 10?

1 Express each of the numbers in Parts a–i as accurately as possible as a power of 10. You can find exact values for some of the required exponents by thinking about the meanings of positive and negative exponents. Others might require some calculator exploration of ordered pairs that satisfy the exponential equation $y = 10^x$.

    a. 100   **b.** 10,000   **c.** 1,000,000

    d. 0.01   **e.** −0.001   **f.** 3.45

    g. −34.5   **h.** 345   **i.** 0.0023

2 Suppose that the sound intensity of a screaming baby was measured as $9.5 \frac{\text{watts}}{\text{cm}^2}$. To calculate the equivalent intensity in decibels, 9.5 must be written as 10^x for some value of x.

    a. Between which two integers does it make sense to look for values of the required exponent? How do you know?

    b. Which of the two integer values in Part a is probably closer to the required power of 10?

    c. Estimate the required exponent to the nearest hundredth. Then use your estimate to calculate a decibel rating for the loudness of the baby's scream.

    d. Estimate the decibel rating for loudness of sound from a television set that registers intensity of $6.2 \frac{\text{watts}}{\text{cm}^2}$.

Common Logarithms As you probably discovered in your work on Problems 1 and 2, it is not easy to solve equations like $10^x = 9.5$ or $10^x = 0.0023$—even by estimation. To deal with this very important problem, mathematicians have developed procedures for finding exponents. If $10^x = y$, then x is called the **base 10 logarithm** of y.

This definition of base 10 or *common logarithm* is usually expressed in function-like notation:

$$\log_{10} a = b \text{ if and only if } 10^b = a.$$

$\log_{10} a$ is pronounced "log base 10 of a." Because base 10 logarithms are so commonly used, $\log_{10} a$ is often written simply as $\log a$. Most scientific calculators have a built-in log function (LOG) that automatically finds the required exponent values.

3 Use your calculator to find the following logarithms. Then compare the results with your work on Problem 1.

    a. log 100   **b.** log 10,000   **c.** log 1,000,000

    d. log 0.01   **e.** log (−0.001)   **f.** log 3.45

    g. log (−34.5)   **h.** log 345   **i.** log 0.0023

4 What do your results from Problem 3 (especially Parts e and g) suggest about the kinds of numbers that have logarithms? See if you can explain your answer by using the connection between logarithms and the exponential function $y = 10^x$.

5 Logarithms can be used to calculate the decibel rating of sounds, when the intensity is measured in $\frac{watts}{cm^2}$.

a. Use the logarithm feature of your calculator to rewrite 9.5 as a power of 10. That is, find x so that $9.5 = 10^x$.

b. Recall that if the intensity of a sound is $10^x \frac{watts}{cm^2}$, then the expression $10x + 120$ can be used to convert the sound's intensity to decibels. Use your result from Part a to find the decibel rating of the crying baby in Problem 2.

6 Assume the intensity of a sound $I = 10^x \frac{watts}{cm^2}$.

a. Explain why $x = \log I$.

b. Rewrite the expression for converting sound intensity readings to decibel numbers using $\log I$.

7 Use your conversion expression from Problem 6 to find the decibel rating of the television set in Problem 2 Part d.

Why Do They Taste Different? You may recall from your study of science that the acidity of a substance is described by its pH rating—*the lower its pH, the more acidic a substance is*. The acidity depends on the hydrogen ion concentration in the substance (in moles per liter). Some sample hydrogen ion concentrations are given below. Since those hydrogen ion concentrations are generally very small numbers, they are converted to the simpler pH scale for reporting.

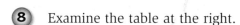

8 Examine the table at the right.

a. Describe how hydrogen ion concentrations [H$^+$] are converted into pH readings.

b. Write an equation that makes use of logarithms expressing pH as a function of hydrogen ion concentration [H$^+$].

Substance	[H$^+$]	pH
Hand soap	10^{-10}	10
Egg white	10^{-9}	9
Sea water	10^{-8}	8
Pure water	10^{-7}	7
White bread	10^{-6}	6
Coffee	10^{-5}	5
Tomato juice	10^{-4}	4
Orange juice	10^{-3}	3

9 Use the equation relating hydrogen ion concentration and pH reading to compare acidity of some familiar liquids.

a. Complete a copy of the table at the right. Round results to the nearest tenth.

b. Explain how your results tell which is more acidic—lemonade, apple juice, or milk.

Substance	[H$^+$] Proportion	pH Reading
lemonade	0.00501	
apple juice	0.000794	
milk	0.000000355	

Summarize
the Mathematics

In work on the problems of this investigation, you learned how physical measurements of sound intensity and acidity of a chemical substance are converted into the more familiar decibel and pH numbers. You also learned how the *logarithm* function is used in those processes.

a How would you explain to someone who did not know about logarithms what the expression $\log y = x$ tells about the numbers x and y?

b What can be said about the value of $\log y$ in each case below? Give brief justifications of your answers.

   i. $0 < y < 1$
   ii. $1 < y < 10$
   iii. $10 < y < 100$
   iv. $100 < y < 1,000$

Be prepared to explain your ideas to the class.

✓ Check Your Understanding

Use your understanding of the relationship between logarithms and exponents to help complete these tasks.

a. Find these common (base 10) logarithms without using a calculator.

   i. $\log 1,000$ **ii.** $\log 0.001$ **iii.** $\log 10^{3.2}$

b. Use the function $y = 10^x$, but not the logarithm key of your calculator, to estimate each of these logarithms to the nearest tenth. Explain how you arrived at your answers.

   i. $\log 75$ **ii.** $\log 750$ **iii.** $\log 7.5$

c. If the intensity of sound from a drag race car is 125 watts per square centimeter, what is the decibel rating of the loudness for that sound?

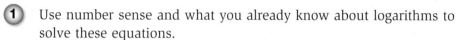

Investigation 2   Solving for Exponents

Logarithms can be used to find exponents that solve equations like $10^x = 9.5$. For this reason, they are an invaluable tool in answering questions about exponential growth and decay. For example, the world population is currently about 6.2 billion and growing exponentially at a rate of about 1.14% per year. To find the time when this population is likely to double, you need to solve the equation

$$6.2(1.0114)^t = 12.4, \text{ or } (1.0114)^t = 2.$$

As you work on the problems of this investigation, look for ways to answer this question:

How can common logarithms help in finding solutions of exponential equations?

1 Use number sense and what you already know about logarithms to solve these equations.

    a. $10^x = 1,000$ **b.** $10^{x+2} = 1,000$

    c. $10^{3x+2} = 1,000$ **d.** $2(10)^x = 200$

    e. $3(10)^{x+4} = 3,000$ **f.** $10^{2x} = 50$

    g. $10^{3x+2} = 43$ **h.** $12(10)^{3x+2} = 120$

    i. $3(10)^{x+4} + 7 = 28$

Unfortunately, many of the functions that you have used to model exponential growth and decay have not used 10 as the base. On the other hand, it is not too hard to transform any exponential expression in the form b^x into an equivalent expression with base 10. You will learn how to do this after future work with logarithms. The next three problems ask you to use what you already know about solving exponential equations with base 10 to solve several exponential growth problems.

2 If a scientist counts 50 bacteria in an experimental culture and observes that one hour later the count is up to 100 bacteria, the function $P(t) = 50(10^{0.3t})$ provides an exponential growth model that matches these data points.

    a. Explain how you can be sure that $P(0) = 50$.

    b. Show that $P(1) \approx 100$.

    c. Use the given function to estimate the time when the bacteria population would be expected to reach 1,000,000. Explain how to find this time in two ways—one by numerical or graphic estimation and the other by use of logarithms and algebraic reasoning.

(3) The world population in 2005 was 6.2 billion and growing exponentially at a rate of 1.14% per year. The function $P(t) = 6.2(10^{0.005t})$ provides a good model for the population growth pattern.

    a. Explain how you can be sure that $P(0) = 6.2$.

    b. Show that $P(1) = 6.2 + 1.14\%(6.2)$.

    c. Find the time when world population would be expected to reach 10 billion if growth continues at the same exponential rate. Explain how to find this time in two ways—one by numerical or graphic estimation and the other by use of logarithms and algebraic reasoning.

Summarize
the Mathematics

In work on the problems of this investigation, you learned how to use logarithms to solve equations related to exponential functions.

  a How can logarithms be used to solve equations in the form $10^{mx + n} = c$?

  b How can logarithms be used to solve scientific problems that lead to equations in the form $a(10^{mx}) = c$?

  c How are the methods you described in Parts a and b related?

Be prepared to explain your ideas to the class.

✓Check Your Understanding

Use logarithms and other algebraic methods as needed to complete the following tasks.

a. Solve these equations.

    i. $5(10)^x = 450$

    ii. $4(10)^{2x} = 40$

    iii. $5(10)^{4x-2} = 500$

    iv. $8x^2 + 3 = 35$

b. The population of the United States in 2006 was about 300 million and growing exponentially at a rate of about 0.7% per year. If that growth rate continues, the population of the country in year $2006 + t$ will be given by the function $P(t) = 300(10^{0.003t})$. According to that population model, when is the U.S. population predicted to reach 400 million? Check the reasonableness of your answer with a table or a graph of $P(t)$.

On Your Own

Applications

1 Find the decibel ratings of these sounds.

   a. A passing subway train with sound intensity reading of $10^{-0.5} \frac{\text{watts}}{\text{cm}^2}$

   b. An excited crowd at a basketball game with sound intensity reading of $10^{1.25} \frac{\text{watts}}{\text{cm}^2}$

2 Find these common (base 10) logarithms without using a calculator and explain your reasoning.

   a. log 100,000   **b.** log 0.001

   c. $\log (10^{4.75})$ **d.** log 1

3 Find the decibel ratings of these sounds.

   a. A door slamming with sound intensity $89 \frac{\text{watts}}{\text{cm}^2}$

   b. A radio playing with sound intensity $0.005 \frac{\text{watts}}{\text{cm}^2}$

4 Pure water has a pH of 7. Liquids with pH less than 7 are called acidic; those with pH greater than 7 are called alkaline. Typical seawater has pH about 8.5, soft drinks have pH about 3.1, and stomach gastric juices have pH about 1.7.

   a. Which of the three liquids are acidic and which are alkaline?

   b. Find the concentration of hydrogen ions in seawater, soft drinks, and gastric juices.

   c. Explain why it is correct to say that the concentration of hydrogen ions in gastric juices is about 25 times that of soft drinks.

   d. If a new soft drink has a hydrogen ion concentration that is one-fifth that of typical soft drinks, what is its pH?

5 Use algebraic reasoning with logarithms to solve the following equations for x.

   a. $\log x = 2$

   b. $15 = 10^x$

   c. $5(10)^{2x} = 60$

   d. $10^{3x-1} = 100,000$

6 The Washington Nationals baseball team was purchased in 2006 for 450 million dollars. If the value of this investment grows at a rate of 5% compounded yearly, the purchase price of the team in year 2006 + t will be given by $V(t) = 450(10^{0.021t})$.

a. Explain how you can be sure that this function gives the correct value of the investment in 2006.

b. Use the function to estimate the value of the investment in 2010.

c. Use logarithms and other algebraic reasoning to estimate the time when the value of the investment will be $1 billion ($1,000 million).

7 Suppose that the average rent for a two-bedroom apartment in Indianapolis is currently $750 per month and increasing at a rate of 8% per year. The function $R(t) = 750(10^{0.033t})$ provides a model for the pattern of expected increase in monthly rent after t more years.

a. Explain how you know that $R(0) = 750$.

b. Find the time when the average rent for a two-bedroom apartment will be $1,000 per month, if inflation continues at the current 8% rate. Show how to find that time in two ways—one by estimation using a table or graph of $R(t)$ and another using logarithms and algebraic reasoning.

8 If an athlete tries to improve his or her performance by taking an illegal drug, the amount of that drug in his or her blood will decline exponentially over time, but tests are quite sensitive to small amounts.
For example, a 200 mg dose of a steroid might decay so that the amount remaining after t days is given by the rule $s(t) = 200(10^{-0.046t})$.

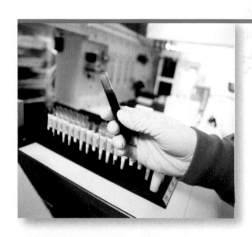

a. Explain how you know that $s(0) = 200$.

b. Estimate the time when only 5 mg of the steroid remains active in the athlete's blood. Show how to find that time in two ways—one by estimation using a table or graph of $s(t)$ and another using logarithms and algebraic reasoning.

9 Suppose that 500 mg of a medicine enters a hospital patient's bloodstream at noon and decays exponentially at a rate of 15% per hour.

a. Use the exponential function $D(t) = 500(10^{-0.07t})$ to predict the amount of medicine active in the patient's blood at a time 5 hours later, where t is time in hours.

b. Find the time when only 5% (25 mg) of the original amount of medicine will be active in the patient's body. Show how to find that time in two ways—one by estimation using a table or graph of $D(t)$ and another using logarithms and algebraic reasoning.

10 The magnitude of an earthquake is often reported using the Richter scale. This rating depends on the amount of displacement recorded by a seismogram and the distance from the epicenter of the earthquake to the device. The table below gives the Richter scale ratings for measurements at a distance of 100 km from the epicenter of an earthquake.

Seismogram Displacement (in meters)	10^{-6}	10^{-5}	10^{-4}	10^{-3}	10^{-2}	10^{-1}	10^{0}	10^{1}	10^{2}
Richter Scale Rating	1	2	3	4	5	6	7	8	9

a. Write a function $R(x)$ which gives the Richter scale rating for an earthquake based upon the displacement x in meters of a seismograph located 100 km away from the epicenter.

b. A scientist noticed a displacement of 0.054 meters on a seismogram located 100 km from the epicenter of an earthquake.

   i. Between what two whole numbers did the Richter scale rating of this quake fall and how do you know from inspecting the table of sample Richter scale values?

   ii. What was the precise Richter scale rating of the earthquake?

c. The earthquake that caused the Indian Ocean Tsunami in December of 2004 reportedly measured 9.15 on the Richter scale. What displacement would be recorded on a seismograph on Simeulue Island, approximately 100 km away from the epicenter?

(Source: en.wikipedia.org/wiki/2004_Indian_Ocean_earthquake)

11 When archaeologists discover remains of an ancient civilization, they use a technique called *carbon dating* to estimate the time when the person or animal died or when the artifact was made from living material.

The amount of radioactive Carbon-14 in such an artifact decreases exponentially according to the function $C(t) = 100(10^{-0.00005255t})$, where t is time in years and $C(t)$ is the percent of the Carbon-14 present in the artifact when it was last living material.

a. What is the half-life of Carbon-14, the time when only 50% remains from an original amount?

b. Suppose that a skeleton is discovered that has only 10% of the Carbon-14 that one finds in living animals. When was that skeleton part of a living animal?

Connections

12 A large number like 2364700 is written in scientific notation as 2.3647×10^6 and a small number like 0.000045382 as 4.5382×10^{-5}.

a. Write each of the following numbers in scientific notation with five significant digits (rounding appropriately where necessary to meet this condition).

   i. 47265   **ii.** 584.73

   iii. 97485302   **iv.** 0.002351

b. Suppose that the only calculator you had was one that could multiply and divide numbers between 1 and 10 (including decimals) but no others. Explain how you could still use this calculator to find these products and quotient.

   i. 584.73×97485302 **ii.** 47265×0.002351

   iii. $47265 \div 584.73$

13 Consider the function $y = \log x$. Study tables and graphs of the function to develop answers for these questions about its properties.

a. What seem to be the domain and range of $\log x$?

b. How does the relationship of logarithms and exponents explain your answer to Part a?

c. How does the pattern of change in $\log x$ over its domain compare with the patterns of change for these functions over the same domain?

   i. $y = x$ **ii.** $y = \dfrac{1}{x}$

   iii. $y = x^2$ **iv.** $y = 2^x$

14 Solve the following exponential equations and then explain how the strategies used are similar to what you use in solving linear equations.

a. $10^{x+2} = 100{,}000$

b. $10^{3x+2} = 10{,}000$

c. $5(10^{3x+2}) + 6 = 506$

15 The time it takes a computer program to run increases as the number of inputs increases. Three different companies wrote three different programs, A, B, and C, to compute the same information. The time, in milliseconds, it takes to run each of the programs when given n inputs is given by the three functions below:

$$A(n) = 10{,}000 + 2 \log n$$
$$B(n) = 100 + 4n^2$$
$$C(n) = (0.00003)(10^n)$$

a. Which program, A, B, or C, takes the least amount of time for 1 input?

b. Which program is most efficient for 300 inputs?

c. Which program is most efficient for 100,000 inputs?

d. Which program would you market to home users? What about business users? Explain your reasoning.

Reflections

16 Explain what is meant by the equation $\log a = b$. Then use your explanation to show why the following statements are true.

   a. $\log 1 = 0$

   b. $\log 10 = 1$

   c. $\log 10^x = x$

17 You are familiar with *linear scales* such as on a ruler or map. On these scales, the *difference* between equally-spaced scale points is a constant.

A Richter scale (see Applications Task 10) is an example of a *logarithmic scale.*

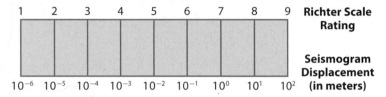

What is constant in the case of equally-spaced scale points on a logarithmic scale?

18 Study a graph of $y = \log x$ for $0 < x \leq 10,000$ and explain how the result justifies inclusion of logarithms in a unit on *nonlinear* functions.

19 With the introduction of logarithmic functions, you are now able to solve exponential equations using exact algebraic reasoning or a CAS, in addition to numeric and graphic estimation. Which method(s) do you prefer for solving equations like the following? Be prepared to explain your choice in each case.

   a. $100 = 4.5x - 885$

   b. $x^2 + x - 2 = 0$

   c. $3x^2 + 7x - 2 = 0$

   d. $5(10^x) = 500$

Extensions

20 Suppose that n is a positive integer.

  a. If $0 < \log n < 1$, what can you say about n?

  b. If $5 < \log n < 6$, what can you say about n?

  c. If $p < \log n < p + 1$, where p is a positive integer, what can you say about n?

21 When you go to the movies, the number of frames that are displayed per second affects the "smoothness" of the perceived motion on the screen. If the frames are displayed slowly, our minds perceive the images as separate pictures rather than fluid motion. However, as the frequency of the images increases, the perceived gap between the images decreases and the motion appears fluid. The frequency f at which we stop seeing a flickering image and start perceiving motion is given by the equation $f = K \log S$, where K is a constant and S is the brightness of the image being projected.

  a. S is inversely proportional to the square of the observer's distance from the screen. What would be the effect on f if the distance to the screen were cut in half? What if the distance to the screen were doubled?

  b. If the image is being projected at a slow frequency and you perceive a flicker, where should you move in the theater: closer to the screen or closer to the rear?

  c. Suppose the show is sold out, and you cannot move your seat. What could you do to reduce the flickering of the image on-screen?

22 Recall that a prime number n is an integer greater than 1 that has only 1 and n as divisors. The first eight primes are 2, 3, 5, 7, 11, 13, 17, and 19. Mathematicians have proved that the number of primes less than or equal to n is approximated by $\dfrac{0.4343n}{\log n}$, quite an accomplishment since the primes appear irregularly among the natural numbers.

  a. Count the actual number of primes less than or equal to n to complete the table below. Plot the (n, *number of primes* $\leq n$) data.

n	10	25	40	55	70	85	100	115	130	145
Number of primes $\leq n$	4			16	19	23	25	30	31	34

  b. Graph $P(n) = \dfrac{0.4343n}{\log n}$, $0 < n \leq 150$. How well does this function model the counts in Part a?

  c. Use the function $P(n)$ to estimate the number of primes less than or equal to 1,000; less than or equal to 100,000; less than or equal to 1,000,000; less than or equal to 10^{18}.

  d. Using the function $P(n)$, about what percent of the numbers up to 10^6 are prime? Up to 10^{18}?

23 Use base 10 logarithms to solve each of these equations for k.

    a. $10^k = 2$ **b.** $10^k = 5$ **c.** $10^k = 1.0114$

24 Use your results from Task 23 and what you know about properties of exponents to show how each of these exponential expressions can be written in equivalent form as $(10^k)^x$ and then 10^{kx}.

    a. 2^x **b.** 5^x **c.** 1.0114^t

25 Use your results from Task 24 and the ideas you developed in Investigation 2 to solve these exponential equations. Then check each solution and be prepared to explain your solution strategy.

    a. $2^x = 3.5$ **b.** $5(2^x) = 35$

    c. $5(2^x) + 20 = 125$ **d.** $5^x = 48$

    e. $3(5^x) + 12 = 60$ **f.** $300(5^x) = 60$

26 Use symbolic reasoning to solve the following equations for x.

    a. $\log(10^x) = 4$ **b.** $2^{2x+2} = 8^{x+2}$

27 The close connection between logarithm and exponential functions is used often by statisticians as they analyze patterns in data where the numbers range from very small to very large values. For example, the following table shows values that might occur as a bacteria population grows according to the exponential function $P(t) = 50(2^t)$:

Time t (in hours)	0	1	2	3	4	5	6	7	8
Population $P(t)$	50	100	200	400	800	1,600	3,200	6,400	12,800

    a. Complete another row of the table with values log (*population*) and identify the familiar function pattern illustrated by values in that row.

    b. Use your calculator to find log 2 and see how that value relates to the pattern you found in the log $P(t)$ row of the data table.

    c. Suppose that you had a different set of experimental data that you suspected was an example of exponential growth or decay, and you produced a similar "third row" with values equal to the logarithms of the population data. How could you use the pattern in that "third row" to figure out the actual rule for the exponential growth or decay model?

Review

28 How are the shape and location of the graph for an exponential function $f(x) = a(b^x)$ related to the values of a and b with $a, b > 0$ and $b \neq 1$?

29 Recall that the volume V of a cone can be found using the formula $V = \frac{1}{3}\pi r^2 h$, where r is the radius of the base and h is the height of the cone.

a. What is the radius of a cone that has a volume of 400 cm³ and a height of 5 cm?

b. Rewrite the formula so that it expresses h as a function of V and r.

c. Rewrite the formula so that it expresses r as a function of V and h.

d. How does the height of a cone change if the volume of the cone is constant but the radius increases?

30 Use algebraic reasoning to solve these equations.

a. $7x^2 + 23 = 100$

b. $x^2 + 13x + 42 = 0$

c. $7x^2 + 23x = 0$

d. $5(x - 8) + 12 = 4 - 7x$

31 Consider the coordinate grid at the right. Describe in words a transformation or composite of transformations that will map the first triangle onto the second. Then provide a coordinate rule for the mapping.

a. $\triangle A$ onto $\triangle B$

b. $\triangle A$ onto $\triangle C$

c. $\triangle A$ onto $\triangle D$

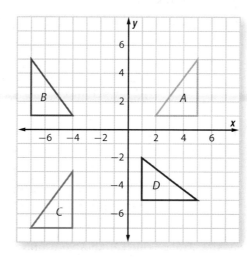

32 For each of the tables of values below:

- Decide if the relationship between x and y can be represented by a linear, exponential, or quadratic function.
- Find an appropriate function rule for the relationship.
- Use your rule to find the y value that corresponds to an x value of 10.

a.

x	-2	-1	0	1	2	3	4	5
y	12	5	0	-3	-4	-3	0	5

b.

x	-2	-1	0	1	2	3	4	5
y	1	2.5	4	5.5	7	8.5	10	11.5

c.

x	-2	-1	0	1	2	3	4	5
y	1	2	4	8	16	32	64	128

33 The table below provides information about the employees at Rosie's Grill. Fill in the remaining parts of the table.

	Full Time	Part Time	Total
Men	2		20
Women		20	
Total			45

Suppose that an employee is chosen at random. Find the probability of each event.

a. The employee is a full-time employee.

b. The employee is a man.

c. The employee is a man and a full-time employee.

d. The employee is a man or is a full-time employee.

34 The diagram below indicates the sidewalks and buildings in one apartment complex. Jamie is responsible for shoveling the sidewalks whenever it snows.

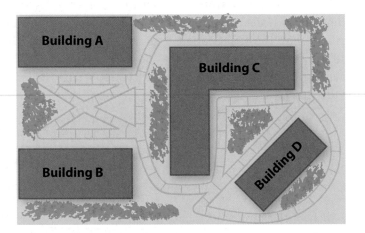

a. Draw a vertex-edge graph that represents this situation. Be sure to identify what the vertices and edges of your graph represent.

b. Explain how you know this graph has an Euler circuit. Then find one circuit.

c. Is the circuit you found in Part b the only Euler circuit for your graph? Explain your reasoning.

Looking
Back

T he lessons in this unit involved many different situations in which quantitative variables were related to each other by linear, quadratic, and inverse variation functions. You also studied *logarithms*, a useful tool in solving problems related to exponential growth and decay. And you learned how to express information and questions about any function using the standard function notation $f(x)$.

As a result of your work on problems in the lessons of this unit, you should be able to better recognize situations in which each type of function is useful. You should also have an extended set of strategies for reasoning about the symbolic expressions and equations that represent those functions and the questions that arise in dealing with them. In particular, you should be more able to write quadratic expressions in useful equivalent forms by expanding products and factoring standard-form expressions. You should be able to solve quadratic equations by using the quadratic formula and by factoring where appropriate. You should also be able to use a variety of strategies to solve systems of equations involving a linear function and an inverse variation function or a linear function and a quadratic function. You should have a beginning understanding of ways to use logarithms to solve equations that involve exponential expressions.

The tasks in this final lesson will help you review and organize your knowledge of nonlinear functions and equations.

(1) **Business Economics** In stores that sell athletic shoes of various kinds, the cost of doing business includes fixed expenses (like rent and pay for employees) and variable expenses (like payments for shoes bought from manufacturers). Operating costs of any store will be a function of those two main factors.

The typical American now owns two or three pairs of athletic shoes, which range in price from a $20 pair of old-fashioned sneakers at a discount store to $135 for top-of-the-line basketball shoes. One big seller has been Nike's Air Pegasus, which, like nearly all athletic shoes, is manufactured by suppliers in Asia. This accounting is based on a sale at an outlet of a large national retailer.

—By Steven Pearlstein

Production labor	$2.75
Materials	9.00
Rent, equipment	3.00
Supplier's operating profit	1.75
Duties	3.00
Shipping	.50
Cost to Nike	**$20.00**
Research/development	$.25
Promotion/advertising	4.00
Sales, distribution, administration	5.00
Nike's operating profit	6.25
Cost to retailer	**$35.50**
Rent	$9.00
Personnel	9.50
Other	7.00
Retailer's operating profit	9.00
COST TO CONSUMER	**$70.00**

Sources: Nike Inc., Reebok International Inc., The Finish Line Inc., Just for Feet Inc., Melville Corp., U.S. Customs Service, Atlantic Footwear Assn., industry consultants and executives, *The Washington Post*

Suppose that at *All Sport Shoes*, the manager estimates the monthly operating cost for the store (in dollars) as a function of the number of pairs of shoes that the store purchases from its suppliers. The rule for that function is $C(x) = 17,500 + 35x$.

a. Calculate and explain the meaning of each of the following:

  i. $C(100)$

  ii. $C(250)$

  iii. $C(0)$

b. What do the numbers 17,500 and 35 tell about the relation between the number of pairs of shoes purchased from the manufacturer and the total cost of doing business at the store for one month?

c. What value of x satisfies $C(x) = 24,500$? What does that value tell about the store's monthly business costs?

d. What table and graph patterns do you expect for this cost function?

e. What is the practical domain of the cost function? What is the practical range?

f. What is the theoretical domain of the function? What is the theoretical range?

2 Developers of Waldo's World amusement park along Interstate I-75 wanted to install a large, illuminated sign that could be seen from a distance. A design firm suggested using a "W" logo made up of parabolas like the graphs below.

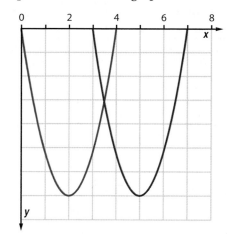

a. What functions will give parabolic graphs with the indicated *x*-intercepts? Write the function rules in two forms—one using a standard $ax^2 + bx + c$ expression and the other using an expression that is the product of two linear factors.

b. What are the minimum points of the graphs of the functions you defined in Part a?

c. If you wanted a W that was the same width as that shown but with minimum points along the line $y = -12$, what adjustment of your function rules in Part a would do the job?

3 The design firm explored a second option using the functions $f(x) = x^2 - 6x$ and $g(x) = x^2 + 2x - 8$ to make the W appear somewhat taller and thinner.

a. Write each of these functions using equivalent factored forms of the expressions $x^2 - 6x$ and $x^2 + 2x - 8$.

b. Use the factored forms to locate the *x*-intercepts and minimum points of each function graph.

c. Will the new W appear somewhat taller and thinner than that in Task 2? Explain your answer.

4 Write each of these quadratic expressions in equivalent expanded or factored form.

a. $(x + 3)(x - 7)$ b. $(x - 6)(x + 6)$

c. $(2x + 3)^2$ d. $x^2 + 8x + 7$

e. $x^2 - 144$ f. $3x^2 + 12x$

5 Solve each of these equations in two ways—by factoring and by use of the quadratic formula. Explain how the number of solutions in each case is shown in the factored form of the quadratic expression and in the results of using the quadratic formula.

a. $x^2 - 5x + 4 = 0$ b. $x^2 - 6x + 9 = 0$

c. $x^2 + 11x + 10 = 0$ d. $3x^2 - 3x - 18 = 0$

6 Consider a quadratic function $f(x) = ax^2 + bx + c$.

   a. Explain the relationship between the graph of $f(x)$ and the values $\dfrac{-b}{2a}$ and $\dfrac{\sqrt{b^2 - 4ac}}{2a}$.

   b. Suppose the quadratic expression $ax^2 + bx + c$ can be written in factored form $(ax - m)(x + n)$. Explain what the linear factors tell about the graph of $f(x)$.

7 The three key variables in any type of racing are *distance*, *speed*, and *time*.

   a. If a runner covers 400 meters in 50 seconds, what is the runner's average speed? What if it takes the runner 60 seconds to cover the same distance? Write a rule that expresses average speed s as a function of distance d and time t.

   b. If a NASCAR driver plans to complete a race of 240 miles at an average speed of 150 miles per hour, how long will the race take? What if the average speed is 180 miles per hour? Write a rule that expresses race time t as a function of distance d and average speed s.

   c. If a participant in a triathlon swims at an average speed of 1.2 meters per second for 40 minutes, how much distance will be covered? What if the average speed drops to 0.9 meters per second and the time increases to 50 minutes? Write a rule that expresses distance d as a function of average speed s and time t.

   d. When Dakota was training for a long-distance race, she made some calculations about time and average speed.

     i. If the race is 24 miles long, how does race time t in hours depend on average speed s in miles per hour?

     ii. Dakota believes that her endurance depends on how fast she runs according to the function $t = 10 - s$. The faster she runs, the shorter the time she can actually keep going. How long can she run at an average speed of 4 miles per hour? How long at an average speed of 8 miles per hour?

     iii. At what speeds can Dakota run that will allow her to run long enough to complete the 24-mile race? Sketch a graph of "endurance time as a function of speed" and "required running time as a function of speed" to illustrate your answer.

8 As is the case with all businesses, the owners of Waldo's World need to make many decisions, including prices to charge for admission to the park.

Market research suggested that the income from admissions I in thousands of dollars would depend on the admission price charged x in dollars according to the rule $I(x) = -0.6x^2 + 28x$. Operating costs C were projected as a function of the admission price charged x according to the rule $C(x) = 250 - 2x$. The owners begin their analysis of profit prospects by first determining the admission price(s) for which income would exceed operating costs.

a. Use algebraic reasoning to determine admission price(s) for which income exceeds operating costs. Illustrate your answer by a sketch of the graphs of the functions involved with key points labeled with their coordinates.

b. Use algebraic reasoning to find the maximum profit possible for Waldo's World under the given assumptions. What ticket price should be charged to maximize profit?

9 Sketch graphs illustrating the number of solution possibilities for systems that link variables with conditions that are given by:

a. two linear equations with two variables.

b. one linear function and one inverse variation function.

c. one linear function and one quadratic function.

10 Evaluate each of these expressions. Give exact values where possible and estimates accurate to two decimal places otherwise.

a. $\log 10,000$

b. $\log 0.0001$

c. $\log (10^{3.72})$

d. $\log (10^{-3.72})$

e. $5 \log (10^7) - 8$

f. $\log 372$

11 Many scientific and business calculations require high degrees of accuracy. So, there is value in not rounding computational results until the end of work on a problem. Early estimates can introduce errors that compound to produce final results that are not accurate enough. When the problem involves exponential expressions, logarithms can be helpful in this regard.

a. Solve the equation $10^{3x + 5} = 100$ algebraically.

b. Write the solution for the equation $10^{3x + 5} = 25$ as an expression involving $\log_{10} 25$.

c. Write a formula for the solution to any equation in the form $10^{ax + b} = c$.

Summarize
the Mathematics

In this unit, you investigated a variety of situations in which linear, quadratic, and inverse variation functions described relationships between variables. You further developed algebraic skills that are useful in writing the expressions for functions in equivalent forms, and you learned how to use logarithms to solve problems involving exponential functions.

a If two variables x and y are related by a function so that $y = f(x)$, what information about the relationship is expressed by statements in the form $f(a) = b$?

b How can knowledge of the x- and y-intercepts of graphs be used to write rules for quadratic functions?

c How do you find the expanded forms of expressions in the following forms?

   i. $(x + a)(x + b)$
  ii. $(x + a)^2$
 iii. $(x + a)(x - a)$

d What strategies do you use to write an expression like $x^2 + mx + n$ in equivalent form as a product of linear factors?

e What are the key steps in solving a quadratic equation:

   i. by factoring?
  ii. by use of the quadratic formula?

How do you decide which of these strategies is most likely to be effective in a particular case?

f What are the number of solution possibilities and solution strategies for systems of equations like:

   i. $f(x) = ax + b$ and $g(x) = \frac{k}{x}$?
  ii. $f(x) = ax + b$ and $h(x) = ax^2 + bx + c$?

g What does it mean to say "$\log b = a$"?

h How can the equation $y = 10^x$ be expressed in equivalent form using common logarithms?

Be prepared to explain your ideas and methods to the class.

✓ Check Your Understanding

Write, in outline form, a summary of the important mathematical concepts and methods developed in this unit. Organize your summary so that it can be used as a quick reference in future units and courses.

NETWORK OPTIMIZATION

Optimization is the process of finding the best. This process is important throughout mathematics and in everyday life. Everywhere you look, you'll see people trying to get the best, whether it's the best deal on a new purchase, the best job, the highest score in a game, or even something as ordinary as the best route from home to school. In your previous study of mathematics, you have often tried to optimize, for example, by finding the best-fitting line for a scatterplot of data, the maximum area of a shape with fixed perimeter, the highest point on the graph of a quadratic function, or the fewest number of colors needed to color the vertices of a graph.

In this unit of *Core-Plus Mathematics*, you will study optimization in the context of networks. For example, you might want to find an optimum road network or an optimum telecommunications network. To help solve network, optimization problems, you will use vertex-edge graphs. The necessary concepts and skills for solving the optimization problems in this unit are developed in two lessons.

Lessons

1 *Optimum Spanning Networks*

Use minimum spanning trees and Hamilton circuits to help find optimum networks that span (reach) all the vertices in a vertex-edge graph.

2 *Scheduling Projects Using Critical Paths*

Use critical path analysis to optimally schedule large projects that are comprised of many smaller tasks, like a building construction project.

Optimum Spanning Networks

Network optimization problems occur in many contexts. You can often represent these problems with *vertex-edge graphs* consisting of points (vertices) and line segments or arcs (edges) between some of the points. Sometimes when solving optimization problems related to vertex-edge graphs, you want to find a network that spans (reaches) all the vertices in the graph and is optimum in some way.

Think about situations involving networks with which you are familiar and what it would mean to find an optimum or best network.

a What kinds of situations in the communications industry might involve finding an optimum network?

b What kinds of situations in the transportation industry might involve finding an optimum network?

c What kinds of situations in manufacturing might involve finding an optimum network?

d What kinds of situations in your own everyday life might involve finding an optimum network?

In this lesson, you will learn about two important types of *optimum spanning networks:* minimum spanning trees and minimum spanning circuits.

Investigation 1 — Minimum Spanning Trees

In this information age, it is important to find the best way to stay informed. You need to have the right information at the right time in order to make the best decisions and take the most effective action. One way to keep informed is through computer networks. Many places, including businesses, schools, and libraries, have computers linked together in networks so that information can be shared among many users. In fact, there is a common saying that "the network *is* the computer."

Suppose that a large school decides to create its own *intranet*, that is, a network of computers within the school. They will create the network by placing 6 *hubs* at different locations around the school, linking nearby computers to each hub, and then connecting all the hubs. Every hub does not need to be connected directly to every other hub, but they must all be connected in some way, directly or indirectly. The problem is to figure out how to connect all the hubs with the least amount of cable. As you work through the problems in this investigation, look for answers to the following questions:

> *How can you find a network that will connect all 6 hubs, directly or indirectly, using the minimum amount of cable?*

> *What are some properties of this network and of similar networks?*

Computer Network Problem Because of the layout of the school, it is not possible to run cable directly between every pair of hubs. The matrix below shows which hubs can be linked directly, as well as how much cable is needed. The hubs are represented by letters and the distances are in meters.

Computer Network Matrix

	A	B	C	D	E	F
A	—	45	—	—	—	15
B	45	—	40	—	40	55
C	—	40	—	15	25	—
D	—	—	15	—	30	55
E	—	40	25	30	—	45
F	15	55	—	55	45	—

1 Examine the computer network matrix.

   a. What does the "25" in the C row mean?

   b. Why is the A-B entry the same as the B-A entry?

   c. Why isn't there a B-B entry? Why isn't there a D-B entry?

   d. Why does it make sense that the matrix is symmetric about the main diagonal?

2 Draw a vertex-edge graph that represents the information in the matrix. Recall that when building a graph model, you must specify what the vertices and edges represent.

3 Use your graph to solve the computer network problem, as follows.

   a. Compare your graph to other students' graphs. Agree on a graph that best represents this problem situation. What do the vertices and edges represent?

   b. What is the least amount of wire needed to connect the hubs so that every hub is linked directly or indirectly to every other hub?

   c. Make a copy of the graph you agreed on in Part a and then darken the edges of a shortest network for this problem.

   d. Compare your shortest network and the minimum amount of cable needed to what other students found. Discuss and resolve any differences.

An Optimum Network An optimum network in this case is a shortest network. Think about how to find a shortest network and the properties of such a network.

4 Write step-by-step instructions for how to find a shortest network in a vertex-edge graph, like the shortest network you found in the computer networking problem. A set of step-by-step instructions like this is called an **algorithm**.

5 Exchange your algorithm for finding a shortest network with the algorithm written by another student or group of students.

   a. Make a new copy of the graph. Carefully follow the steps in each other's algorithm for finding a shortest network. Compare results. Does each algorithm work? Is each algorithm written carefully enough so that anyone can follow the directions and find a shortest network?

   b. Work together to refine the algorithms that work.

6 Think about the properties of the shortest wiring networks you have been investigating. State whether each of the following statements is *true* or *false*. In each case, give a reason justifying your answer. Compare your answers to those of other students and resolve any differences.

   Statement I   There is only one correct answer possible for the minimum amount of cable needed to connect all 6 hubs.

   Statement II   There can be more than one shortest network for a given situation.

   Statement III   There is more than one algorithm for finding a shortest network.

   Statement IV   A shortest network must be all in one piece; that is, the network must be **connected**.

   Statement V   All vertices must be joined by the network.

   Statement VI   A shortest network cannot contain any circuits. (Recall that a **circuit** is a path that starts and ends at the same vertex and does not repeat any edges.)

7 The type of shortest network that you have been studying is an important vertex-edge graph model, which can be described with some useful mathematical terms.

   a. A connected graph that has no circuits is called a **tree**. Why does it make sense to call such a graph a *tree*?

   b. A tree in a connected graph that reaches (that is, includes or connects) all the vertices in the graph is called a **spanning tree**. Why does it make sense to say that such a tree is *spanning*?

   c. A graph with numbers on its edges is called a **weighted graph**. The numbers on the edges, whatever they represent, are called **weights**. The computer network graph that you have been working with is a weighted graph. What do the weights represent?

   d. A **minimum spanning tree** in a weighted connected graph is a tree that has minimum total weight and *spans* the graph—that is, it includes every vertex. Explain why the shortest networks you have found in the computer network graph are minimum spanning trees.

More Algorithms As you may have concluded in Problem 6, there are several possible algorithms for finding a minimum spanning tree in a connected graph.

8 Study the algorithm below.

Step 1: Draw all the vertices but no edges.

Step 2: Add an edge with the smallest weight that will not create a circuit. If there is more than one such edge, choose any one. The edge you add does not have to be connected to previously-added edges, and you may use more than one edge of the same weight.

Step 3: Repeat Step 2 until it is no longer possible to add an edge without creating a circuit.

a. Follow the steps of this algorithm to construct a minimum spanning tree for the computer network graph.

b. Explain why this algorithm could be called a *best-edge algorithm*.

c. Compare the minimum spanning tree you get using this best-edge algorithm to the one you found in Problem 3. How do the total weights of the minimum spanning trees compare?

9 Examine the graph below.

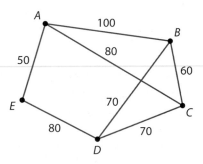

a. Use the best-edge algorithm from Problem 8 to find a minimum spanning tree for this graph. Calculate its total weight.

b. Explain why the algorithm can produce different minimum spanning trees.

c. Find all possible minimum spanning trees for the graph. Compare their total weights.

d. How is the best-edge algorithm similar to, or different from, the algorithms you produced in Problems 4 and 5?

10 Students in one class claimed that the following algorithm will produce a minimum spanning tree in a given graph.

Step 1: Make a copy of the graph with the edges drawn lightly.

Step 2: Choose a starting vertex.

Step 3: For the vertex where you are, darken the shortest edge from that vertex that will not create a circuit. (If there is more than one such edge, choose any one.) Then move to the end vertex of that edge.

Step 4: Repeat Step 3 until it is not possible to add another edge.

Complete Parts a–f to test this algorithm. First, make four copies of the graph below.

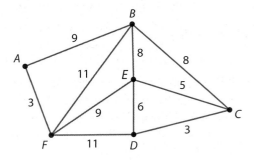

a. Apply the algorithm to the graph starting with vertex *E*. What is the total weight of the network you get?

b. Explain why this algorithm could be called a *nearest-neighbor algorithm*.

c. Apply the algorithm starting with vertex *F*. Record the total weight of the resulting network.

d. Apply the algorithm starting with vertex *A*. What happens?

e. Now use the best-edge algorithm described in Problem 8 to find a minimum spanning tree for this graph.

f. Do you think the nearest-neighbor algorithm is a good algorithm for finding a minimum spanning tree? Write a brief justification of your answer.

11 The best-edge algorithm from Problem 8 was first published by American mathematician Joseph Kruskal, and so it is often called **Kruskal's algorithm**. Kruskal discovered the algorithm while he was still a graduate student in the 1950s. How are the nearest-neighbor algorithm and Kruskal's algorithm similar? How are they different?

12 Two important questions about any algorithm are "Does it always work?" and "Is it efficient?".

a. Do you think Kruskal's algorithm (described in Problem 8) will always work to find a minimum spanning tree in any connected graph? Use vertex-edge graph software to investigate your conjecture. Create several connected weighted graphs. Use these test graphs and the software to help answer the following questions.

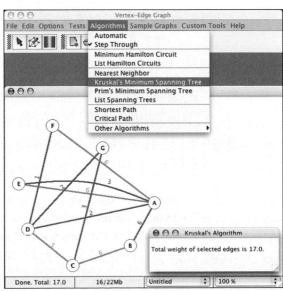

- Does Kruskal's algorithm produce a spanning tree in a connected graph? (Use the software to run the algorithm on several graphs; check to see if the result is always a spanning tree.)

- Is it possible that different runs of Kruskal's algorithm could produce different spanning trees in the same graph? (Try different runs on the same graph for several different graphs. If Kruskal's algorithm produces just one spanning tree for each graph, create a new graph for which you think Kruskal's algorithm can generate different spanning trees, then test it.)

- If Kruskal's algorithm produces several different spanning trees for a given graph, do they all have the same total weight?

- Will any spanning tree produced by Kruskal's algorithm have the smallest possible total weight? (Use the software to generate and examine all possible spanning trees for a given graph. Check to see if a spanning tree generated by Kruskal's algorithm has smallest weight among all possible spanning trees.)

b. Suppose you run all these tests on many graphs, and each time Kruskal's algorithm produces a minimum spanning tree. Is this enough evidence to conclude that Kruskal's algorithm will always work to find a minimum spanning tree in any connected graph? Explain your reasoning.

c. The two questions—Does it always work? and Is it efficient?—should be considered for all algorithms. So far, you have only investigated the first question. For Kruskal's algorithm, mathematicians have proven that the answer to both questions is "yes." Kruskal's algorithm will efficiently find a minimum spanning tree for any connected graph. How would you answer these questions for the nearest-neighbor algorithm?

Summarize the Mathematics

In this investigation, you learned how to find and interpret minimum spanning trees.

a Does every connected graph have a minimum spanning tree? Explain your reasoning.

b Is it possible for a given graph to have more than one minimum spanning tree? Can different minimum spanning trees for the same graph have different total weights?

c What information does the total weight of a minimum spanning tree give you?

d Describe in your own words the basic strategy of Kruskal's best-edge algorithm. Do the same for the nearest-neighbor algorithm. Which of these two algorithms is guaranteed to produce a minimum spanning tree for any connected graph?

Be prepared to share your ideas and reasoning with the class.

✓ Check Your Understanding

A landscape architect has been contracted to design a sprinkler system for a large lawn. There will be six sprinkler heads that must be connected by a buried network of pipes to the main water source. The possible connections and distances in yards are shown in the diagram below. The main water source is represented by vertex *B*. What is the least amount of pipe needed to construct the sprinkler network? Draw a landscape plan showing the optimum sprinkler network.

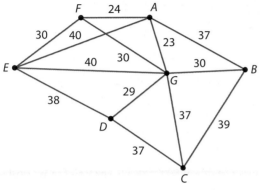

Main Water Source

Investigation 2 — The Traveling Salesperson Problem (TSP)

In the last investigation, you learned about a particular type of optimum spanning network—a minimum spanning tree. The vertex-edge graphs you investigated were *weighted graphs*, since they had *weights* (numbers) on the edges. Such numbers often represent distance, but they can also represent other quantities such as time or cost. In this investigation, you will study another type of optimum spanning network in weighted graphs, related to the following problem.

The **Traveling Salesperson Problem**, often simply referred to as the **TSP**, is one of the most famous problems in mathematics. Here is a statement of the problem.

A sales representative wants to visit several different cities, each exactly once, and then return home. Among the possible routes, which will minimize the total distance traveled?

This problem is historically known as the Traveling Salesman Problem since the original context for the problem referred to salespeople, almost all men at the time the problem was named in the 1950s, who traveled from city to city selling their products. As is common in the present day, we will typically refer to this problem simply by its initials—TSP. As you work through this investigation, look for answers to this question:

What are some ideas and methods that are helpful
as you try to solve the TSP?

1 Although the TSP is classically stated in terms of salesmen, cities, and distance, it also refers to similar problems in other contexts. Consider the TSP in the context of this airfare graph.

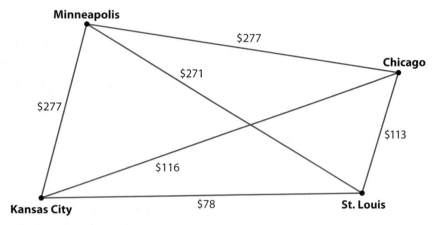

a. Solve the TSP for this weighted graph. What does "weight" represent in this case?

b. Compare your solution to those of other students. Resolve any differences.

c. How do you know that there is no circuit less expensive?

2 A route through a graph that starts at one vertex, visits all the other vertices exactly once, and finishes where it started is called a **Hamilton circuit**.

a. Describe the TSP in terms of Hamilton circuits.

b. How many different Hamilton circuits are there in the graph in Problem 1? For the purpose of finding the total weight of circuits, two circuits are different only if they have different edges. It does not matter where you start or which direction you go around the circuit.

3 Think about the method you used to find the optimum circuit in Problem 1.

a. Write a description of the method you used.

b. Could you generalize your method to find an optimum circuit for traveling to the capital cities of all 48 contiguous states in the United States? Explain your reasoning.

Best-Edge Algorithm In Investigation 1, you used Kruskal's best-edge algorithm to find a minimum spanning tree. One group of students devised the following best-edge algorithm for the TSP. They claim it will solve the TSP. Here is their algorithm:

> Step 1: Make a copy of the graph with the edges drawn lightly.
> Step 2: Darken the shortest edge not yet used, provided that:
> • you do not create a circuit of darkened edges, unless all of the vertices are included.
> • no vertex is touched by 3 darkened edges.
> (The edge you darken does not have to be connected to previously darkened edges.)
> Step 3: Repeat Step 2 as long as it is possible to do so.

4 Analyze this best-edge algorithm.

   a. Why do you think the algorithm requires that you do not create a circuit of darkened edges unless all of the vertices are included?

   b. Why do you think the algorithm requires that no vertex is touched by 3 darkened edges?

5 Apply the algorithm to the airfare graph in Problem 1. Does this algorithm produce a solution to the TSP? Explain.

Brute-Force Method One method that certainly will work to solve the TSP is to list all possible Hamilton circuits, compute the weight of each one, and choose the minimum. This approach of checking all possibilities is sometimes called a *brute-force method*. With computers available to do all the calculations, you might naturally think this is the way to proceed.

6 Use vertex-edge graph software to try the brute-force method for a weighted complete graph with 5 vertices. That is, consider a graph with 5 vertices that has exactly one edge between each pair of vertices, and the edges are weighted. Do the following.

   a. Create a weighted complete graph with 5 vertices.

   b. Use a method of your choice to try to find a solution to the TSP for this 5-vertex graph, using software or not.

   c. Use a brute-force method to find a solution to the TSP for this graph. That is, find all possible Hamilton circuits, compute the total weight of each, and select a circuit with the minimum total weight. Compare to what you found in Part b.

In Problem 6, you were able to use a computer to carry out a brute-force method to solve the TSP for a 5-vertex graph. However, even though computers are fast, the amount of time required for a brute-force computer solution must be taken into account. For example, in the next two problems, you will determine how long it will take a computer to use a brute-force method to solve the TSP for the 26 cities shown in the following map. Assume each city is connected directly to all the others, and the tour starts at Atlanta.

Seattle
Portland
Minneapolis ○
Milwaukee ○
Detroit   Boston ○
Chicago ○
Cleveland ○
Omaha ○
Newark
Denver ○
Indianapolis ○ ○ Pittsburgh ○
Philadelphia
San Francisco
Kansas City ○
Cincinnati   D.C.
St. Louis
Tulsa ○
Los Angeles
Memphis ○
Atlanta
Dallas ○
New Orleans
Houston ○

7 Find the number of possible circuits for the 26-city TSP using the following reasoning.

a. Starting from Atlanta, how many cities could be the first stop?

b. Once you choose a city for this first stop, how many cities could be the second stop in the circuit? Remember that every city is connected directly to every other city, and each city is visited exactly once.

c. How many different first-stop/second-stop possibilities are there? Justify your answer.

d. How many cities could be the third stop of the circuit? How many different routes of first-stop/second-stop/third-stop are there?

e. How many different circuits are possible using all the cities?

8 Suppose you could use the fastest computer in the world to solve the 26-city TSP.

a. Using the brute-force method running on the world's fastest supercomputer, how long do you think it would take to solve the 26-city TSP? Make a quick guess.

b. The TOP500 project, which began in 1993, keeps track of the world's fastest computers. Twice a year, they release a list of the 500 most powerful computer systems. In June 2007, they announced that the BlueGene/L System, developed jointly by IBM and the U.S. Department of Energy, was the world's fastest. The BlueGene/L shown at the left was officially rated at 280.6 TFLOPS (teraFLOPS). Thus, roughly speaking, it can carry out 280.6 trillion calculations per second. Suppose you could use the BlueGene/L System to solve the 26-city TSP. Consider how fast the BlueGene/L could compute the total weight of all the possible different Hamilton circuits. (It would take additional time to find all the circuits.) Assume that the BlueGene/L can compute the weight of 280 trillion circuits each second. How many seconds will it take to compute the weight of all the circuits? How many years? (Source: www.top500.org)

c. How fast do you think a computer would need to be to provide a brute-force solution to this problem in a reasonable amount of time?

More with Hamilton Circuits Early work with Hamilton circuits involved graphs that were not weighted. For example, around 1857, Sir William Rowan Hamilton, a famous Irish mathematician for whom Hamilton circuits are named, invented a game called the *Traveller's Dodecahedron, A Voyage Round the World.* Recall that a dodecahedron has 20 vertices and 12 faces which are regular pentagons.

9 The *Traveller's Dodecahedron* game includes a wooden compressed dodecahedron with a peg at each vertex and a silk cord. The vertices represent cities from around the world, like Canton, Delhi, and Zanzibar. The object of the game is to start at one city, visit the other 19 cities exactly once, and finish back where you started. The silk cord is wound around the pegs to keep track of the journey.

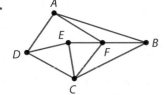

a. Explain how the wooden figure shown can be thought of as a "compressed" dodecahedron. Then think about completely flattening the compressed dodecahedron. The flattened figure looks like a vertex-edge graph with 20 vertices and 11 pentagonal regions. Draw this vertex-edge graph.

b. Instead of using string, use your pencil to trace a path in this graph that will win the game.

c. Explain why a winning path is a Hamilton circuit.

d. Does it matter where you start? Explain why or why not.

10 For the following two graphs, find a Hamilton circuit if one exists. If there is no Hamilton circuit, explain why not.

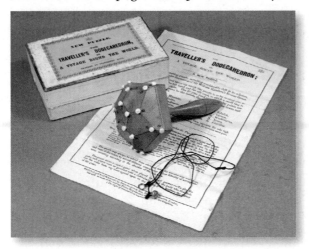

a.

b.

In this investigation, you learned about Hamilton circuits and explored algorithms for solving the TSP.

a What is the relationship between a Hamilton circuit and a solution to the TSP?

b Describe in your own words the best-edge algorithm for the TSP discussed in Problems 4 and 5. Can you always solve the TSP using this algorithm?

c Describe the brute-force method for solving the TSP. In theory, will this method always solve the TSP? Will this method provide a practical solution to the TSP? Explain.

Be prepared to share your descriptions and thinking with the class.

✓Check Your Understanding

The matrix at the right shows the mileage between four cities.

$$\begin{array}{c} & \begin{array}{cccc} A & B & C & D \end{array} \\ \begin{array}{c} A \\ B \\ C \\ D \end{array} & \left[\begin{array}{cccc} 0 & 20 & 25 & 40 \\ 20 & 0 & 35 & 45 \\ 25 & 35 & 0 & 30 \\ 40 & 45 & 30 & 0 \end{array}\right] \end{array}$$

a. Represent the information in the matrix with a weighted graph.

b. Trace all the different Hamilton circuits starting at *A*. List the vertices in each circuit. Record the total length of each circuit.

c. Would you get different answers in Part b if the starting vertex was *B*?

d. Is there a difference between circuit *A-B-C-D-A* and circuit *A-D-C-B-A*? Explain your reasoning.

e. What is the solution to the TSP for this graph?

Investigation 3   Comparing Graph Topics

So far in this lesson, you have studied minimum spanning trees, Hamilton circuits, and the TSP. In this investigation, you will compare different graph topics. As you work through the investigation, look for answers to this question:

> *What are similarities and differences among different graph topics that you have studied?*

1 **TSP versus Minimum Spanning Trees** There are some interesting connections between these two graph topics.

   a. Describe how a solution to the TSP is similar to, and yet different from, a minimum spanning tree.

b. The title of this lesson is "Optimum Spanning Networks."

    i. Explain how a minimum spanning tree can be viewed as an optimum spanning network.

    ii. Explain how a solution to the TSP can be viewed as an optimum spanning network.

c. A best-edge algorithm can be used to try to solve the TSP and to find a minimum spanning tree. For which of these two problems does this algorithm always work?

2 **Hamilton versus Euler Circuits** In *Core-Plus Mathematics* Course 1, you may have studied *Euler circuits*—circuits that use each edge of a graph exactly once.

a. State the definition of a Hamilton circuit. Then describe the key difference between an Euler circuit and a Hamilton circuit.

b. Consider the graph at the right.

    i. Find an Euler circuit.

    ii. Find a Hamilton circuit.

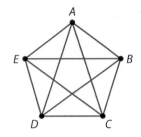

3 The graph in Problem 2 has both an Euler and a Hamilton circuit. Draw the following graphs (if possible).

a. A graph that does not have an Euler circuit or a Hamilton circuit

b. A graph that does not have an Euler circuit but does have a Hamilton circuit

c. A graph that does have an Euler circuit but does not have a Hamilton circuit

4 Some graphs have Euler or Hamilton circuits and some do not. You may recall that for Euler circuits there is a theorem that gives a simple, easily testable condition for whether or not an Euler circuit exists. This is not the case for Hamilton circuits. That is, no one knows a nice theorem that states, "A graph has a Hamilton circuit if and only if ...." However, there are some graph properties that give you some information about Hamilton circuits.

a. Look back at your work on Problem 3 above and on Problem 10 in Investigation 2 (page 411). Can you think of a property of a graph that will guarantee that it does *not* have a Hamilton circuit? Make a conjecture and give an argument to support it.

b. Using vertex-edge graph software, generate several graphs that have the property you conjectured in Part a, then test them to see if they have a Hamilton circuit. Does your conjecture still hold or do you need to revise it?

c. Generate and test other graphs to help you propose another property that you think will guarantee that a graph does not have a Hamilton circuit.

⑤ Matrices and Graphs You learned about adjacency matrices for graphs in previous units. In the next few problems, you will use a similar type of matrix called a *distance matrix*. Consider the following road network. There are seven small towns in Johnson County that are connected to each other by gravel roads, as in the diagram below. (The diagram is not drawn to scale and the roads are often curvy.) The distances are given in miles.

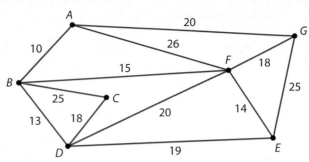

The county wants to pave some of the roads so that people can get from every town to every other town on paved roads, either directly or indirectly. To save money, they also want the total number of miles paved to be as small as possible.

a. Copy the vertices of this graph on a sheet of paper. Then draw a network of paved roads that will fulfill the county's requirements.

b. Each entry in a **distance matrix** for a graph is the length of a shortest path between the corresponding pair of vertices. For example, each entry in the partially completed distance matrix at the right is the shortest distance between the corresponding pair of towns on the paved road network that you found in Part a above.

$$
\begin{array}{c c}
 & \begin{array}{ccccccc} A & B & C & D & E & F & G \end{array} \\
\begin{array}{c} A \\ B \\ C \\ D \\ E \\ F \\ G \end{array} &
\left[\begin{array}{ccccccc}
\_ & \_ & 41 & \_ & 39 & \_ & \_ \\
\_ & \_ & \_ & \_ & 29 & 15 & \_ \\
\_ & 31 & \_ & \_ & \_ & \_ & \_ \\
23 & 13 & 18 & \_ & 42 & 28 & 46 \\
\_ & 29 & \_ & \_ & \_ & \_ & \_ \\
\_ & \_ & \_ & \_ & \_ & \_ & \_ \\
43 & \_ & \_ & \_ & 32 & \_ & \_
\end{array} \right]
\end{array}
$$

   i. Explain why the *A-C* entry is 41 and the *D-E* entry is 42.

   ii. Why will this matrix have symmetry about its main diagonal?

   iii. Fill in the remaining entries of the matrix. Divide the work among some of your classmates.

⑥ Use the distance matrix to further analyze the road network in Problem 5.

a. Which two towns are farthest apart on the paved-road network?

b. Compute the row sums of the distance matrix. What information do the row sums give about distances on the paved-road network?

c. Which town seems to be most isolated on the paved-road network? Which town seems to be most centrally located? Explain how these questions can be answered by examining the distance matrix.

Which towns might be dissatisfied with this paved-road network? Why? What are some other considerations that might be taken into account when planning an optimum paved-road network?

Solved and Unsolved Problems In this lesson, you examined four fundamental vertex-edge graph topics: minimum spanning trees, the TSP, Hamilton circuits, and Euler circuits. For two of these topics—the TSP and Hamilton circuits—the key problems are currently unsolved! New applications and new mathematics have been developed as researchers continue to work on these problems.

For the TSP, the key problem is to find an efficient solution method that will work in all situations. You have seen one method, a best-edge algorithm, that is efficient but does not guarantee a solution. You have seen another method, the brute-force method, that guarantees a solution but is not efficient. No one knows a method that is both efficient and works in all situations.

For Hamilton circuits, the key problem is to find some testable condition(s) that will completely describe graphs that have a Hamilton circuit. In Problem 4, you made some conjectures about properties of graphs that guarantee that a Hamilton circuit does *not* exist. In Extensions Task 22, you are asked to examine some special types of graphs for which it is possible to deduce whether or not a Hamilton circuit exists. However, there are no known general results for arbitrary graphs. That is, no one knows a theorem that completely characterizes graphs that have a Hamilton circuit; no one knows an efficient algorithm for determining if a graph has a Hamilton circuit; and no one knows an efficient algorithm that is guaranteed to find a Hamilton circuit if it exists.

In contrast, the general problems of finding a minimum spanning tree and finding an Euler circuit are well solved. There are known efficient algorithms that are proven to work in both cases. In addition, the problem of characterizing graphs that have Euler circuits is also solved (a connected graph has an Euler circuit if and only if all the vertices of the graph have even degree).

In this investigation, you have compared several graph topics and problems.

a Describe similarities and differences between the problem of finding a minimum spanning tree and the TSP.

b Describe similarities and differences between Hamilton circuits and Euler circuits.

c What information does the distance matrix for a minimum spanning tree give you?

Be prepared to share your ideas with the class.

✔Check Your Understanding

Consider the road network below, where distances shown are in miles.

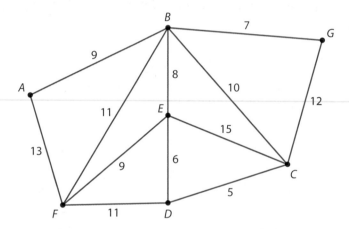

a. Suppose some of the roads need to be plowed after a snowstorm. Find the shortest possible network of plowed roads that will allow cars to drive from every town to every other town on plowed roads.

b. What is the shortest distance from *A* to *F* on the plowed-road network?

c. When there is no snow, all the roads can be used. Find the shortest distance from *A* to *F* when all the roads are clear.

d. Suppose you want to take a tour of all seven towns in the summer when all the roads are clear. Find a route that will visit all the towns without visiting any town more than once, except that you finish the tour where you started. Then, find the shortest such route.

e. Which of Parts a–d above involve finding:

  i. a minimum spanning tree?

  ii. a Hamilton circuit?

  iii. a solution to the TSP?

Applications

1 A restaurant has opened an outdoor patio for evening dining. The owner wants to hang nine decorative light fixtures at designated locations on the overhead latticework. Because of the layout of the patio and the latticework, it is not possible to install wiring between every pair of lights. The matrix below shows the distances in feet between lights that can be linked directly. The main power supply from the restaurant building is at location X. The owner wants to use the minimum amount of wire to get all nine lights connected.

	X	A	B	C	D	E	F	G	H	I
X	—	18	—	—	11	—	—	13	17	—
A	18	—	16	—	—	15	15	—	—	—
B	—	16	—	16	12	—	—	—	—	—
C	—	—	16	—	—	—	—	12	—	—
D	11	—	12	—	—	—	—	10	—	—
E	—	15	—	—	—	—	7	—	—	—
F	—	15	—	—	—	7	—	—	—	—
G	13	—	—	12	10	—	—	—	18	—
H	17	—	—	—	—	—	—	18	—	8
I	—	—	—	—	—	—	—	—	8	—

a. What is the minimum amount of wire needed to connect all nine lights?

b. Suppose the electrician decides to start at the power supply X, then go to the closest light, then go to the closest light from there, and so on. What algorithm does she seem to be using?

c. Apply the electrician's algorithm to the graph, starting at X. Describe what happens.

2 There are many situations in which it is useful to detect *clustering*. For example, health officials might want to know if outbreaks of the flu are spread randomly over the country or if there are geographic clusters where high percentages of people are sick. Geologists might want to know if the distribution of iron ore is spread evenly through an ore field or if high densities of ore are clustered in particular areas. Economists might want to know if small business start-ups are more common, that is, clustered, in some areas. There are several techniques that have been devised to detect clustering. A technique involving minimum spanning trees is illustrated in the following copper-ore mining context.

Great Lakes Mining Company would like to know if copper ore is evenly distributed throughout a particular region or if there are clusters of ore. The company drills a grid of nine test holes in each of two ore fields. The following diagrams show the grid of test holes in each ore field along with the percentage of copper, expressed as a decimal, in the sample from each test hole.

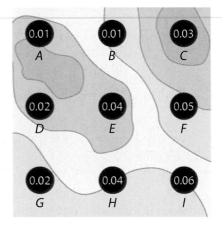

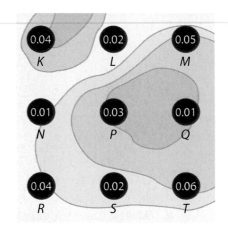

a. Construct a graph model for the grid in each ore field. Represent each test hole as a vertex, and connect two vertices with an edge if the test holes are next to each other (vertically, horizontally, or diagonally).

b. For each edge, compute the absolute value of the difference between the concentrations of copper at the two vertices on the edge. Label the edge with this number. For example, consider the grid on the left. Since the concentration at test hole *A* is 0.01 and the concentration at test hole *E* is 0.04, label the edge connecting *A* and *E* with |0.01 − 0.04| or 0.03.

c. Find a minimum spanning tree for each of the two graphs. What is the total weight of each minimum spanning tree? (Note that "total weight" here refers to the sum of the concentration differences on each edge.)

d. Now consider the connection between the length of a minimum spanning tree and clusters of ore concentrations.

   i. In general, if there is a cluster of test holes with similar concentrations of copper, will the numbers on the edges in that cluster be large or small? Why?

   ii. If there is more clustering in one of the ore fields, will the length of the minimum spanning tree for that ore field be larger or smaller than the other one? Why?

   iii. Which of the two ore fields in this example has greater clustering of concentrations of copper? Explain in terms of minimum spanning trees.

3 The graph to the right shows a road network connecting six towns (not shown to scale). The distances shown are in miles. The Highway Department wants to plow enough roads after a snowstorm so that people can travel from any town to any other town on plowed roads. However, because of the time and cost involved, officials want to plow as few miles of road as possible.

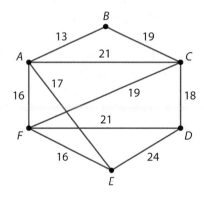

a. Find and draw a network that will meet the Highway Department's requirements. What is the total number of miles that must be plowed?

b. As you know from this lesson, there may be several networks that satisfy the Highway Department's requirements. Find all the plowed-road networks that will work. Check the total length of each such network and make sure that you get the same total mileage for each that you got in Part a.

c. Construct a distance matrix for each shortest network (that is, for each minimum spanning tree) that you found in Parts a and b. (List the vertices in alphabetical order in your matrix.)

d. For each plowed-road network, which town is most centrally located? On what quantitative (numerical) information did you base your decision?

e. Each plowed-road network that you found has the same total length. Despite this, do you think one is better than another? Justify your answer by using information from the graphs and the distance matrices.

4 Three different features appear in a local newspaper every day. The features are scheduled to be printed in three jobs, all on the same printing press. After each job, the press must be cleaned and reset for the next job. After the last job, the press is reset for the first job to be run the next morning. The time, in minutes, needed to set up the press between each pair of jobs is shown in the matrix below.

$$
\begin{array}{c c c c}
 & A & B & C \\
A & - & 25 & 15 \\
B & 30 & - & 25 \\
C & 20 & 20 & -
\end{array}
$$

The newspaper production manager wants to schedule the jobs so that the total set-up time is minimum.

a. Model this situation with a *weighted digraph*.

b. Show how a solution to the TSP will tell you how to schedule the jobs so that the total set-up time is minimum.

c. In what order should the jobs be scheduled, and what is the minimum total set-up time?

5 Integrated circuit boards are used in a variety of electronic devices, including kitchen appliances, video games, automobile ignition systems, and the guidance systems in commercial airliners. To manufacture a circuit board, a laser must drill as many as several million holes on a single board. This is usually done with a laser in a fixed position; the circuit board is turned to the positions that must be drilled. For maximum efficiency, the board must end up in its original position, no hole should pass under the laser more than once, and the total distance that the board is moved should be as small as possible.

To see how this problem is solved using graphs, consider a simple situation in which there are just four holes to be drilled. The distance, in millimeters, that the board must be moved from one hole to another is given in the matrix below.

$$
\begin{array}{c c c c c}
 & A & B & C & D \\
A & - & 0.02 & 0.02 & 0.01 \\
B & 0.02 & - & 0.04 & 0.02 \\
C & 0.02 & 0.04 & - & 0.05 \\
D & 0.01 & 0.02 & 0.05 & -
\end{array}
$$

a. Represent the information in the matrix with a weighted graph.

b. Explain why solving the circuit board problem is the same as solving the TSP for this graph.

c. Find the order for drilling the holes that will minimize the total distance that the board has to be moved.

6 Information is processed by computers as strings of 1s and 0s—called **binary strings**. Suppose the information comes from a rotary mechanical controller like what is found on a motorized wheelchair. How can the rotary position of a controller be converted into a binary string so that a computer can process the information? One method, patented in 1953 by Frank Gray and still used today, is based on a set of binary strings called a *Gray code*.

A **Gray code** is an ordered set of binary strings with the following properties.

- Every string of a given length is in the list.
- Each string in the list differs from the preceding one in exactly one position.
- The first and last strings in the list differ in exactly one position.

a. Here is a Gray code using binary strings of length two:

10   00   01   11

Verify that the three properties of a Gray code are satisfied.

When the strings are short and there are so few of them, it is possible to find a Gray code by trial and error. Using Hamilton circuits is one way to find Gray codes with longer strings. Consider strings of length three. A binary string of length three has a 1 or a 0 in each of the three positions. For example, 100 and 011 are binary strings of length three.

b. Build a graph model by letting the vertices be the eight binary strings of length three. Two vertices are connected with an edge if the two strings differ in exactly one position. For example, 010 differs in exactly one position (the 3rd position) from 011, so the vertices representing 010 and 011 should be connected by an edge.

c. Find a Hamilton circuit in the graph. List all the distinct vertices in the circuit in order.

d. Is the list you made in Part c a Gray code? Why or why not?

e. The diagram at the right represents a rotary controller with 8 positions. A movable crank is centered at the inner circle of the diagram and extends to the outermost circle. As you rotate this crank, it passes over the circular sectors comprised of blue and yellow regions. For each sector to which the crank is moved, an electrical device converts the blue regions into 1s and the yellow regions into 0s. Thus, each of the 8 angular positions is converted into a 3-digit binary string. (The white inner circle is not part of the coding, since it represents the shaft of the crank.) On a copy of the diagram, moving clockwise, label the remaining sectors with binary strings. Is this a Gray code? If so, trace a Hamilton circuit in the graph model of Part b that corresponds to this code.

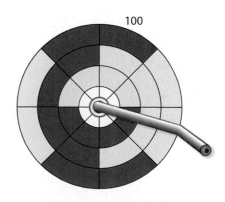
100

f. Why do you think a Gray code is desirable in the situation of a rotary controller as described in Part e?

7. A family with seven members in different parts of the world wants to keep informed about family news at home in the United States. But international calling can be expensive. The family wants to set up a telephone-calling network so everyone will know the latest news for the least total cost. A family member in the United States will call Felix, and then Felix will start the message through the network. The table below shows the cost for a 30-minute phone call between each pair of family members who are abroad.

Phone Call Costs (in dollars)

	Amy	Felix	Eliza	Kit	Owen	Ruby	Raquel
Amy		3.50	4.75	3.80	4.10	2.85	5.10
Felix	3.50		3.75	2.50	4.50	4.10	3.40
Eliza	4.75	3.75		2.95	3.15	4.40	3.50
Kit	3.80	2.50	2.95		4.25	3.30	3.40
Owen	4.10	4.50	3.15	4.25		2.95	3.25
Ruby	2.85	4.10	4.40	3.30	2.95		3.60
Raquel	5.10	3.40	3.50	3.40	3.25	3.60	

a. What is the total cost of the least expensive calling network they can set up?

b. Write a description of who should call whom in this least expensive calling network.

8. Martin and William have invited friends, who do not all know each other, for dinner. All eight people will be seated at a round table, and the hosts want to seat them so that each guest will know the people sitting on each side of him or her. In the graph below, two people who know each other are connected by an edge.

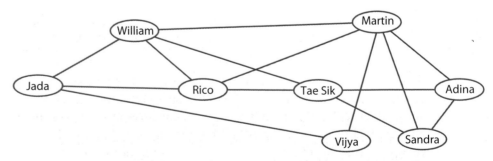

a. Find a Hamilton circuit. Explain how you can use the Hamilton circuit to seat the people according to Martin and William's requirement.

b. Sketch a diagram of the round table and show how the people should be seated.

Connections

9 In this lesson, you found optimum networks by finding minimum spanning trees. Graphs for problems involving minimum spanning trees always have numbers associated with their edges. Now consider connected graphs that do not have numbers on the edges. In these cases, you might still be interested in finding a *spanning tree*.

a. Find a spanning tree for each graph below. Describe the method you used to find the spanning trees.

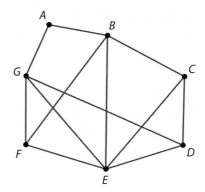 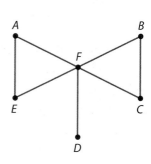

b. Find three different spanning trees for the following graph.

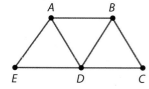

c. Write a rule relating the number of vertices *V* in a connected graph and the number of edges *E* in a spanning tree for the graph.

10 If you make a rectangular frame, like framing used for scaffolding, it is necessary to brace it with a diagonal strip. Without such a strip, it can deform, as illustrated below.

A shape like this  will deform under a load to a shape like this, unless it is braced like this.

Buildings and bridges are often constructed of rectangular steel grids, such as those shown below. To make grids rigid, you do not have to brace each cell with a diagonal, but you do have to brace some of them.

a. One of the two grids at the right is rigid, and the other is not.

   i. Which is the rigid grid? Explain your choice.

   ii. For the rigid grid, remove some of the braces without making the grid nonrigid. (You may want to make a physical model to help.)

   iii. For the nonrigid grid, add some diagonal braces to make the grid rigid.

Grid A   **Grid B**

b. You can use vertex-edge graphs to help solve problems like those on the previous page. The first step is to model the grid with a graph. Let the vertices represent the rows and columns of the grid. Then draw an edge between a row-vertex and a column-vertex if the cell for that row and column is braced. The graph for Grid A is drawn below.

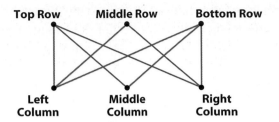

    i. Explain why there is an edge from the top-row vertex to the middle-column vertex.

    ii. Explain why there is no edge from the middle-row vertex to the middle-column vertex.

    iii. Construct the graph for Grid B.

c. Compare the graphs for Grid A and Grid B.

    i. Is the graph for Grid A connected? Is Grid A rigid?

    ii. Is the graph for Grid B connected? Is Grid B rigid?

    iii. Do you think connectedness of a graph model of a grid will ensure the corresponding grid is rigid? Draw another connected graph and another nonconnected graph, each of which could model a grid. What is true about the rigidity of the corresponding grids? Do these examples support your conjecture?

d. A rigid grid may have "extra" bracings. For example, in Part a, you discovered that it was possible to remove some of the cell bracings of Grid A and still maintain the rigidity. Investigate what this means in terms of the corresponding graph.

    i. On a copy of the graph for Grid A, eliminate "extra" edges, one at a time. Stop when you think that removing another edge will result in a graph that represents a nonrigid grid. How is your final "subgraph" related to the original graph?

    ii. Add the minimum number of bracings to Grid B to make it rigid. Draw the corresponding graph. Is there anything special about the graph?

    iii. Describe as completely as you can what you think are the properties of a vertex-edge graph that represents a grid that is rigid and has the minimum number of bracings. Draw a graph that has the properties, then check to see if the graph corresponds to a grid that is rigid with the minimum number of bracings. Draw a rigid grid with the minimum number of bracings, then draw the graph that represents that grid. If necessary, use these and other examples to refine your graph description.

11 For vertex-edge graphs, the position of the vertices, the lengths of the edges, and the shape of the graph are not essential. All that really matters is how the vertices are connected by the edges. So, for example, when drawing a graph that corresponds to a distance matrix, the graph does not need to be drawn to scale. On the other hand, in other areas of geometry, like coordinate geometry as studied in the *Coordinate Methods* unit, position, size, and shape *are* essential. Investigate this key difference between graph theory and other areas of geometry by considering the following distance matrix. Each entry is the shortest distance, in miles, between the two corresponding towns.

$$\begin{array}{c}\begin{array}{ccc} & \text{W} & \text{R} & \text{T}\end{array}\\\begin{array}{c}\text{Woebegone (W)}\\\text{Rivendell (R)}\\\text{Troy (T)}\end{array}\left[\begin{array}{ccc} - & 60 & 100 \\ 60 & - & 80 \\ 100 & 80 & - \end{array}\right]\end{array}$$

a. Draw a vertex-edge graph that represents the information in the matrix.

b. Use a compass and ruler to draw a scale diagram showing the distances between the three towns. Assume straight-line roads between the towns.

c. State a question involving these three towns that is best answered using the scale diagram.

d. State a question that could be answered using either model—the scale diagram or the vertex-edge graph.

12 A **Hamilton path** is a route that uses each vertex of a graph exactly once. (Thus, a Hamilton circuit may be thought of as a special type of Hamilton path; it is a circuit that uses each vertex exactly once.) Hamilton paths can be used to analyze tournament rankings.

Consider a round-robin tennis tournament involving four players. The matrix below shows the results of the tournament. Recall that the matrix is read from row to column with a "1" indicating a win. For example, the "1" in the Flavio-Simon entry means that Flavio beat Simon.

Tournament Results

$$\begin{array}{c}\begin{array}{cccc} & \text{J} & \text{S} & \text{F} & \text{B}\end{array}\\\begin{array}{c}\text{Josh (J)}\\\text{Simon (S)}\\\text{Flavio (F)}\\\text{Bill (B)}\end{array}\left[\begin{array}{cccc} 0 & 0 & 0 & 1 \\ 1 & 0 & 0 & 1 \\ 1 & 1 & 0 & 1 \\ 0 & 0 & 0 & 0 \end{array}\right]\end{array}$$

a. Draw a digraph representing the information in the matrix.

b. Find all the Hamilton paths in the graph.

c. Use the Hamilton path to rank the players in the tournament. Explain the connection between your ranking and the Hamilton path.

d. In the *Matrix Methods* unit, you used row sums and powers of matrices to rank tournaments. Rank the tournament as you did in that unit and compare your ranking to that in Part c above.

13 In Investigation 2, you considered a Hamilton circuit on a dodecahedron. In fact, many other polyhedra permit Hamilton circuits. For example, all of the Platonic solids (regular polyhedra) permit Hamilton circuits. Consider the cube.

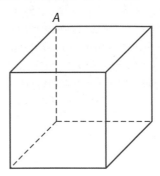

a. How many edges will be in a Hamilton circuit on a cube? Draw a diagram showing one such circuit.

b. Will all Hamilton circuits on a cube be congruent figures? Explain.

c. If a circuit is described by the sequence of vertices in the order visited, how many different Hamilton circuits begin at vertex *A* in the diagram above?

14 The Internet can be thought of as a huge vertex-edge graph (see figure on page 399). Such large graphs are sometimes called *massive graphs*. The degrees of vertices in massive graphs often obey an inverse power law. That is, the number of vertices of degree d is proportional to $\frac{1}{d^b}$, for some constant $b > 0$. In particular, for many Internet graphs, b is between 2 and 3. Explain what this inverse power law relationship tells you about the numbers of vertices with very large degree and those with very small degree. (You can read more about this at www.mathaware.org/mam/04/essays/graphs.html.)

Reflections

15 Think about the characteristics of those graphs which are also trees.

a. Explain why a tree can be considered a *minimum* connected graph.

b. Explain why a tree can be considered a *maximum* graph with no circuits.

16 Look back at Problem 8 (page 410) of Investigation 2. Find news about the world's fastest computer today. Using the power of that computer, how long will it take to solve the 26-city TSP using the brute-force method? How long will it take to solve a 27-city TSP problem?

17 With the rapid development of more and more powerful computers, do you think that any problem can eventually be solved with a brute-force method by having a computer check all possibilities? Or do you think that there is some fundamental limitation to the ability of computers to solve problems? Explain your thinking.

18 There is a story that composer Igor Stravinsky (1882–1971) was asked how he would describe his music pictorially. Pointing to the diagram below, he replied, "This is my music:"

 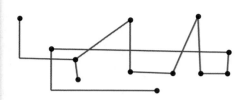

a. What do you think Stravinsky meant?

b. One of Stravinsky's most well-known compositions is *The Rite of Spring*. Listen to *The Rite of Spring*. Why do you think Stravinsky used a tree graph to describe his music? (Incidentally, because *The Rite of Spring* sounded so unexpectedly different and unusual, the premiere performance caused a riot in the audience.)

19 Kruskal's algorithm for minimum spanning trees was developed in the 1950s, and new results related to the TSP are discovered almost every year. It is estimated that more new mathematics has been developed in the last 20 years than in all the past history of mathematics. In fact, most of the mathematics you have investigated in this unit has been developed in the last few decades or even more recently. Do some research to find out about some recent development in mathematics. Write a brief essay on what you find. You might describe how current mathematics is used in the plot of a recent novel or television show, or you could explain a recent news story about mathematics from the radio, newspaper, or Internet. (One good source is Math in the Media from the American Mathematical Society: www.ams.org/mathmedia/.)

Extensions

20 The nearest-neighbor algorithm you investigated in Problem 10 on page 404 did not always produce a minimum spanning tree in a connected graph. Below is a modified version of that algorithm called *Prim's algorithm*.

Step 1: Make a copy of the graph with the edges drawn lightly.

Step 2: Choose a starting vertex. This is the beginning of the tree.

Step 3: Find all edges that have one vertex in the tree constructed so far. Darken the shortest such edge that does not create a circuit. If there is more than one such edge, choose any one.

Step 4: Repeat Step 3 until all vertices have been reached.

a. Test Prim's algorithm using three copies of the following graph from Problem 10.

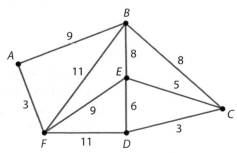

  i. Apply the algorithm starting at vertex *E*.

  ii. Apply the algorithm starting at vertex *C*.

  iii. Apply the algorithm starting at vertex *A*.

b. Compare Prim's algorithm with the nearest-neighbor algorithm in Problem 10 on page 404.

   i. How are they similar? How are they different?

   ii. Compare the results in Part a above to those you got with the nearest-neighbor algorithm in Problem 10.

c. Prim's algorithm is sometimes called a tree-growing algorithm. Explain why a tree is constructed at each stage of Prim's algorithm.

d. Compare Prim's algorithm to Kruskal's best-edge algorithm from Problem 8 on page 404.

   i. Using vertex-edge graph software, generate several weighted connected graphs. Then apply Kruskal's and Prim's algorithms to find a minimum spanning tree in each graph. Apply each algorithm in automatic mode and in user-activated edge-by-edge mode, if available.

   ii. Describe how Kruskal's and Prim's algorithms are similar. Describe how they are different.

   iii. Do you prefer one algorithm over the other to find a minimum spanning tree? Why?

e. Do you think Prim's algorithm is a good procedure for finding a minimum spanning tree? Write a brief justification of your answer.

21 To find a minimum spanning tree in a graph, you look for a network of existing edges that joins all the vertices and has minimum length. In some situations, you may want to create a minimum spanning network by adding new vertices and edges to the original graph. Such a network is called a **Steiner tree** (named after Jacob Steiner, a nineteenth century mathematician at the University of Berlin). You can find Steiner trees using geometry.

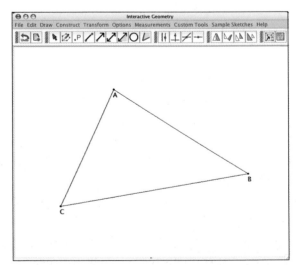

a. Using interactive geometry software, draw a triangle, *ABC*, in which all the angles are less than 120°. You can think of this triangle as a vertex-edge graph.

Your goal in Parts b and c is to find a shortest network that joins all three of the triangle's vertices.

b. A minimum spanning tree for this triangular graph is just the network consisting of the two shortest sides. Find the sum of the lengths of the two shortest sides. Now you have the length of a minimum spanning tree.

c. In Part b, you found a minimum spanning tree and its total length. Now consider a network that spans all three vertices, A, B, and C, and you are allowed to insert a new vertex and new edges to create this spanning network. You will investigate whether this new spanning network is shorter than the minimum spanning tree you found in Part b. Begin by using the software to perform the following construction.

  i. Insert a new point (vertex) inside the triangle; label it D.

  ii. Construct segments from the inside point D to each of the vertices of the triangle. This network, consisting of segments $\overline{DA}$, $\overline{DB}$, and $\overline{DC}$, is a network that spans all three of the triangle's vertices (but it uses a vertex and edges that are not part of the original triangle).

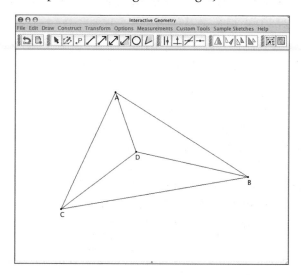

  iii. Measure the length of this network by measuring each segment and adding the three lengths.

  iv. Use the software to grab the inside point and drag it around. Note that the network length changes as the point is moved. Drag the point around until the network length is as small as possible.

Is this length smaller than the minimum spanning tree length from Part b?

d. Find the measure of the three central angles that surround the inside point. Make a conjecture about the measures of these angles when the inside point is moved to a position giving the shortest connected network. Test your conjecture on some other triangles.

22 You have learned in this lesson that there is no known theorem that provides complete and efficiently testable conditions for whether or not a graph has a Hamilton circuit. However, for some graphs, you can easily decide if they have a Hamilton circuit.

a. Consider *complete graphs*. A **complete graph** is a graph that has exactly one edge between every pair of vertices. Complete graphs with three and five vertices are shown below.

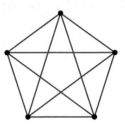

Draw the complete graphs with four and six vertices. Which complete graphs have a Hamilton circuit? Explain.

b. Consider *complete bipartite graphs*. A **complete bipartite graph**, denoted $K_{n,m}$, is a graph consisting of two sets of vertices, one with n vertices and the other with m vertices. There is exactly one edge from each vertex in the one set to each vertex in the other set. There are no edges between vertices within a set.

- Draw these complete bipartite graphs: $K_{2,2}$, $K_{2,3}$, $K_{3,3}$, $K_{3,4}$.
- Which complete bipartite graphs have a Hamilton circuit? Justify your answer.

23 In this lesson, you have learned that there is no known efficient algorithm for solving the TSP. However, there are efficient algorithms for finding a minimum spanning tree. Since both minimum spanning trees and solutions to the TSP are optimum spanning networks (an optimum spanning tree and optimum spanning circuit, respectively), perhaps a minimum spanning tree could be used to find an approximate solution to the TSP.

a. To test this idea, start with a weighted graph where the weights are regular Euclidean distance. Use one of the efficient algorithms you know (like Kruskal's algorithm) to find a minimum spanning tree.

b. Suppose you have a circuit that is a solution to the TSP for the graph in Part a. Delete one edge from the circuit.

   i. Explain why the resulting network is a spanning tree for the graph.

   ii. Explain why the total weight of this spanning tree is greater than or equal to the total weight of the minimum spanning tree you have found.

   iii. Explain why the total weight of this spanning tree is less than the total weight of the circuit which is a solution to the TSP.

   Thus, the total weight of the minimum spanning tree is less than the total weight of a circuit that is a solution to the TSP.

c. Now consider the minimum spanning tree, and convert it into a Hamilton circuit, as follows:

- Duplicate each edge of the minimum spanning tree.

- Choose a starting point and begin traveling on the duplicated edges of what used to be the minimum spanning tree. Because of the duplicated edges you can backtrack and make a circuit out of what used to be the minimum spanning tree. Some vertices will get visited more than once in this circuit. Remember, you really want a Hamilton circuit; so eventually, you only want to use each vertex once. As you move along the circuit of duplicated edges, when you reach a vertex on the circuit you have visited before, look beyond it to the next unvisited vertex on the circuit. Take a shortcut to that unvisited vertex by simply "cutting the corner" and avoiding the vertex you have already visited. Explain why the "cutting the corner" distance is shorter than the distance that goes through the skipped vertex.

- Continue in this way, cutting the corners to skip vertices you have already visited, until you end up back at the starting vertex.

d. Using the Hamilton circuit you created in Part c, justify the following statements.

- The weight of this Hamilton circuit is less than the weight of the duplicated-edges circuit.

- The weight of the duplicated-edges circuit is twice the weight of the minimum spanning tree.

- Thus, the weight of the Hamilton circuit you have created is less than twice the weight of the minimum spanning tree.

e. By putting the concluding statements together from Parts b and d, explain why you have found a Hamilton circuit with weight that is within twice the optimum weight of a solution to the TSP. Thus, you have a reasonable approximate solution to the TSP.

24 You have seen in this lesson that there is no known efficient method for solving the TSP. The same is true for finding Hamilton circuits and paths. Most experts believe that efficient solutions for these problems will never be found, at least not by using traditional electronic computers. In 1994, computer scientist Leonard M. Adleman of the University of Southern California in Los Angeles opened up the possibility of using nature as the computer to solve these problems. Dr. Adleman successfully carried out a laboratory experiment in which he used DNA to do the computations needed to solve a Hamilton path problem. Dr. Adleman stated, "This is the first example, I think, of an actual computation carried out at the molecular level." This method has not been shown to solve all Hamiltonian problems, and the particular problem solved was quite small, but it opens up some amazing possibilities for mathematics and computer science.

a. The graph that Dr. Adleman used in his experiment is shown below. All of the information in the graph was encoded using strands of DNA, and then the computations needed to find a Hamilton path were carried out by biochemical processes. Of course, this graph is small enough that the Hamilton path can also be found without gene splicing or conventional computers. Find the Hamilton path for this graph.

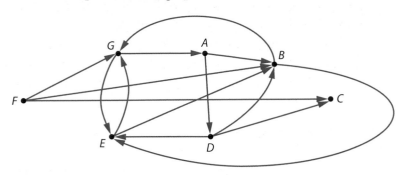

b. Find out more about this groundbreaking experiment in "molecular computation" by reading some of the articles below and conducting an Internet search. Write a short report incorporating your findings.

- Adleman, Leonard M. Molecular Computation of Solutions to Combinatorial Problems. *Science*, November 11, 1994.

- Delvin, Keith. Test Tube Computing with DNA. *Math Horizons*, April 1995.

- Kolata, Gina. Scientist At Work: Leonard Adleman: Hitting the High Spots of Computer Theory. *The New York Times*, Late Edition, December 15, 1994.

- Do an Internet search to find recent information.

25 Since solving the TSP has applications for so many different kinds of networks, like telephone and transportation networks, mathematicians are always looking for better algorithms that will solve the problem for larger graphs. In 1954, a 49-city instance of the TSP was solved. The record in 2004 was 24,978 cities—all cities in Sweden. Find the latest news about the TSP and write a brief report on the history of the TSP, the current record, and recent developments. A good source as of 2006 was the TSP site at the Georgia Institute of Technology: www.tsp.gatech.edu.

Review

TSP solved for 24,978 cities in Sweden

26 Rewrite each expression in the shortest equivalent form without parentheses.

a. $3(15 - 8x) - 2x(x + 9)$ **b.** $\frac{2}{5}(3x) + \frac{1}{5}(4 + 7x) - 1$

c. $-6 + 5(-3 + 4x) - 8x$ **d.** $(x + 4)(2x + 7)$

27 Estimate the answer to each question. Then calculate the exact answers and compare to your estimates.

a. 15 is what percent of 75? **b.** What is 35% of 168?

c. 45 is 40% of what number?

28 Trapezoid *ABCD* has vertex matrix $\begin{bmatrix} 1 & 1 & 3 & 8 \\ 1 & 5 & 5 & 1 \end{bmatrix}$.

    a. Find the area of trapezoid *ABCD*.

    b. Find the perimeter of trapezoid *ABCD*.

    c. Trapezoid *A′B′C′D′* is the image of trapezoid *ABCD* after applying the transformation $(x, y) \rightarrow (x + 3, y - 2)$. Find the area and perimeter of trapezoid *A′B′C′D′* by using properties of the transformation. Explain your reasoning.

    d. Trapezoid *WXYZ* is the image of trapezoid *ABCD* after applying the transformation $(x, y) \rightarrow (5x, 5y)$. Find the area and perimeter of trapezoid *WXYZ* by using properties of the transformation. Explain your reasoning.

29 The registration cost per person on a softball team varies inversely with the number of people on the roster for the team.

    a. If only 10 people are on the roster, each person must pay $37. How much would each person have to pay if there were 15 people on the roster?

    b. Write a rule that expresses cost per person as a function of number of people. Then identify the constant of proportionality and explain what it means in this context.

    c. Describe how the cost per person changes as the number of people on the roster increases.

30 The histograms below show the ages of patients who were seen by two different doctors during one week in October. The scales on the Number of Patients axes are the same for both graphs.

Dr. Cabala

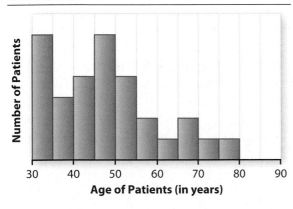

Dr. Dimas

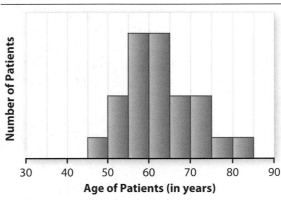

    a. For which office was the mean age of the patients seen greater? Explain your reasoning.

    b. For which office was there greater variation in the ages of the patients seen? Explain your reasoning.

31 Solve each equation algebraically—without use of the quadratic formula, if possible.

    a. $x^2 - 15x + 26 = 0$ **b.** $x^2 = -8x + 20$

    c. $48 = -x(x + 16)$ **d.** $2x^2 + 5x = 1$

Scheduling Projects Using Critical Paths

Careful planning is important to ensure the success of any project. This is particularly true in the case of planning large projects such as the construction of a new high school, shopping mall, or apartment complex. Even smaller projects such as house remodeling, organizing a school dance, or hosting a party can profit from careful planning. Network optimization using critical paths can help you plan well.

In this lesson, you will learn how to use *critical paths* in vertex-edge graphs to schedule projects that involve many tasks. This technique is called the *Program Evaluation and Review Technique* (*PERT*) or the *Critical Path Method* (*CPM*). Critical path analysis is one of the most frequently used mathematical management techniques.

Investigation 1 — Building a Model

As you have seen before, a first step in modeling a situation is often to make a diagram. In the case of scheduling large projects like a school dance project, it is important to consider this question:

> *How can you create a vertex-edge graph model that will help in the analysis and scheduling of the project?*

Listed here are some of the tasks that you may have found necessary in planning a spring dance. These are the tasks that will be used for the rest of this investigation. The order in which these tasks would need to be completed may vary from school to school.

Tasks

Book a Band or DJ (*B*)
Design the Poster (*D*)
Choose and Reserve the Location (*L*)
Display the Posters (*P*)
Choose a Theme (*T*)
Arrange for Decorations (*DC*)

At Marshall High School, the prerequisites for the various tasks are as follows:

- The tasks that need to be done just before booking the band are choosing the location and choosing a theme.
- The tasks that need to be done just before designing the poster are booking the band and arranging for decorations.
- There are no tasks that need to be done just before choosing and reserving the location.
- The task that needs to be done just before displaying the posters is designing the posters.
- There are no tasks that need to be done just before choosing a theme.
- The task that needs to be done just before arranging for decorations is choosing a theme.

Tasks to be done *just* before a particular task are called **immediate prerequisites**.

 Using the prerequisite information for Marshall High School, complete a table like the one below showing which tasks are immediate prerequisites for others. Such a table is called an **immediate prerequisite table**.

Dance Plans

Task	Immediate Prerequisites
Book a Band or DJ (B)	L, T
Design the Poster (D)	
Choose and Reserve the Location (L)	
Display the Posters (P)	
Choose a Theme (T)	
Arrange for Decorations (DC)	

 Using the procedure below, complete a diagram showing how all of the tasks are related to each other.

Step 1: Place the circle labeled S at the far left of your paper. Put a new circle at the far right of your paper, where the diagram will end. Label it F for "Finish." The circles labeled S and F do not represent actual tasks. They simply indicate the start and finish of the project.

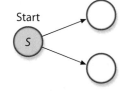

Step 2: Draw two empty circles to the right of S representing tasks that do not have any immediate prerequisites. Which tasks in the immediate prerequisite table from Problem 1 should be represented by the two empty circles? Label the two circles with those tasks.

Step 3: Moving to the right again, draw a circle for each remaining task. Draw an arrow between two circles if one task (at the tail of the arrow) is an immediate prerequisite for the other task (at the tip of the arrow).

Step 4: Finish the diagram by drawing connecting arrows from the final task or tasks to the circle marked F.

Step 5: If necessary, redraw the diagram so that it looks orderly.

3 Compare your diagram with the diagrams of others.

   a. Do the diagrams look different? Explain the differences.

   b. Does everyone's diagram accurately represent the information in the immediate prerequisite table?

   c. Decide on one organized, orderly diagram that best represents the situation. Make a copy of that diagram.

4 A diagram like the one you drew in Problem 3 is a vertex-edge graph where the edges are *directed*, that is, the edges are arrows. Graphs like this are called **directed graphs**, or **digraphs**. Use your digraph and the immediate prerequisite table to help answer the following questions.

   a. Which of the following pairs of tasks can be worked on at the same time by different teams? Explain your reasoning.

      i. Tasks L and B

     ii. Tasks L and T

    iii. Tasks L and DC

    iv. Tasks L and D

   b. Find one other pair of tasks that can be worked on at the same time.

     Explain, in terms of the school dance project and the individual tasks, why it is reasonable for these tasks to be worked on at the same time.

Summarize
the Mathematics

In this investigation, you learned how to use a digraph to show how tasks involved in the dance project are related to each other.

a What do the vertices of the project digraph represent?

b How are prerequisite tasks represented in the project digraph?

c How are tasks that can be worked on at the same time represented in the project digraph?

Be prepared to share your ideas with the class.

✔Check Your Understanding

"Turning around" a commercial airplane at an airport is a complex project that happens many times every day.

Suppose that the tasks involved are unloading arriving passengers, cleaning the cabin, unloading arriving luggage, boarding departing passengers, and loading departing luggage. The relationships among these tasks are as follows:

- Unloading the arriving passengers must be done just before cleaning the cabin.

- Cleaning the cabin must be done just before boarding the departing passengers.

- Unloading the arriving luggage must be done just before loading the departing luggage.

- All activities in the cabin of the airplane (unloading and boarding passengers and cleaning the cabin) can be done at the same time as loading and unloading luggage.

Construct an immediate prerequisite table and project digraph for this situation.

Investigation 2 Critical Paths and the Earliest Finish Time

You have seen that a large project, like a school dance or "turning around" a commercial airplane, consists of many individual tasks that are related to each other. Some tasks must be done before others can be started. Other tasks can be worked on at the same time. A digraph is a good way to show how all the tasks are related to each other.

The real concern in a large project is to get all the tasks done most efficiently. In particular, it is important to know the least amount of time required to complete the entire project. This minimum completion time is called the **earliest finish time** (**EFT**). As you work through the problems in this investigation, look for answers to this question:

How can you use the project digraph to find the earliest finish time for a project?

1 There are many reasonable estimates that you and your classmates might make for how long it will take to complete each task of the school dance project. Experience at one school suggested the task times and prerequisites displayed in the following table. These task times will be used for the rest of this lesson.

Planning a Dance

Task	Task Time	Immediate Prerequisites
Choose and Reserve a Location (L)	2 days	none
Choose a Theme (T)	3 days	none
Book a Band or DJ (B)	7 days	L, T
Arrange for Decorations (DC)	5 days	T
Design the Poster (D)	5 days	B, DC
Display the Poster (P)	2 days	D

Put these task times into the project digraph you constructed in the last investigation by entering the task times into the circles (vertices) of the digraph.

2 Assume that you have plenty of qualified people to work on the project. Use the immediate prerequisite table and the project digraph to help you figure out how to complete the project most efficiently.

   a. What is the least amount of time required to complete the whole project (that is, what is the EFT for the project)? Explain.

   b. Compare answers and explanations with some of your classmates. Resolve any differences.

   c. Is the earliest finish time for the whole project equal to the sum of all the individual task times? Explain why or why not.

3 Think about the EFT in terms of paths through the project digraph.

   a. Which path through the graph corresponds to the earliest finish time for all the tasks? Write down your answer and an explanation. Compare answers and explanations with some of your classmates. Resolve any differences.

   b. How many paths are there through the project digraph, from S to F? List in order the vertices of all the different paths. For each path, compute the total time of all tasks on the path. Explain how this is a brute-force method for finding the EFT.

4 A path through the project digraph that corresponds to the earliest finish time is called a **critical path**.

   a. Mark the edges of the critical path so that it is easily visible.

   b. Compare your critical path from Part a to the critical paths found by others. If the paths are different, discuss the differences and decide on the correct critical path.

   c. Describe the connections among the critical path, the EFT, and the longest path.

5 If all the posters are to be displayed 30 days before the dance, how many days before the dance should work on the project begin? Justify your answer based on your work on the previous problems.

6 Now consider what happens to the EFT and critical path if certain tasks take longer to complete than expected.

a. The project digraph below represents the school dance project you have been working on. Compare this digraph to the one you have constructed and used in the previous problems. If the graphs look different in some way, explain why this digraph and your digraph both accurately represent the project.

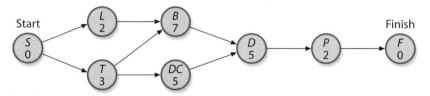

b. What happens to the earliest finish time if it takes 6 days, instead of 5 days, to design the poster (task *D*)?

c. What happens to the earliest finish time if it takes 9 days, instead of 7 days, to book the band (task *B*)?

d. A task on a critical path is called a **critical task**. What happens to the earliest finish time if one of the critical tasks takes longer than expected to complete?

e. What happens to the earliest finish time if it takes 6 days, instead of 5 days, to arrange for the decorations (task *DC*)?

f. Suppose it takes 3 days, instead of 2 days, to choose and reserve a location (task *L*).

   i. What happens to the EFT?

   ii. What is the critical path?

g. Suppose it takes 6 days, instead of 2 days, to choose and reserve a location (task *L*).

   i. What happens to the EFT?

   ii. What is the critical path?

h. What can happen to the earliest finish time and the critical path if one of the tasks that is *not on the critical path* takes longer than expected to complete?

Critical Paths, PERT, CPM, and Technology Finding a critical path and the EFT for a project are essential parts of the Program Evaluation and Review Technique (PERT) and the Critical Path Method (CPM). These methods are among the most widely-used mathematical tools in business, industry, and government. Of course, most projects are much larger than the school dance project you have been working on. Computer software is used to help solve large critical path problems. To explore how that might be done, use vertex-edge graph software or project management software to help solve the following problem.

Suppose a car manufacturer is considering a new fuel-efficient car. Since this new car will cost a lot of money to produce, the company wants to be sure that it is a good idea before they begin production. So, they do a feasibility study. (In a feasibility study, you look at things like estimated cost and consumer demand to see if the idea is practical.) The feasibility study itself is a big project that involves many different tasks. The table below shows tasks, times, and prerequisites for the feasibility study project.

Feasibility Study Project

Task	Task Time	Immediate Prerequisites
Design the Car (A)	7 weeks	none
Plan Market Research (B)	2 weeks	none
Build Prototype Car (C)	10 weeks	A
Prepare Advertising (D)	4 weeks	A
Prepare Initial Cost Estimates (E)	2 weeks	C
Test the Car (F)	5 weeks	C
Finish the Market Research (G)	3.5 weeks	B, D
Prepare Final Cost Estimates (H)	2.5 weeks	E
Prepare Final Report (I)	1 week	F, G, H

a. Generate the feasibility study project digraph. What is a critical path and the EFT for this project?

b. Use the software to experiment with different task times and see the effect on the EFT and critical path. For example:

   i. What happens to the EFT if you double all the task times? Explain why this is so.

   ii. Modify some task times so that the critical path does not change but the EFT increases. State a general rule for when this will happen.

   iii. Modify some task times so that neither the critical path nor the EFT changes. State conditions for when this will happen.

   iv. Modify some task times so that the critical path changes, and there is only one critical path.

   v. Modify some task times so there is more than one critical path.

   vi. *Optional:* Experiment with adding new vertices, times, and edges to the graph. Describe the effects on the EFT and critical path(s).

In this lesson, you have learned mathematical concepts and methods to help efficiently schedule projects. In particular, you have learned about critical paths and how to find the EFT for a project.

a How can you find the EFT by examining a digraph for a project?

b What is a critical path for a project, and why is it "critical"?

Be prepared to share your ideas with the entire class.

✓Check Your Understanding

Examine the digraph below.

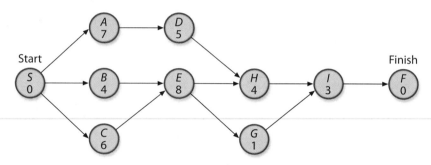

a. How many paths are there through this digraph, from *S* to *F*? List all the different paths, and compute the length of each path.

b. Find the critical path and the EFT.

c. Are there any tasks that can have their task times increased by 3 units and yet not cause a change in the EFT for the whole project? If so, which tasks? If not, why not?

Applications

1 Shown below is the immediate prerequisite table for preparing a baseball field.

Preparing a Baseball Field

Task	Task Time	Immediate Prerequisites
Pick up Litter (*L*)	4 hours	none
Clean Dugouts (*D*)	2 hours	*L*
Drag the Infield (*I*)	2 hours	*L*
Mow the Grass (*G*)	3 hours	*L*
Paint the Foul Lines (*P*)	2 hours	*I, G*
Install the Bases (*B*)	1 hour	*P*

a. Find at least two tasks that can be worked on at the same time.

b. Draw a digraph for this project.

c. Mark the critical path.

d. What is the EFT for the whole project? (Assume that you have enough people to work on the project and the task times shown are the times needed, given the people who will be working on the tasks.)

e. Do you think that the task times given in the table are reasonable? Change at least one task time. Use your new time(s) to find a critical path and the EFT.

2 Suppose that your school is planning to organize an Earth Day. You will have booths, speakers, and activities related to planet Earth and its environment. Such a project will require careful planning and coordination among many different teams that will be working on it.

Here are six tasks that will need to be done as part of the Earth Day project and estimates for the time to complete each task.

Planning Earth Day

Task	Task Time
Decide on Topics	6 days
Get Speakers	5 days
Choose Date and Location	3 days
Design Booths	2 weeks
Build Booths	1 week
Make Posters	6 days

a. Decide on immediate prerequisites for each of the tasks and construct an immediate prerequisite table.

b. Draw a project digraph.

c. Find a critical path and the EFT.

3 On a large construction project, there is usually a general contractor (the company that coordinates and supervises the whole project) and smaller contractors (the companies that carry out specific parts of the project, like plumbing or framing).

Suppose that the company responsible for putting in the foundation for a building estimates the times shown in the following prerequisite table. The general contractor wants the foundation done in 13 days. Can the foundation crew meet this schedule? If so, explain. If not, propose a plan for what they should do in order to shorten task times and finish on schedule.

Putting in a Foundation

Task	Task Time	Immediate Prerequisites
Measure the Foundation (A)	1 day	none
Dig Foundation (B)	4 days	A
Erect Forms (C)	6 days	B
Obtain Reinforcing Steel (D)	2 days	A
Assemble Steel (E)	3 days	D
Place Steel in Forms (F)	2 days	C, E
Order Concrete (G)	1 day	A
Pour Concrete (H)	3 days	F, G

4 Shown below is the immediate prerequisite table for building a house. Assume that three specialists are working on each task.

Build a House

Task	Task Time	Immediate Prerequisites
Clear Land (C)	2 days	none
Build Foundation (BF)	3 days	C
Build Upper Structure (U)	15 days	BF
Electrical Work (EL)	9 days	U
Plumbing Work (P)	5 days	U
Complete Exterior Work (EX)	12 days	U
Complete Interior Work (IN)	10 days	EL, P
Landscaping (H)	6 days	EX

a. Find at least two tasks that can be worked on at the same time.

b. Draw the digraph for this project.

c. Mark the critical path.

d. What is the EFT for the project?

e. Suppose that each specialist works 8 hours per day and is paid an average of $20 per hour. What will the total labor costs be?

f. Suppose that some plumbing supplies will be late in arriving, so installing the plumbing will take 10 days. How does this affect the EFT and critical path?

5 A prerequisite table for putting on a school play is shown below.

Planning a School Play

Task	Task Time (in days)	Immediate Prerequisites
Choose a Play (A)	7	none
Tryouts (B)	5	A
Select Cast (C)	3	B
Rehearsals (D)	25	C
Build Sets and Props (E)	20	A
Create Advertising (F)	4	C
Sell Tickets (G)	10	F
Make/Get Costumes (H)	20	C
Lighting (I)	7	E, H
Sound and Music (J)	9	E
Dress Rehearsals (K)	5	D, I, J
Opening Night Prep (L)	1	K

a. You must report to the principal how long it will take before the play is ready to open. What will you report?

b. Suppose that because of a conflict with another special event, you find out that you must complete the project in 6 fewer days than the EFT. In order to meet this new timetable, you decide to shorten the time for some of the tasks.

   i. Describe which tasks you will shorten and how you will shorten them.

   ii. Show how this will result in a new EFT that is 6 days less than the original EFT. (Shortening task times like this is sometimes called **crashing** the task times.)

c. Another way to attempt to shorten the EFT is to figure out a way to change some of the prerequisites. Suppose that you decide to change the prerequisites for setting up the lighting by doing that task whether or not the set and props are built. Thus, task *E* is no longer an immediate prerequisite for task *I*. How much time will this save for the EFT?

d. Describe at least one other reasonable rearrangement of prerequisites to shorten this EFT. By how many days does your rearrangement change the EFT?

6 Suppose that you and two friends are preparing a big dinner to serve 20 people.

a. List 4–8 tasks that must be done as part of this project.

b. Decide how long each task will reasonably take to complete.

c. Decide on the immediate prerequisites for each task and construct the immediate prerequisite table.

d. Draw the project digraph.

e. Find a critical path and the EFT.

7 Reproduced below is the digraph from the Check Your Understanding task on page 442. The critical path is shown by the red arrows. Verify that the EFT is 21.

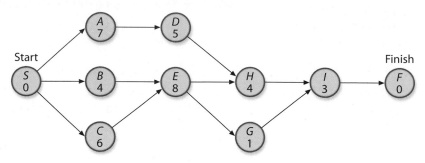

a. How do the critical path and EFT change if:

   i. the time for task *C* decreases by 5?

   ii. the time for task *C* increases by 5?

   iii. the time for task *D* increases by 2?

   iv. the time for task *D* increases by 3?

   v. the time for task *D* decreases by 4?

b. Write a summary describing how changes in times for tasks on and off the critical path affect the EFT and the critical path.

c. Construct the immediate prerequisite table for the project digraph.

8 It is possible for a project digraph to have more than one critical path. Consider a modified version of the school dance project digraph shown below.

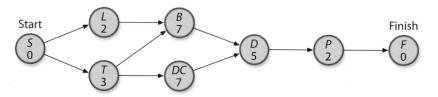

a. Find the EFT for the project.

b. How many critical paths are there?

c. List all the critical tasks.

9 Shown below is a modified version of the project digraph from the Check Your Understanding task on page 442.

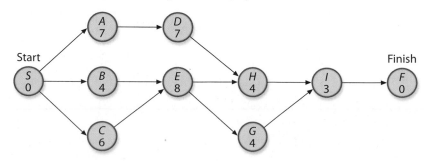

a. Find the EFT for the project.

b. How many critical paths are there?

c. List all the critical tasks.

Connections

10 The adjacency matrix for a graph can be used to enter the graph into a computer. You also can get information about the graph and the project just by looking at the adjacency matrix. Consider the digraph below.

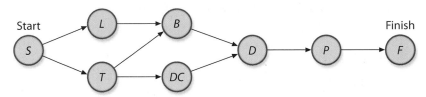

a. Construct the adjacency matrix for the digraph with vertices *L, T, B, DC, D,* and *P*.

b. Add up all the numbers in row *B*. What does this sum mean in terms of prerequisites?

c. Add up all the numbers in column *B*. What does this sum mean in terms of prerequisites?

d. What does a row of all zeroes mean in terms of prerequisites?

e. What does a column of all zeroes mean in terms of prerequisites?

11 The **indegree** of a vertex in a digraph is the number of arrows coming into it. The **outdegree** is the number of arrows going out of it. Consider the digraph above in Connections Task 10.

a. What are the indegree and outdegree of *B*?

b. What is the outdegree of *T*?

c. Ignoring *S* and *F*, what is the indegree of *T*? The outdegree of *P*?

d. What do indegree and outdegree mean in terms of prerequisites?

e. How can you compute indegree and outdegree by looking at the rows and columns of the adjacency matrix for a digraph?

12 Explain why the digraphs below could *not* be used to model a simple project.

a.

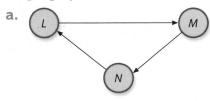

b.

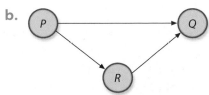

c. Are there any four-vertex configurations that can never occur in a project digraph? Explain and illustrate your answer.

13 Interview some adults in business or government who use PERT or the Critical Path Method to manage projects. Based on what you find out, describe one or two recent real-world examples of how critical paths are used to schedule projects.

Reflections

14 The title of this unit is *Network Optimization*. Optimize means find the best. The "best" may be different in different situations.

  a. Give an example in which the shortest path is not the best path.

  b. Explain why the best solution method for a small graph may not be the best solution method for a large graph. Give an example. (You may choose an example from this lesson or from Lesson 1.)

15 Initially, did you find something particularly difficult or confusing in finding critical paths and EFTs for projects? If so, how would you explain that idea to a friend to help her or him avoid the confusion?

16 The weighted graphs in Lesson 1 are sometimes called *edge-weighted graphs*. Explain why the project digraphs you worked with in Investigation 2 of Lesson 2 could be called *vertex-weighted graphs*.

17 What are similarities and differences among critical paths, Hamilton paths (Connections Task 12, page 425), and Euler paths?

Extensions

18 A leading contemporary researcher in discrete mathematics is Fan Chung Graham, a mathematician who earned her doctorate in 1974 and has worked at Bell Labs, Harvard University, the University of Pennsylvania, and the University of California at San Diego. Although some of the problems that Fan Chung works on look like games, they often have important applications in areas like communication networks and design of computer hardware and software. This task is an example of such a problem.

Fan Chung Graham

   In the following graph, suppose a person is standing at each vertex. The first letter of each vertex label is the name of the vertex; the second letter is the destination that each person must reach by walking along edges of the graph. The goal is for all of the people to walk to their destinations without overusing any edge. By trying to solve a problem like this, you can find out how accessible a network is—that is, if it has any "bottlenecks" where there is excessive traffic.

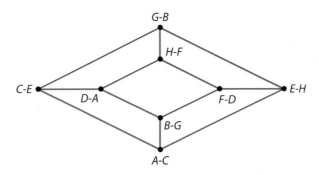

a. Assume that an edge is "overused" if it is used more than twice. Find routes for all walkers so that everyone reaches his or her destination but no edge is overused.

b. Are other routes possible? Describe the strategy you used to find the routes.

c. State and try to solve at least one other problem related to this situation.

19 In an immediate prerequisite table, the task times given are the times needed to complete each task using the expected available resources. You might need to figure out what resources are needed (like how many workers are needed) so that you can maintain the task times. Or, if the available resources change, then the task times could change. You will explore such situations in this problem.

Reproduced below is the immediate prerequisite table for the project of preparing a baseball field for play from Applications Task 1. In that problem, you assumed that there are plenty of people to work on the project and that the task times shown are the times needed given the people who will be working on each task. In this problem, suppose the task times shown are how long it takes to complete each task when one person works on the task.

Preparing a Baseball Field

Task	Task Time	Immediate Prerequisites
Pick up Litter (L)	4 hours	none
Clean Dugouts (D)	2 hours	L
Drag the Infield (I)	2 hours	L
Mow the Grass (G)	3 hours	L
Paint the Foul Lines (P)	2 hours	I, G
Install the Bases (B)	1 hour	P

a. Suppose only one person will work on any given task, and you have plenty of people available to work on the whole project. In this situation, what is the EFT for the project?

b. Suppose again that only one person will work on any given task. Each person can work on as many tasks as needed. In order to finish by the EFT in Part a, what is the fewest number of people needed to work on the project?

c. Suppose you have to complete the whole project by yourself. What is the EFT in this situation?

d. Suppose you and one friend are hired to prepare the field. You and your friend decide that you might work together or separately to complete any task. If you work together for an entire task, you can reduce the task time by half from that shown in the immediate prerequisite table. The only task that must be done by someone working alone is mowing the grass, since there is only one mower. How would you assign duties in order to complete the project in the least amount of time?

20 In this lesson, you have found critical paths in many different digraphs. In this task, you will explore several methods that might be used to try to find a critical path.

a. Recall the nearest-neighbor algorithm that you tried for minimum spanning trees in Lesson 1, on page 404. When using that algorithm, you move from vertex to vertex always choosing the "nearest-neighbor" vertex, with the goal of finding an overall *shortest* network. Although this type of algorithm has some appeal, it did not work for finding a minimum spanning tree. In the case of a critical path, you want a *longest* path. Thus, modify the nearest-neighbor algorithm so that it is a "farthest-neighbor" algorithm. Apply this algorithm to the project digraph in the Check Your Understanding task on page 442. Does it work to find a critical path?

b. Think about a method for finding the EFT for the whole project by starting at the Start vertex and working forward, finding the EFT for each task, until you reach the Finish vertex. Then once you have the task EFTs, work backwards to find a critical path. Try out this method and provide a more detailed description.

c. If you tried a method other than those in Parts a and b to find a critical path, write a step-by-step description of your method for how to find a critical path. Test your algorithm using several digraphs. Modify as needed based on your test results. You might use vertex-edge graph software or project management software to help test your algorithm.

21 Besides finding the earliest finish time (EFT) for the whole project, it is often useful to find the EFT for each task. To see how to do this, consider the simple project digraph below.

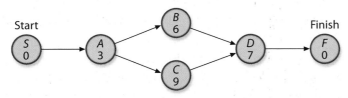

a. The EFT for a given task is the least amount of time required to finish that task. Keep in mind that in order to finish the task, all of its prerequisites must be finished as well.

  i. What is the least amount of time needed to finish task *A*?

  ii. What is the least amount of time needed to finish task *C*?

  iii. Find the EFT for the rest of the tasks in this project.

b. Make a copy of the project digraph. Write the EFT for each task just above the vertex representing the task.

c. How does the EFT for the last task compare to the EFT for the whole project?

d. Describe the method you used to figure out the task EFTs.

Review

22 Solve each equation by reasoning with the symbols themselves.

a. $x^2 - 7x = 44$

b. $x(x + 8) = -6$

c. $\dfrac{12}{x} = x + 1$

d. $(x + 3)(x + 4) = 20$

23 Triangle *ABC* has vertices $A(-2, 3)$, $B(1, 7)$, and $C(2, 0)$.

a. What type of triangle is $\triangle ABC$? Be as specific as you can.

b. Draw $\triangle ABC$ and its reflection image across the *x*-axis. Label the coordinates of the image. What type of triangle is the image? Why?

c. Draw $\triangle ABC$ and its image after a 90° clockwise rotation about the origin. Label the coordinates of the image. What type of triangle is the image? Why?

24 Rewrite each radical expression in simplest form.

a. $\sqrt{24}$

b. $\sqrt{350}$

c. $\sqrt{\dfrac{80}{36}}$

d. $\dfrac{\sqrt{45}}{6}$

e. $\sqrt{6}\sqrt{72}$

f. $\sqrt{36 + 100}$

25 Find the values for x and y in the triangles below.

a.
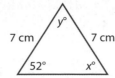

b.
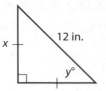

c.
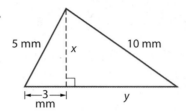

d.
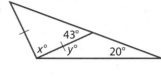

26 Would you expect each pair of variables below to have a positive, negative, or no correlation? Explain your reasoning.

a. The number of minutes that Grant has been running and the distance he has left to run if he plans on running three miles

b. The age of a person and the last digit of his or her zip code

c. The length of the grass on a baseball field during the summer and the number of days since it was last mowed

d. The number of phones in a house and the number of people living in the house

Looking Back

In this unit, you used vertex-edge graphs to optimize a variety of networks. For example, you used minimum spanning trees to optimize computer networks and road networks; you used Hamilton circuits and solutions to the TSP to optimize travel networks; and you used critical paths to find the optimum (earliest) finish time for large projects. Tasks in this final lesson will help you pull together and review what you have learned.

1 Consider the map below of a region in Kentucky.

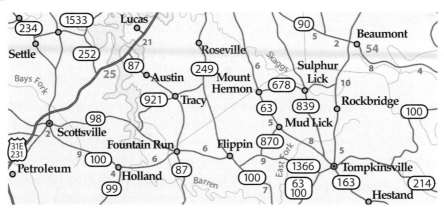

Map © 1997 by Rand McNally, R.L. 97-S-79

a. State a problem related to this map that could be solved by finding a minimum spanning tree. Then describe how you would go about solving the problem.

b. State a problem related to this map that could be solved by finding a solution to an instance of the TSP. Describe how you would go about solving the problem.

2 Suppose that you are the editor for a school newspaper. Study the following background information about the publishing process.

It takes 10 days for the reporters to research all the news stories. It takes 12 days for other students, working at the same time as the reporters, to arrange for the advertising. The photographers need 8 days to get all the photos. However, they cannot start taking photos until the research and the advertising arrangements are complete. The reporters need 15 days to write the stories after they have done the research. They can write while the photographers are getting photos. It takes 5 days to edit everything after the stories and the photos are done. Then it takes another 4 days to lay out the newspaper and 2 more days for printing.

Write a report to your teacher-advisor explaining how long it will take to turn out the next edition of the paper. State which steps of the publishing process will need to be monitored most closely. Include diagrams and complete explanations in your report.

3 You have used vertex-edge graphs to model a variety of situations. You can also use vertex-edge graphs to represent and analyze relationships among the new concepts that you are learning. This is done using a type of graph called a *concept map*. In a concept map, the vertices represent ideas or concepts, and edges illustrate how the concepts are connected. The edges may or may not have labels. The first step in building a concept map for some area of study is to list all the concepts you can think of. Here is the beginning of such a list for this unit on network optimization:

- network
- spanning tree
- minimum spanning tree
- algorithms
- best-edge algorithm

a. Add to this list. Include all the concepts from this unit that you can recall.

The next step in building a concept map is to let the concepts in the list be the vertices of a graph, and then draw edges between vertices to show connections between concepts. The edges should be labeled to show *how* the vertex concepts are connected. There are many different concept maps that can be drawn. The beginning of one concept map for this unit is drawn below.

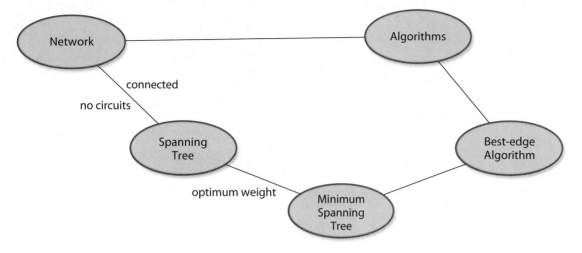

b. Interpret the sample concept map above.

   i. Why is there an edge between network and spanning tree?

   ii. There are two labels on the edge between network and spanning tree. Explain both of those labels. What other label could be put on this edge?

   iii. Why is there an edge between best-edge algorithm and minimum spanning tree? Put an appropriate label on this edge.

c. Complete the concept map by adding all the concepts (vertices) that you listed in Part a along with appropriate edges and labels that show connections among the concepts.

d. Compare your concept map with those of other classmates. Discuss similarities and differences.

Summarize the Mathematics

In this unit, you have studied important concepts and methods related to network optimization.

a) Optimization or "finding the best" is an important theme throughout this unit. Describe three problem situations from the unit in which you "found the best." In each case, explain how you used a vertex-edge graph to solve the problem.

b) The major topics you studied are minimum spanning trees, the TSP, and critical paths. For each topic, briefly describe what it is and what kinds of problems it can be applied to.

c) You examined a variety of algorithms and methods for solving network optimization problems, including best-edge algorithms and brute-force methods. For each of these two solution procedures:

   i. describe the basic strategy,

   ii. give examples of problems that can be solved using the procedure, and

   iii. list some advantages and disadvantages of the procedure.

Be prepared to share your examples and descriptions with the entire class.

✓ Check Your Understanding

Write, in outline form, a summary of the important mathematical concepts and methods developed in this unit. Organize your summary so that it can be used as a quick reference in future units and courses.

TRIGONOMETRIC METHODS

Trigonometry, or triangle measure, is an important tool used by surveyors, navigators, engineers, builders, astronomers, and other scientists. Triangulation and trigonometry provide methods to indirectly determine otherwise inaccessible distances and angle measures. The same tools are useful in the design and analysis of mechanisms involving triangles in which the lengths of two sides of a triangle are fixed while the length of the third side is allowed to vary.

Through your work on the investigations in this unit, you will develop the understanding and skill needed to solve problems using trigonometric methods. Key ideas will be developed in two lessons.

Lessons

1 Trigonometric Functions

Use angles in standard position in a coordinate plane to define the trigonometric functions sine, cosine, and tangent. Interpret and apply those functions in the case of situations modeled with right triangles.

2 Using Trigonometry in Any Triangle

Develop the Law of Sines and the Law of Cosines, and use those relationships to find measures of sides and angles in triangles. Solve equations involving several variables for one of the variables in terms of the others.

Trigonometric Functions

In Course 1 of *Core-Plus Mathematics,* you learned that triangles are rigid figures whose shapes and sizes are completely determined by conditions involving as few as three parts. In this unit, you will learn methods for determining the measures of the remaining parts of a triangle.

An automobile jack that comes with some makes of automobiles is shown below. It can be modeled by a pair of triangles with a common variable-length side. Points *A* and *C* are connected by a threaded rod, and the lengths *AB*, *BC*, *AD*, and *CD* are all 1 foot. The distance between points *A* and *C* increases or decreases depending on the direction the rod is turned.

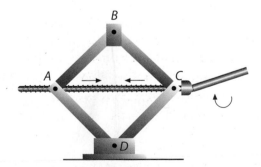

The jack mechanism is a good example of how the measures of the six parts of a triangle (three sides and three angles) are interconnected. Altering the size of one part changes the size of other parts; and if you know measures of some parts, you can determine the measures of others.

Think about the design and function of this automobile jack. Use the "Auto Jack" custom tool to test your ideas.

a About how long would the threaded rod need to be if the jack is to be stored with points *B* and *D* as close together as possible?

b As the distance *AC* decreases, how do the angle measures of △*ABC* change? How does the distance between point *B* and the threaded rod change?

c How does the height of the jack, *BD*, compare to the length of the altitude of △*ABC* drawn from point *B*? Explain your thinking.

d Suppose the jack is set so that *AC* is as long as possible. As the threaded rod is turned at a constant rate, the distance *AC* decreases at a constant rate. How would you describe the rate at which the height *BD* of the jack changes?

In this lesson, you will use angles in a coordinate plane to define trigonometric functions and use those functions to determine measures of unknown parts in right triangles. These functions also form the basis for finding unknown parts of any triangle, as you will learn in the second lesson. Throughout this unit, you will see the practical power of trigonometric methods for solving a wide range of applied problems.

Investigation 1 Connecting Angle Measures and Linear Measures

One of the most effective strategies for calculating distances that cannot be measured directly is to represent the situation with a triangle in which the length of one side is the desired distance and other sides and/or angles can be measured. Trigonometry provides methods for using the known parts of a triangle to calculate those that are unknown. The key connections are the trigonometric ratios sine, cosine, and tangent defined as functions of angles in standard position in a coordinate plane. As you work on the problems in this investigation, look for answers to the following questions:

How are the sine, cosine, and tangent functions defined?
How can their values be estimated?

1. Study the diagram below of a radio transmitter tower with two support wires attached to it at *A* and *E* and to the ground at *B* and *D*. First, focus on the triangle formed by the tower, the ground, and the shorter support wire.

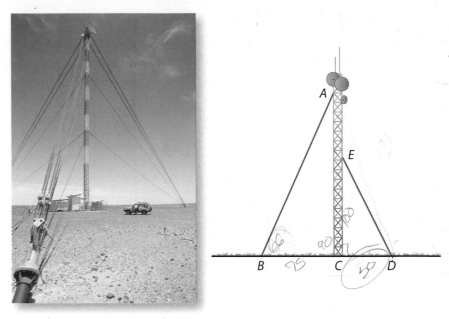

a. If *CD* = 50 feet and *CE* = 100 feet, is the size and shape of right △*DCE* completely determined? Why or why not?

b. How long is support wire $\overline{DE}$?

c. If you were asked to attach a support wire, 125 feet long, from point *D* to the tower, how far up the tower would you have to climb?

d. What relationship among sides of a right triangle have you used in Parts b and c?

2. Now focus on the right triangle formed by the tower, $\overline{BC}$, and the longer support wire $\overline{AB}$.

a. If *BC* = 75 feet and m∠*ABC* = 66°, is the size and shape of right △*ABC* completely determined? Why or why not?

b. Explain why the size and shape of a right triangle are completely determined when the measures of two sides are known or the measure of one acute angle and the length of one side are known.

c. It should be possible to use the information given in Part a to find *AC*, the distance above the ground at which support wire $\overline{AB}$ is attached to the tower. Will the Pythagorean Theorem help? Explain.

To solve problems like Problem 2 Part c, you need to find a connection between angle measures and segment lengths in a right triangle. In this case, it is helpful to think of an angle as being formed by rotating a ray about its endpoint from an *initial position* to a *terminal position*.

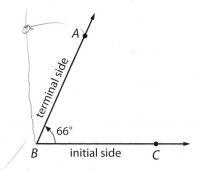

The point about which the ray is rotated is the **vertex** of the angle. The initial position of the ray is the **initial side** of the angle. The terminal position of the ray is the **terminal side** of the angle. To indicate the direction of rotation from initial side to terminal side, it is customary to say the angle has *positive measure* if the rotation is counterclockwise and has *negative measure* if the rotation is clockwise.

3 The diagram at the right shows four points on the terminal side of an angle in **standard position** in a coordinate plane. The vertex of the angle is at the origin and its initial side coincides with the positive *x*-axis.

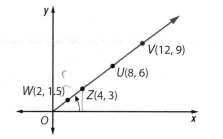

a. For each point (x, y) shown on the terminal side, find the ratios $\frac{y}{x}$.

   i. $U(8, 6)$ **ii.** $V(12, 9)$

   iii. $W(2, 1.5)$ **iv.** $Z(4, 3)$

b. How do the ratios $\frac{y}{x}$ compare in each case? Why does that make sense?

c. For each point (x, y) in Part a, suppose r is the distance from the origin to the point. How do you think the ratios $\frac{x}{r}$ would compare in each case? The ratios $\frac{y}{r}$? Check your conjectures.

4 Your discoveries about the ratios of lengths in Problem 3 apply to any angle in standard position. Consider the diagram below that shows an angle with degree measure θ (Greek letter "theta") in standard position.

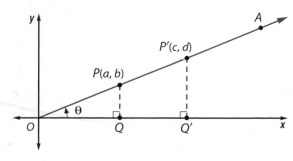

Let $P(a, b)$ and $P'(c, d)$ be any two points on the terminal side $\overrightarrow{OA}$, other than the origin O.

  a. Find the slope of $\overleftrightarrow{OA}$ using points O and P. Find the slope of $\overleftrightarrow{OA}$ using points O and P'.

  b. How are $\dfrac{PQ}{OQ}$ and $\dfrac{P'Q'}{OQ'}$ related? Why?

  c. Explain why $\triangle OP'Q'$ is the image of $\triangle OPQ$ under a size transformation with center at the origin and magnitude $k = \dfrac{c}{a}$. Use the following questions to guide your thinking.

   i. What is the image of point O under this transformation?

   ii. Why is point Q' the image of point Q under this transformation?

   iii. Why is point P' the image of point P under this transformation?

  d. Use your work in Part c to help explain each step in the reasoning below.

   i. $OQ' = k(OQ)$ and $OP' = k(OP)$.

   So, $\dfrac{OQ'}{OQ} = \dfrac{OP'}{OP}$.

   So, $OQ' \cdot OP = OQ \cdot OP'$.

   So, $\dfrac{OQ}{OP} = \dfrac{OQ'}{OP'}$.

   ii. $OP' = k(OP)$ and $P'Q' = k(PQ)$.

   So, $\dfrac{OP'}{OP} = \dfrac{P'Q'}{PQ}$.

   So, $OP' \cdot PQ = OP \cdot P'Q'$.

   So, $\dfrac{PQ}{OP} = \dfrac{P'Q'}{OP'}$.

  e. Write a statement describing how your work in Parts b and d support your discoveries in Problem 3.

  f. Draw diagrams similar to that above for cases where $\theta > 90°$. Does reasoning similar to that in Parts b–d hold in these cases? Explain.

Your work in Problem 4 shows that if $P(x, y)$ is a point (not the origin) on the terminal side of an angle in standard position and $r = \sqrt{x^2 + y^2}$, then the ratios $\dfrac{y}{x}$, $\dfrac{y}{r}$, and $\dfrac{x}{r}$ do not depend on the choice of P; they depend only on the measure of $\angle POQ$. That is, these ratios are functions of the measure θ of the angle. These functions are called **trigonometric functions** and are given special names as indicated below.

tangent of $\theta = \tan \theta = \dfrac{y}{x}$ $(x \neq 0)$

sine of $\theta = \sin \theta = \dfrac{y}{r}$

cosine of $\theta = \cos \theta = \dfrac{x}{r}$

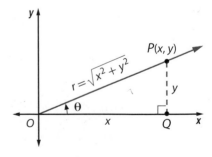

5 The terminal side of an angle in standard position with measure θ contains the given point. In each case, draw the angle on a coordinate grid. Then find $\cos \theta$, $\sin \theta$, and $\tan \theta$.

  a. $P(12, 5)$

  b. $P(-6, 4)$

  c. For any angle with measure θ, is it possible for $\sin \theta > 1$? For $\cos \theta > 1$? For $\tan \theta > 1$?

6 The diagram below shows a portion of a circle with radius 10 cm, drawn on a 2-mm grid. Angles are marked off in 10° intervals, so $m\angle AOP_1 = 10°$, $m\angle AOP_2 = 20°$, $m\angle AOP_3 = 30°$, and so on. You can use this diagram to calculate approximate values of $\cos\theta$, $\sin\theta$, and $\tan\theta$ for angles with measure θ between 0° and 90°.

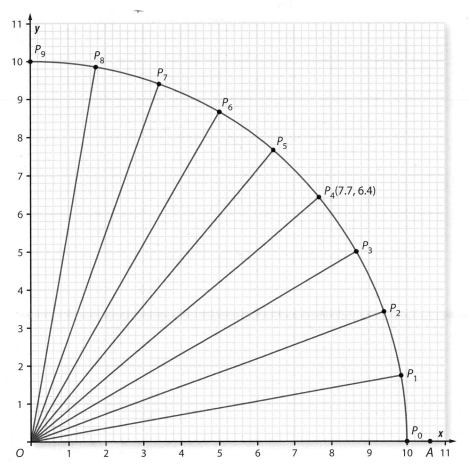

a. Verify the entries in the table below for the case of $\angle AOP_4$.

b. Make a copy of the table and then use the diagram to find the missing entries. Share the workload.

	P_n				
$m\angle AOP_n$	x	y	$\cos\theta$	$\sin\theta$	$\tan\theta$
0°					
20°	9.4	3.4			
40°	7.7	6.4	0.77	0.64	0.84
60°			0.50		
80°					
90°					

7 In Problem 6, you were able to find approximate values of the trigonometric functions of 20°, 40°, 60°, and 80°. In the case of angles with measure 0° and 90°, you were able to determine *exact* values (with the exception of tan 90°). You can use geometric reasoning to find exact values of tangent, sine, and cosine of 45°.

a. Draw an angle of 45° in standard position.

b. What is an equation for the line that makes an angle of 45° with the positive *x*-axis?

c. What do you know about the coordinates of any point on the terminal side of this angle?

d. Choose a point (x, y) on the terminal side of the angle and then find tan 45°, sin 45°, and cos 45°. Give exact values, not decimal approximations.

8 Now return to Problem 2 about the radio transmitter tower. Suppose $BC = 75$ feet and m∠ABC = 66°.

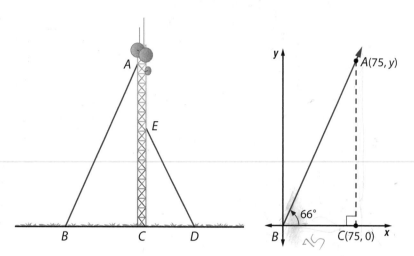

To determine *AC*, the distance above the ground at which support wire *AB* is attached to the tower:

a. Draw ∠ABC in standard position as in the diagram above on the right. Explain why the coordinates of points *A* and *C* are labeled as shown.

b. Write an expression for tan 66° in terms of the given information.

c. Use the diagram in Problem 6 to calculate an approximate value of tan 66°.

d. Using your results from Parts b and c, find the approximate height *AC*.

e. How long is support wire $\overline{AB}$?

f. Show how you could use a trigonometric function to find the approximate length of $\overline{AB}$ without first finding the height *AC*. Compare your answer to that obtained in Part e.

9 The values of the trigonometric functions you found by using the grid on page 463 were rough approximations of the function values. Several centuries ago, mathematicians spent years calculating values of the trigonometric functions by hand to several decimal places so they could be used in surveying and astronomy. Today, more advanced mathematical methods are used. You will study those methods in a later course. A portion of a table of trigonometric function values with four-digit accuracy is shown below.

Angle	sin	cos	tan
45°	0.7071	0.7071	1.0000
46°	0.7193	0.6947	1.0355
47°	0.7314	0.6820	1.0724
48°	0.7431	0.6691	1.1106
49°	0.7547	0.6561	1.1504
50°	0.7660	0.6428	1.1918
51°	0.7771	0.6293	1.2349
52°	0.7880	0.6157	1.2799
53°	0.7986	0.6018	1.3270
54°	0.8090	0.5878	1.3764
55°	0.8192	0.5736	1.4281
56°	0.8290	0.5592	1.4826
57°	0.8387	0.5446	1.5399
58°	0.8480	0.5299	1.6003
59°	0.8572	0.5150	1.6643
60°	0.8660	0.5000	1.7321

Angle	sin	cos	tan
61°	0.8746	0.4848	1.8040
62°	0.8829	0.4695	1.8807
63°	0.8910	0.4540	1.9626
64°	0.8988	0.4384	2.0503
65°	0.9063	0.4226	2.1445
66°	0.9135	0.4067	2.2460
67°	0.9205	0.3907	2.3559
68°	0.9272	0.3746	2.4751
69°	0.9336	0.3584	2.6051
70°	0.9397	0.3420	2.7475
71°	0.9455	0.3256	2.9042
72°	0.9511	0.3090	3.0777
73°	0.9563	0.2924	3.2709
74°	0.9613	0.2756	3.4874
75°	0.9659	0.2588	3.7321

a. Compare the values of sine, cosine, and tangent of 45° that you found in Problem 7 with the values in the table above.

b. Compare the values of tan 66° and cos 66° that you used in Problem 8 with the values in the table above.

c. As the measure of an angle increases from 45° to 75°,

   i. how does the sine of the angle change?

  ii. how does the cosine of the angle change?

 iii. how does the tangent of the angle change?

d. Why do the patterns of change in Part c make sense in terms of the diagram on page 463?

e. Use the table to determine each of these trigonometric function values.

   i. sin 58°   **ii.** cos 48°   **iii.** tan 72°

f. Given the function values below, use the table to determine the measure of the angle θ whose terminal side is in the first quadrant.

   i. sin θ = 0.8910

   ii. cos θ = 0.5878

   iii. tan θ = 1.1106

Summarize
the Mathematics

In this investigation, you explored the sine, cosine, and tangent, three members of a new family of functions called trigonometric functions.

a How do each of these functions provide a connection between angle measure and linear measure?

b Suppose the terminal side of an angle in standard position with measure θ lies on the line with equation $y = 2x$, $x \geq 0$.

   i. Find tan θ.   **ii.** Find cos θ.   **iii.** Find sin θ.

c Suppose θ is the measure of an angle in standard position whose terminal side is in the first quadrant and tan $\theta = \frac{4}{5}$.

   i. Find cos θ.   **ii.** Find sin θ.

d Describe the pattern of change for each function as the measure of an angle in standard position increases from 0° to 90°.

   i. cos θ   **ii.** sin θ   **iii.** tan θ

Be prepared to explain your responses to the class.

✓ Check Your Understanding

The terminal side of an angle in standard position with measure θ contains the point $P(4, 7)$.

a. Draw a sketch of the angle.

b. Find sin θ, cos θ, and tan θ.

c. Use the table on page 465 to estimate θ to the nearest degree.

The highest point on the Earth's surface is the peak of Mount Everest in the Himalaya mountain range along the Tibet-Nepal border in Asia. The most recent calculations indicate that Mount Everest rises 8,872 meters (29,108 feet) above sea level. As early as 1850, surveyors had estimated the height of that peak with error of only 0.4%. The first climbers known to actually reach the summit were Tenzing Norgay and Edmund Hillary in 1953.

In Investigation 1, you explored how the tangent and cosine functions could be used to calculate a height and distance that could not be measured directly. As you work on the problems of this investigation, make note of answers to the following question:

> *How can trigonometric functions be used to calculate heights*
> *like that of Mount Everest and other distances*
> *that cannot be measured directly?*

1 The following sketch shows the start of one surveyor's attempt to determine the height of a tall mountain without climbing to the top herself.

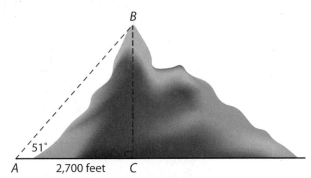

a. Use the given information to calculate the lengths of $\overline{AB}$ and $\overline{BC}$. (Use the table on page 465 to calculate approximate values of trigonometric functions as needed.)

b. Suppose that a laser ranging device allowed you to find the length of $\overline{AB}$ and the *angle of elevation* $\angle BAC$, but you could not measure the length of $\overline{AC}$. How could you use this information (instead of the information from the diagram) to calculate the lengths of $\overline{AC}$ and $\overline{BC}$?

2 The trigonometric functions are often used in problems modeled with right triangles, as in Problem 1. It is helpful to be able to use these functions without first placing an acute angle of the triangle in standard position in a coordinate plane. Examine the diagram of right △*ABC* with ∠*C* a right angle.

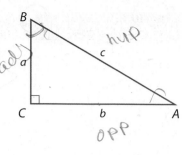

a. Explain why the following **right triangle definitions** of sine, cosine, and tangent make sense.

$$\text{tangent of } \angle A = \tan A = \frac{a}{b} = \frac{\text{length of side } opposite \angle A}{\text{length of side } adjacent \text{ to } \angle A}$$

$$\text{sine of } \angle A = \sin A = \frac{a}{c} = \frac{\text{length of side } opposite \angle A}{\text{length of } hypotenuse}$$

$$\text{cosine of } \angle A = \cos A = \frac{b}{c} = \frac{\text{length of side } adjacent \text{ to } \angle A}{\text{length of } hypotenuse}$$

b. Write expressions for tan *B*, sin *B*, and cos *B*.

3 Chicago's *Bat Column*, a sculpture by Claes Oldenburg, is shown below.

a. About how tall do you think the column is? What visual clues in the photo did you use to make your estimate?

b. In the diagram at the right, what lengths and angles could you determine easily by direct measurement (and without using high-powered equipment)?

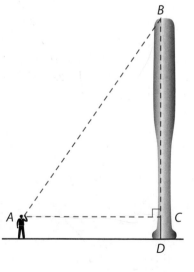

c. Which trigonometric functions of ∠*A* involve side $\overline{BC}$? Of these, which also involve a measurable length?

d. Which of the trigonometric functions of ∠*B* involve side $\overline{BC}$ and a measurable length? If you know the measure of angle of elevation ∠*A*, how can you find the measure of ∠*B*?

e. To find the height of *Bat Column*, Krista and D'wan proceeded as follows. First, Krista chose a spot (point *A*) 20 meters from the sculpture (point *C*). D'wan estimated the angle of elevation at *A* by sighting the top of the sculpture along a protractor and using a weight as shown. He measured $\angle A$ to be 55°. What is the measure of $\angle B$?

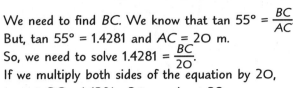

f. They next used the following reasoning to find the height *BC*.

> We need to find *BC*. We know that $\tan 55° = \dfrac{BC}{AC}$.
> But, $\tan 55° = 1.4281$ and $AC = 20$ m.
> So, we need to solve $1.4281 = \dfrac{BC}{20}$.
> If we multiply both sides of the equation by 20,
> we get $BC = 1.4281 \cdot 20$, or about 29 m.

i. Why did they decide to use the tangent function rather than the sine function?

ii. How did they know that $\tan 55° = 1.4281$?

iii. Check that each step in their reasoning is correct.

iv. How do you think Krista and D'wan used this information to calculate the height of *Bat Column*?

g. Ken said he could find the length *AB* (the line of sight distance) by solving $\cos 55° = \dfrac{AC}{AB}$.

i. Use Ken's idea to find the length *AB*.

ii. What is another way you could find *AB* by using a different trigonometric function?

iii. Could you find *AB* without using a trigonometric function? Explain your reasoning.

④ As you have seen in Problems 1 and 3, an important part of solving problems using trigonometric methods is to decide on a trigonometric function that uses given information. For each right triangle below, write two equations involving trigonometric functions of acute angles that include *s* and the indicated length. Then rewrite each equation in an equivalent form "$s = \ldots$."

a.

b.

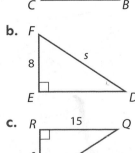

c. R   15   Q

s

P

⑤ Rather than using tables, today it is much easier to find values of the trigonometric functions with a calculator.

    a. To calculate a trigonometric function value for an angle measured in degrees, first be sure your calculator is set in *degree* mode. Then simply press the keys that correspond to the desired function. For example, to calculate $\sin 27.5°$ on most graphing calculators, press ⌨SIN **27.5** ⌨) ⌨ENTER. Try it. Then calculate $\cos 27.5°$ and $\tan 27.5°$. Although your calculator displays 10 digits, you should report estimates of trigonometric function values to the nearest ten-thousandth.

    b. Use your calculator to find the sine, cosine, and tangent of 66°. Of 54°. Compare these results with those in the table on page 465.

    c. Use your calculator to find the sine, cosine, and tangent of 45°. Compare these results with the exact values you found in Problem 7 of Investigation 1.

⑥ Each part below gives angle measure and side length information for right $\triangle ABC$ with $\angle C$ a right angle. For each, sketch and label the triangle. Then find the lengths of the remaining two sides and find the measure of the third angle.

    a. $\angle B = 52°$, $a = 5$ m   **b.** $\angle A = 48°$, $a = 15$ mi

    c. $\angle A = 31°$, $b = 8$ in.   **d.** $\angle A = 70°$, $c = 14$ cm

Summarize
the Mathematics

The trigonometric functions sine, cosine, and tangent are useful in calculating lengths in situations modeled with right triangles. Refer to the right triangle below in summarizing your thinking about how to use trigonometric functions in the situations described.

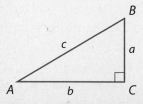

a If you knew b and the measure of $\angle A$, how would you use that information to find a? How could you find $m\angle B$? How could you find c?

b If you knew c and the measure of $\angle B$, how would you use that information to find a? How could you find $m\angle A$? How could you find b?

c If you knew a and the measure of $\angle A$, how would you use that information to find c?

Be prepared to explain your methods to the entire class.

✔ Check Your Understanding

Terri is flying a kite and has let out 500 feet of string. Her end of the string is 3 feet off the ground.

a. If ∠*KIT* has a measure of 40°, approximately how high off the ground is the kite?

b. As the wind picks up, Terri is able to fly the kite at a 56° angle with the horizontal. Approximately how high is the kite?

c. What is the highest Terri could fly the kite on 500 feet of string? What would be the measure of ∠*KIT* then?

d. Your answers in Parts a–c are estimates. When you actually fly a kite, what are some factors that might cause your answers to be somewhat inaccurate?

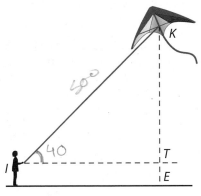

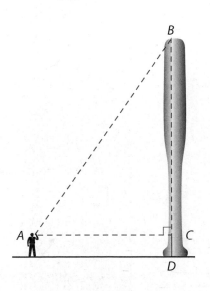

Investigation **3** What's the Angle?

In Investigation 2, you used trigonometric functions to determine an unknown or inaccessible distance in situations that could be modeled with right triangles. In such situations, trigonometric functions can also be used to determine the measure of an angle when you know the lengths of two sides of a triangle. As you work on this investigation, look for answers to the following question:

> *How can trigonometric functions be used to determine the measure of an acute angle in a right triangle when the lengths of two sides are known?*

1 In Investigation 2, you explored ways to use trigonometric functions to indirectly measure the height of Chicago's *Bat Column* sculpture. Cal read some literature that said the *Bat Column's* height *BD* is about 30 meters.

a. Cal sighted to the top of the sculpture from a point *A*, 25 meters from point *C* and 1.5 meters above the ground. Experiment with your calculator to estimate, to the nearest degree, the measure of the angle of elevation that Cal should expect at point *A*.

b. Suppose Cal sighted from a point 18 meters from point *C* and 1.5 meters above the ground. Use the same method as in Part a to estimate the measure of the angle of elevation from this point.

c. Check your answers in Parts a and b using the table on page 465.

2 Using your calculator, estimate (to the nearest degree) the measure of acute $\angle B$ for each of the following trigonometric functions. Check your estimate in each case using the table on page 465.

   a. $\sin B = \frac{3}{4}$ **b.** $\cos B = \frac{1}{2}$ **c.** $\tan B = 2$

3 You can also use the *inverse function* capabilities of your calculator to produce the acute angle measure when you know a trigonometric function value as in Problem 2. The function $\sin^{-1}$ (read "inverse sine") is related to the sine function for acute angles in a way similar to the way the square root and squaring operation are related. One function "undoes" the other. Thus, $\sin^{-1} x$ is the acute angle whose sine is x. The inverse cosine (denoted $\cos^{-1}$) and inverse tangent (denoted $\tan^{-1}$) are similarly related to the cosine and tangent functions.

   a. Suppose you know $\sin A = \frac{4}{5}$. Use the $\sin^{-1}$ function of your calculator to compute the measure of $\angle A$. (Make certain your calculator is set in degree mode.)

   b. Use inverse trigonometric functions to find the measure of $\angle B$ that corresponds to each of the function values given in Problem 2. Compare these values to the values you obtained in that problem.

   c. Use the calculator display at the right to find sin 37.58950296°.

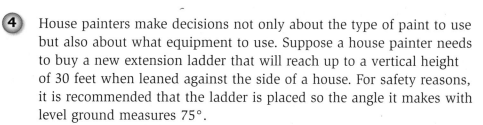

   d. Use your calculator to find the degree measure of the acute angle in each of the following cases, where possible.

   i. $\tan B = 1.84$

   ii. $\sin A = 0.852$

   iii. $\sin A = 2.15$

   iv. $\cos B = 0.213$

4 House painters make decisions not only about the type of paint to use but also about what equipment to use. Suppose a house painter needs to buy a new extension ladder that will reach up to a vertical height of 30 feet when leaned against the side of a house. For safety reasons, it is recommended that the ladder is placed so the angle it makes with level ground measures 75°.

   a. What is the minimum fully-extended length of a ladder that meets these requirements?

   b. Suppose the painter's assistant extends the new ladder to 28 feet and leans it against a house from a point 11.5 feet from the house.

   i. What angle does the ladder make with the ground?

   ii. At what vertical height from the ground does the ladder meet the side of the house?

   iii. To what length should the ladder be adjusted if it is to reach the same point on the side of the house but make the recommended 75° angle with the ground?

   iv. How far will the ladder be placed from the house after it is adjusted as in part iii?

Summarize
the Mathematics

The sine, cosine, and tangent functions are useful in finding the measures of acute angles of a right triangle using the lengths of two sides. Refer to the right triangle below to summarize your thinking about such situations.

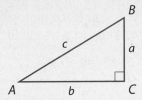

a If you knew a and c, how would you use that information to find the measure of $\angle B$? How could you then find m$\angle A$? How could you find b?

b If you knew b and c, how would you use that information to find the measure of $\angle B$? How could you then find m$\angle A$?

c If you knew a and b, how would you use that information to find the measure of $\angle A$?

Be prepared to explain your methods to the entire class.

✔Check Your Understanding

A person on an oil-drilling ship in the Gulf of Mexico sees a semi-submersible platform with a tower on top of it. The tower stands 130 meters above the platform floor.

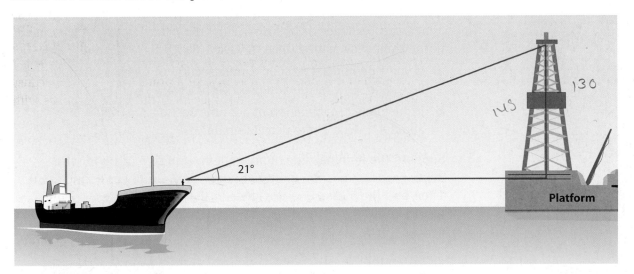

a. If the observer's position on the boat is 15 meters under the floor of the platform and the angle of elevation to the top of the rig is 21°, what three distances can you find? Find them.

b. Suppose the boat moves so the observer is 200 meters from the center line of the tower. What is the angle of elevation now?

On Your Own

1 More people these days are exercising regularly. Exercise scientists measure the amount of work done by people in various forms of exercise so they can learn more about its effect. One popular form of exercise is walking on a treadmill.

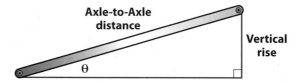

a. What features of a treadmill do you think would increase or decrease the amount of work done by the walker?

b. One index that exercise scientists use is the *percent grade* of the treadmill. Percent grade is computed as 100 multiplied by the sine of the measure of the angle of elevation θ of the treadmill. Suppose θ is in standard position. Compute the percent grade of a two-meter (axle-to-axle) treadmill with a vertical rise of 0.25 meters. Of 0.33 meters.

c. How do you think the percent grade is related to the amount of work a person does on a treadmill?

2 Steep hills on highways are the scourge of long-distance bikers. To measure the percent grade of a section of highway, surveyors use transits to estimate the average angle of elevation (or inclination) over a measured distance of highway. Then the percent grade is computed in the same way as described in Task 1 Part b for a treadmill.

a. If you ride down a straight 3-mile section of highway that has an 8% grade, how far do you drop vertically?

b. If the angle of inclination of a 2-mile section of straight highway is about 4°, what is the percent grade?

3 Suppose the terminal side of an angle in standard position with measure θ contains the indicated point. Find sin θ, cos θ, and tan θ. Then find the measure of the angle to the nearest degree.

a. $P(3, 4)$

b. $P(5, 12)$

c. $P(0, -10)$

d. $P(-5, 5)$

4 In Fort Recovery, Ohio, there is a monument to local soldiers who died in battle. Mr. Knapke, a teacher at the local high school, challenged his class to find as many ways as they could to measure the height of the monument indirectly.

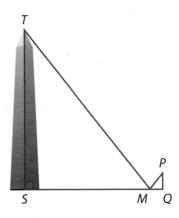

Pedro, whose eye level P is 5.8 feet, proposed a novel solution. He placed a mirror M on the ground 45 feet from the center of the monument's base and then moved to a point 2.6 feet further from the monument where he could just see to the top of the monument in the mirror. He recalled from his earlier studies that the angle of incidence and the angle of reflection are congruent.

a. On a copy of the diagram above, show all of the given information.

b. Figure out how Pedro found the height of the monument. What is the height?

c. Describe another method to find the height of the monument.

5 A survey team was to measure the distance across a river over which a bridge is to be built. They set up a survey post on their side of the river directly across from a large tree on the other side. Then they walked downstream a distance that they measured to be 400 meters. From the downstream position, they sighted the survey post and then rotated their calibrated transit to the tree to find the sighting angle to be 31°.

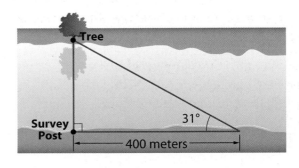

a. Determine the distance directly across the river, that is, from the survey post to the tree on the opposite bank.

b. Determine the distance from the surveyors' sighting point to the tree on the opposite bank.

6 In each right triangle below, you are given two measures. Find the remaining side lengths and angle measures.

a.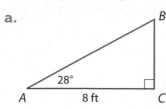

b.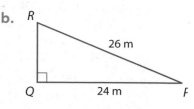

c.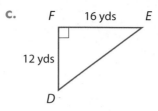

7 From the eye of an observer at the top of a cliff 125 m from the surface of the water, the *angles of depression* to two sailboats, both due west of the observer, are 16° and 23°. Calculate the distance between the sailboats.

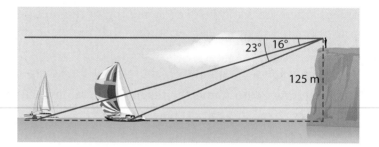

8 Commercial aircraft usually fly at an altitude between 9 and 11 kilometers (about 29,000 to 36,000 feet). When an aircraft is landing, its gradual descent to an airport runway occurs over a long distance. Assume the path of descent is a line.

a. Suppose a commercial airliner begins its descent from an altitude of 9.4 km with an angle of descent of 2.5°. At what distance from the runway should the descent begin?

b. Suppose a commercial airliner flying at an altitude of 11 km begins its descent at a horizontal distance 270 km from the end of the runway. What is its angle of descent?

c. The *cockpit cutoff angle* of an airliner is the angle formed by the pilot's horizontal line of sight and her line of sight to the nose of the plane. Suppose a pilot is flying an aircraft with a cockpit cutoff angle of 14° at an altitude of 1.5 km. In her line of sight along the nose of the plane, she sights the near edge of a lake.

   i. How far is she from the edge of the lake, measuring along her line of sight?

   ii. What is the horizontal distance to the near edge of the lake?

Connections

9 The diagram below extends the diagram on page 463 to show several other angles, $\angle AOP_n$, in standard position in a coordinate plane. Angles are marked off in 10° intervals.

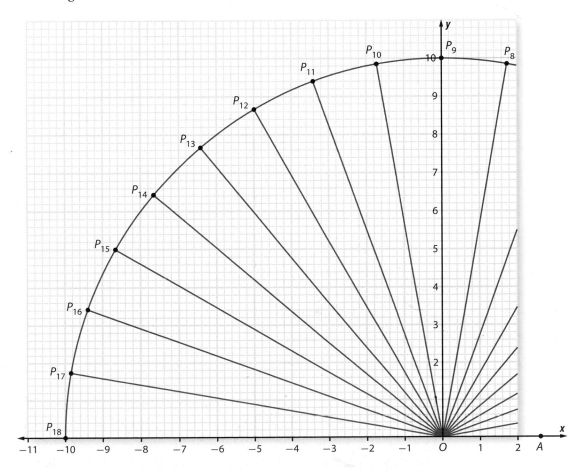

a. Use the diagram to calculate approximate values of the following:

   i. cos 120°   ii. sin 120°   iii. tan 120°

   iv. cos 160°   v. sin 160°   vi. tan 160°

b. What is cos 180°? sin 180°? tan 180°?

c. How could you use symmetry of the semicircle and your completed copy of the table on page 463 to determine each of the trigonometric function values in Part a?

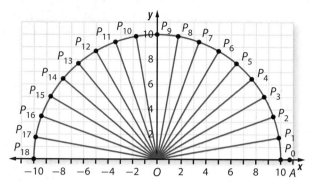

d. Refer to the diagram above. As the measure of $\angle AOP_n$ increases from 90° to 180°,

   i. how does the sine of the angle change?

   ii. how does the cosine of the angle change?

   iii. how does the tangent of the angle change?

---

10 The diagram at the right shows a portion of a circle with radius 1 and center at the origin. For acute angle θ in the diagram, the values sin θ, cos θ, and tan θ can all be seen as segment lengths. Explain why:

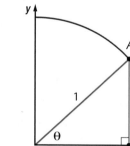

a. cos $\theta = BC$

b. sin $\theta = AC$

c. tan $\theta = DE$

11 A portion of a circle with radius 1 and center at the origin is shown below. Point P_1 is the image of point A under a 45° counterclockwise rotation about the origin. Point P_2 is the image of point A under a counterclockwise rotation of θ degrees about the origin.

a. What is the length of $\overline{OP_1}$? Of $\overline{OP_2}$? Explain your reasoning.

b. Why does $a = \cos 45°$? Why does $b = \sin 45°$?

c. What is the x-coordinate c of point P_2 written as a trigonometric function of θ? What is the y-coordinate d?

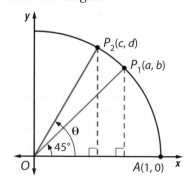

d. Based on your work in Parts a and b, what must be true about the expression $(\sin 45°)^2 + (\cos 45°)^2$? Evaluate the expression by calculating the function values.

e. What must be true about the expression $(\sin \theta)^2 + (\cos \theta)^2$? Explain your reasoning.

12 In Investigation 1, you used geometric reasoning to determine *exact* values of sin 45°, cos 45°, and tan 45°. You also determined exact values of these functions for angles of measure 0° and 90° (with the exception of tan 90°). Use equilateral triangle *ABC* to help you determine the *exact* trigonometric function values below. $\overline{AM}$ is a median of the triangle. (The side length 2 is chosen for ease of computation.)

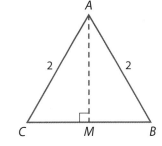

a. cos 60°   b. sin 60°   c. tan 60°

d. cos 30°   e. sin 30°   f. tan 30°

13 The diagram below shows a portion of a circle with radius 1 and center at the origin. Point *A′* is the image of point *A* and point *B′* is the image of point *B* under a 30° counterclockwise rotation about the origin.

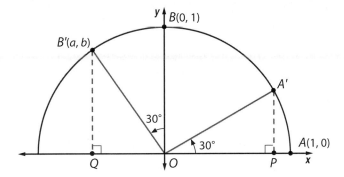

a. Explain why the coordinates of point *A′* are (cos 30°, sin 30°).

b. What is the length of $\overline{OB'}$? Why?

  i. Write the *x*-coordinate *a* of point *B′* as a trigonometric function of 30°.

  ii. Write the *y*-coordinate *b* of point *B′* as a trigonometric function of 30°.

c. In the *Coordinate Methods* unit, you learned that you could build a matrix representation of a rotation about the origin by knowing what happens to the points with coordinates (1, 0) and (0, 1). In that unit, you found the entries of the 30° counterclockwise rotation matrix to be

$$\begin{bmatrix} \dfrac{\sqrt{3}}{2} & -\dfrac{1}{2} \\ \dfrac{1}{2} & \dfrac{\sqrt{3}}{2} \end{bmatrix}.$$

Express the entries of this matrix in terms of cos 30° and sin 30°.

d. Transform the point *B*(0, 1) by multiplying by the matrix in Part c. Do you obtain the coordinates of *B′*?

e. How could you use the sine and cosine functions to build a matrix for a counterclockwise rotation of θ degrees with center at the origin? Check your answer for the case of a 90° counterclockwise rotation.

14 At the opening of this lesson, you may have explored the operation of an automobile jack using the interactive geometry "Auto Jack" custom tool. Recall that points *A* and *C* are connected by a long threaded rod, and the lengths *AB*, *BC*, *AD*, and *CD* are all 1 foot. The distance *AC* is longer or shorter depending on the direction the rod is turned.

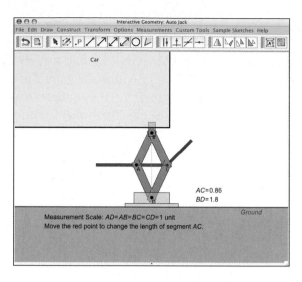

a. If *AC* = *x*, use the Pythagorean Theorem to write a rule that expresses the height *BD* as a function of *x*.

b. What is the domain of the function in Part a?

c. Use a graph of the function to explore how the height *BD* of the jack changes as *AC* changes. Sketch the graph and describe the shape of the graph as precisely as you can.

d. How is the pattern of change you saw in the graph reflected in the table of function values? Start at 0 and use increments of 0.1.

15 In the *Coordinate Methods* unit, Connections Task 18 (page 188) and Connections Task 21 (page 225), you prepared a table summarizing connections between important geometric ideas and their coordinate representations. Review the table you completed and then add to it coordinate representations and examples for the ideas of "angle in standard position" and "trigonometric functions of angle measure."

16 An angle in standard position has its terminal side in the first quadrant on a line whose equation is given. For each of the following lines, determine the measure of the angle.

   a. $y = 20x$ **b.** $y = 8x$ **c.** $y = 0.5x$

17 The length of a side of a rectangle is 8 inches and the length of its diagonal is 9.5 inches.

   a. What is the perimeter of this rectangle? What is its area?

   b. What is the measure of an angle formed by a diagonal and an 8-inch side?

   c. What is the measure of an angle formed by a diagonal and a side that is not 8 inches long?

18 The Pyramid of Cheops in Egypt is a square pyramid. The base edge measures 230 meters, and each face makes an angle of 51.8° with the horizontal floor.

a. Make a sketch of the Pyramid of Cheops.

b. Determine the vertical height of the pyramid.

c. If you were climbing the pyramid, what would be the shortest route to the top? What is the length of this route?

d. Determine the dimensions and angle measures of the triangular faces of the pyramid. Then determine the area of each face.

e. Determine the volume of the Pyramid of Cheops.

Reflections

19 A line through the origin forms an acute angle with the positive x-axis of degree measure θ called the *angle of inclination* of the line. Express the slope of the line as a trigonometric function of θ. What is the equation of the line?

20 Suppose $\triangle ABC$ is a right triangle with $\angle C$ a right angle.

a. If $m\angle A = 50°$, what is $m\angle B$? If $m\angle B = 10°$, what is $m\angle A$?

b. If you know the measure of one acute angle of a right triangle, explain why you can always determine the measure of the other acute angle.

c. Two angles whose measures sum to 90° are called **complementary angles**. In what sense do they complement each other?

d. The term "cosine" suggests that the cosine of an acute angle is the sine of the complement of that angle. Does $\cos 50° = \sin 40°$? Does $\cos 80° = \sin 10°$?

e. Suppose $\angle D$ and $\angle E$ are acute angles of a right triangle.

 i. Use a diagram to explain why $\cos D = \sin E$.

 ii. Why can the equation in part i also be written as $\cos D = \sin (90° - D)$? Why can it also be written as $\sin E = \cos (90° - E)$?

21 What combinations of known facts about parts of a right triangle enable you to find information about other parts?

22 The label on a jar of creamy peanut butter claims that each serving (2 tablespoons) contains 16 grams of fat, 6 grams of carbohydrate, and 8 grams of protein. A recipe for peanut butter cookies calls for 5 tablespoons of peanut butter.

   a. Write proportions whose solutions would tell the number of grams of fat f, carbohydrate c, and protein p from peanut butter in the cookie recipe.

   b. Solve each of the equations in Part a.

   c. Explain how the equations solved in Part b are similar to equations with trigonometric functions used to find side lengths in right triangles.

23 When using trigonometric methods to solve a problem represented by a right triangle, how do you decide whether to use the sine, cosine, or tangent function?

24 If the angle of depression from point A at the top of a mountain to point B at a lower elevation is 20°, what is the angle of elevation from point B to point A? Explain why, using an appropriate diagram.

25 Do a search on "surveying" on the Internet. Look for ways in which surveyors use trigonometry, the instruments they use, and the training and skills that are required. Write a brief report describing your findings.

Extensions

26 As you learned in this lesson, two different points on the terminal side of the same angle in standard position determine two different right triangles, but ratios of corresponding sides of the two triangles are equal. In this task, you will examine another way of justifying this fact.

   a. Refer to Diagram I below. What is the equation of $\overleftrightarrow{OP}$?

Diagram I

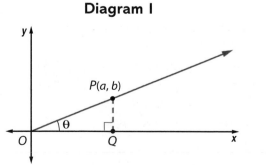

Diagram II

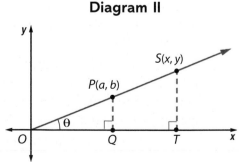

   b. Use the coordinates of P to write each of the following lengths as an algebraic expression:

    i. OQ **ii.** QP **iii.** OP

c. In Diagram II on page 482, $S(x, y)$ is any other point on $\overleftrightarrow{OP}$ other than the origin, $(0, 0)$. Explain why point S has coordinates of the form $\left(x, \dfrac{b}{a}x\right)$.

d. Use the coordinates of S to write each of the following lengths as an algebraic expression:

   i. OT **ii.** TS **iii.** OS

e. Compare the following pairs of ratios:

   i. $\dfrac{QP}{OQ}$ and $\dfrac{TS}{OT}$ **ii.** $\dfrac{QP}{OP}$ and $\dfrac{TS}{OS}$ **iii.** $\dfrac{OQ}{OP}$ and $\dfrac{OT}{OS}$

   What can you conclude?

27 In $\triangle ABC$ shown here, $\overline{BD}$ is an altitude.

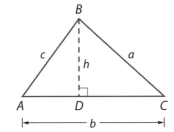

a. Express $\sin C$ in terms of h and a.

b. Explain why the following is true:

   area of $\triangle ABC = 0.5 \cdot ab \cdot \sin C$.

c. Write another formula for the area of $\triangle ABC$, this one in terms of $\sin A$.

d. Using Parts b and c, show that $a \cdot \sin C = c \cdot \sin A$.

28 A surveyor made a sighting to the top of a mountain peak from point A, located on a flat plateau. The angle of elevation from point A was 28°. He then moved 200 feet directly toward the mountain to point B on the same plateau. The angle of elevation from point B was 39°. Find the height of the mountain peak above the plateau. Draw a figure, and explain your steps.

29 While camping by the Merced River in Yosemite Valley, a group of friends were admiring a particular tree on the opposite bank. Maria claimed that the height of the tree could be determined from the group's side of the river by the following method:

- Measure a 50-meter segment, $\overline{AB}$, on this shore.
- Consider the base of the tree to be located at point C and use a sighting device to measure $\angle BAC$ and $\angle ABC$.
- From A, measure the angle of elevation to the top of the tree, located at point D.
- Use the measurements above with some trigonometric functions to calculate the height.

The friends measured $\angle BAC$ to be 54° and $\angle ABC$ to be 74°. The angle of elevation from point A to the treetop was 21°. Draw a sketch of this situation and determine whether Maria was correct. If she was, compute the height; if she was not, explain why not.

30 Find the tallest object on your school campus that cannot be easily measured directly. Prepare a plan to determine the height of the object. Carry out the plan, if it is plausible. Prepare a brief report of your method and findings.

31 Modern satellite communication systems make it possible for us to determine our exact (within a meter or so of accuracy) location on the Earth's surface by sending a signal from a handheld Global Positioning System (GPS) device. Perhaps you have GPS capability on your cell phone.

The GPS sends signals to satellites which relate the signal angle and distance to known reference sites. This technology makes it possible to have a system built into an automobile that shows a dashboard map giving both car and selected destination locations as well as how to reach the destination.

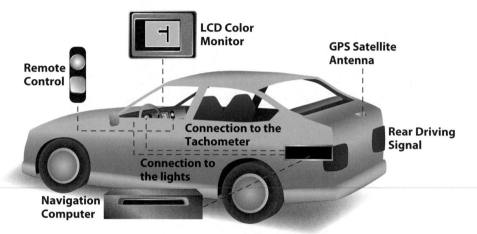

An actual GPS uses *spherical trigonometry* because of the spherical shape of the Earth. Consider a simpler problem of locating objects on a large flat surface, using sensors atop very tall towers.

Devise a system that you think would allow the sorts of calculations required to locate a GPS transmitter on this flat surface relative to a coordinate system.

Review

32 Solve the following equations for x.

a. $\dfrac{5x}{12} = \dfrac{17}{23}$

b. $\dfrac{8 - x}{3} = \dfrac{x + 5}{10}$

c. $\dfrac{x + 6}{2x - 3} = \dfrac{7}{10}$

d. $\dfrac{10}{2x - 3} = \dfrac{7}{3x}$

33 Rewrite each expression with the smallest possible integer under the radical.

a. $\sqrt{60}$

b. $\sqrt{75}$

c. $3\sqrt{27}$

d. $\sqrt{25 + 144}$

34 Carl and Lisa are both drawing triangles *ABC* that meet each set of criteria below. Must the triangles they draw be congruent? Explain your reasoning.

a. The lengths of the three sides are 3 cm, 8 cm, and 7 cm.

b. The measures of the three angles are 120°, 20°, and 40°.

c. $AB = 5$ in., $m\angle B = 60°$, and $m\angle A = 50°$

d. $m\angle C = 100°$, $BC = 10$ cm, and $AC = 8$ cm

e. $m\angle B = 35°$, $BC = 10$ in., $AC = 8.5$ in.

35 Find the area of $\triangle ABC$ in each diagram.

a.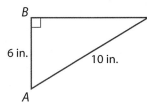

b.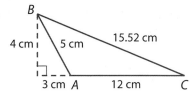

c.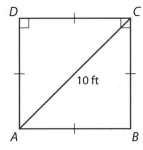

d.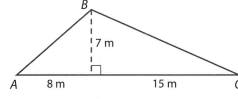

36 Solve each of these quadratic equations by factoring, if possible. If you don't see a way to factor, use the quadratic formula. Check your work by substituting your proposed solutions in the original equation.

a. $x^2 - 3x - 10 = 0$

b. $x^2 + 9x + 11 = 7$

c. $2x^2 - 12x + 16 = 0$

d. $3x^2 - x = 7$

37 Evaluate each of the following expressions, given $a = 4$, $b = 3$, $c = -3$, and $d = \frac{1}{3}$.

a. $2c + a^2 - (b + a)$

b. $a^2 + c^2 - 2abd$

c. $6d + 2(b - c)^2$

d. $(b - a) + abc$

e. $\sqrt{a^2 + b^2}$

38 Tim collected data about the amount of time he spent reading and the number of pages he read. His data are displayed in the scatterplot below.

(scatterplot: x-axis "Time (in minutes)" from 10 to 100; y-axis "Number of Pages" from 0 to 110)

a. Estimate the correlation coefficient for the data. Explain your reasoning.

b. The regression equation is $y = 0.85x + 0.69$. Explain the meaning of the slope of this line in terms of the context.

c. One of the points on the graph has coordinates (92, 106). Find the residual for this point.

d. Tim has 50 pages left in the book he is currently reading. If he reads for 45 minutes before he goes to sleep, do you expect he will finish his book? Explain your reasoning.

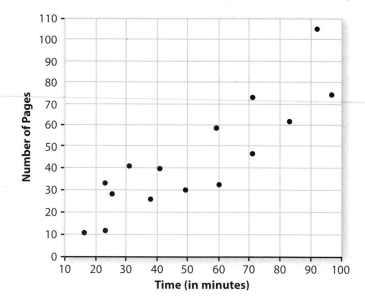

39 The matrix below gives the cost of building trails between five scenic areas in Baldy Mountain State Park.

$$
\begin{array}{c c c c c c}
 & A & B & C & D & E \\
A & \begin{bmatrix} 0 \\ 1{,}250 \\ 459 \\ 3{,}465 \\ 1{,}524 \end{bmatrix} & \begin{matrix} 1{,}250 \\ 0 \\ 875 \\ 1{,}300 \\ 1{,}600 \end{matrix} & \begin{matrix} 459 \\ 875 \\ 0 \\ 2{,}379 \\ 1{,}437 \end{matrix} & \begin{matrix} 3{,}465 \\ 1{,}300 \\ 2{,}379 \\ 0 \\ 889 \end{matrix} & \begin{matrix} 1{,}524 \\ 1{,}600 \\ 1{,}437 \\ 889 \\ 0 \end{matrix} \end{bmatrix}
\end{array}
$$

a. Create a vertex-edge graph that could represent this situation.

b. The State Park rangers want to build trails so hikers can get from any one of these scenic areas to another. Which trails should they build to minimize the cost of building the trails? What is the minimum cost?

c. Use the best-edge algorithm to find a set of trails that would allow hikers to start at *A* and make a loop that visits all five of these attractions. Find the cost of building your loop trail. Are you guaranteed that you have found the cheapest way to build a loop trail that starts at *A*? Explain your reasoning.

Using Trigonometry in Any Triangle

I n Lesson 1, you learned several strategies for using the Pythagorean Theorem and the trigonometric functions (sine, cosine, and tangent) to calculate unknown side lengths and angle measures in situations represented by right triangles. But often, problem situations are modeled by triangles that are not right triangles.

Consider, for example, the problem of developing an accurate map of the floor of the Grand Canyon.

Suppose a surveyor sights point *C*, the tip of a pointed spur deep in the canyon, from triangulation points *A* and *B* on the south rim. $\overline{AB}$ was measured to be 2.68 miles. Using a transit, m∠*CAB* was found to be 64°; m∠*CBA* was found to be 34°.

a Draw and label a triangle representing this situation.

b Is the triangle a right triangle? How can you be sure of your answer?

c What side lies opposite the 34° angle in your triangle? Is sin 34° equal to a ratio of lengths of two sides of your triangle? If so, which ones? If not, why not?

d How might you go about determining the distances *AC* and *BC*?

In this lesson, you will investigate two important properties of any triangle that relate angle measures and side lengths known as the Law of Sines and the Law of Cosines. These properties will add to the trigonometric methods available to you for making indirect measurements and for analyzing mechanisms in which the lengths of two sides of a triangle are fixed but the length of the third side varies.

Investigation 1 The Law of Sines

If the triangle that models a situation involving unknown distances is not (or might not be) a right triangle, then it is not so easy to determine the distances; but it can be done. One method that is sometimes helpful uses the *Law of Sines*. As you work on the problems of this investigation, look for answers to the following question:

> *What is the Law of Sines, and how can it be used to find side lengths or angle measures in triangles?*

Suppose that two park rangers who are in towers 10 miles apart in a national forest spot a fire that is far away from both of them. Suppose that one ranger recognizes the fire location and knows it is about 4.9 miles from that tower.

With this information and the angles given in the diagram below, the rangers can calculate the distance of the fire from the other tower.

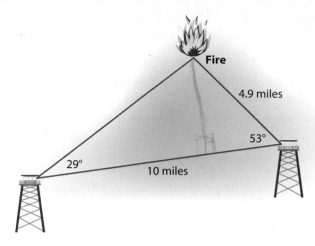

One way to start working on this problem is to divide the obtuse triangle into two right triangles as shown below:

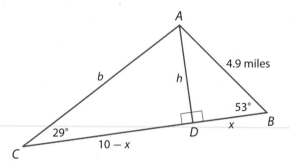

At first, that does not seem to help much. Instead of one segment of unknown length, there are now four! On the other hand, there are now three triangles in which you can see useful relationships among the known sides and angles.

1. Use trigonometry or the Pythagorean Theorem to find the length of $\overline{AC}$. When you have one sequence of calculations that gives the desired result, see if you can find a different approach.

2. In one class in Seattle, Washington, a group of students presented their solution to Problem 1 and claimed that it was the quickest method possible. Check each step in their reasoning and explain why each step is or is not correct.

   (1) $\dfrac{h}{b} = \sin 29°$ (3) $\dfrac{h}{4.9} = \sin 53°$

   (2) $h = b \sin 29°$ (4) $h = 4.9 \sin 53°$

   (5) $b \sin 29° = 4.9 \sin 53°$

   (6) $b = \dfrac{4.9 \sin 53°}{\sin 29°}$

   (7) $b \approx 8.1$ miles

   Compare your solution from Problem 1 with this reported solution.

3 The approach used in Problem 2 to calculate the unknown side length of the given triangle illustrates a very useful general relationship among sides and angles of *any* triangle.

  a. Explain why each step in the following derivation is correct for the acute $\triangle ABC$ below.

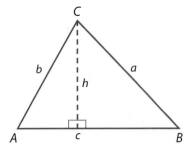

$$\frac{h}{b} = \sin A \qquad (1)$$
$$h = b \sin A \qquad (2)$$
$$\frac{h}{a} = \sin B \qquad (3)$$
$$h = a \sin B \qquad (4)$$
$$b \sin A = a \sin B \qquad (5)$$
$$\frac{\sin A}{a} = \frac{\sin B}{b} \qquad (6)$$

  b. How would you modify the derivation in Part a to show that $\frac{\sin B}{b} = \frac{\sin C}{c}$?

The relationship derived in Problem 3 for acute angles A, B, and C holds in any triangle, for all three of its sides and their opposite angles. It is called the **Law of Sines** and can be written in two equivalent forms. The cases for a right triangle or an obtuse triangle are derived in Extensions Tasks 22 and 23.

In any triangle ABC with sides of lengths a, b, and c opposite $\angle A$, $\angle B$, and $\angle C$, respectively:

$$\frac{\sin A}{a} = \frac{\sin B}{b} = \frac{\sin C}{c}$$

or equivalently,

$$\frac{a}{\sin A} = \frac{b}{\sin B} = \frac{c}{\sin C}.$$

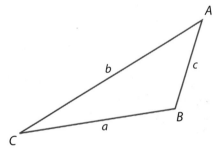

You can use the Law of Sines to calculate measures of angles and lengths of sides in triangles with even less given information than the fire-spotting problem at the beginning of this investigation. In practice, you only use the equality of two of the ratios at any one time.

4 A class in San Antonio, Texas, agreed on the following representation of the surveying problem in the Think About This Situation (page 489). Use what you know about angles in a triangle and the Law of Sines to determine the distances *AC* and *BC*.

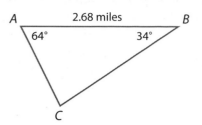

5 Suppose that two rangers spot a forest fire as indicated on the diagram below. Find the distances from each tower to the fire.

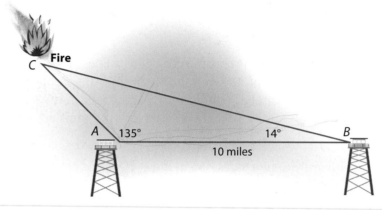

Summarize
the Mathematics

The Law of Sines states a relation among sides and angles of any triangle. It can often be used to find unknown side lengths or angle measures from given information. Suppose you have modeled a situation with △*PQR* as shown below.

a What minimal information about the sides and angles of △*PQR* will allow you to find the length of $\overline{QR}$ using the Law of Sines? How would you use that information to calculate *QR*?

b What minimal information about the sides and angles of △*PQR* will allow you to find the measure of ∠*Q*? How would you use that information to calculate m∠*Q*?

Be prepared to explain your thinking and methods to the entire class.

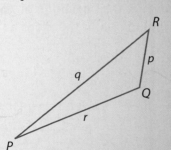

✓ Check Your Understanding

A commuter airplane off course over the Atlantic Ocean reported experiencing mechanical problems around 9:15 P.M. The pilot sent two calls, one to Boston Logan International Airport and one to the regional airport in nearby Beverly. Air traffic controllers at the two airports reported the angles shown in the diagram below. How far was the plane from the closer airport?

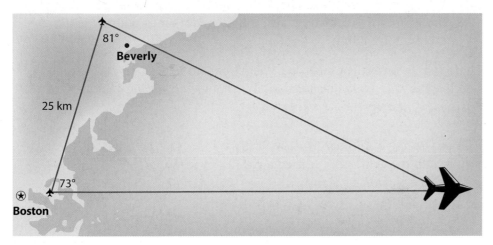

Investigation ② The Law of Cosines

In the previous investigation, you examined how the Law of Sines is helpful in solving problems in which the measures of two angles and the length of one side of a triangle are known. For problems in which other combinations of side lengths or angle measures of a triangle are known, a second property of triangles called the *Law of Cosines* can be helpful. As you work on the following problems, look for answers to this question:

> *What is the Law of Cosines, and how can it be used to find side lengths or angle measures in triangles?*

In a right triangle, one angle is always known (the 90° angle), and the Pythagorean Theorem shows how the lengths of the two legs and the hypotenuse are related to each other. When that relationship is expressed as an equation, it is possible to solve for any side length in terms of the others. In Problems 1 and 2, you will investigate how the relationship among the sides changes as the right angle changes to an acute or obtuse angle.

① Consider a linkage with two sides of fixed length: 12 cm and 16 cm. Here, $AC = 12$ cm and $BC = 16$ cm.

   a. What is the distance from A to B when the angle at C is a right angle?

   b. How does the distance from A to B change as $\overline{AC}$ is rotated to make smaller and smaller angles at C? How does that distance change if $\overline{AC}$ is rotated to make larger angles at C?

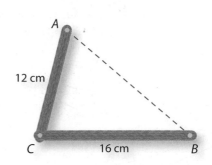

2. Using an actual physical linkage, interactive geometry software, or careful drawings, test your answers to Problem 1 by carefully measuring the distance from point A to point B in each case. Record your measures for further use in Problem 7.

 a. $m\angle C = 30°$ **b.** $m\angle C = 70°$

 c. $m\angle C = 130°$ **d.** $m\angle C = 150°$

3. Why is it impossible to check the measured distances from A to B in Problem 2 by calculations using the Law of Sines without getting more information?

There is a second trigonometric principle for finding relationships among side lengths and angle measures of any triangle. It is called the **Law of Cosines**.

In any triangle ABC with sides of length a, b, and c opposite ∠A, ∠B, and ∠C, respectively:

$$c^2 = a^2 + b^2 - 2ab \cos C$$

In Extensions Task 25, you are asked to provide a justification of this important relationship.

4. The Law of Cosines states a relationship among the lengths of three sides of a triangle and the cosine of an angle of the triangle. If you know the lengths of two sides and the measure of the angle between the two sides, you can use the Law of Cosines to calculate the length of the third side.

 a. Write the Law of Cosines to calculate the length a in $\triangle ABC$ if you know the lengths b and c and the measure of $\angle A$.

 b. Write a third form of the Law of Cosines for $\triangle ABC$ for when you know the measure of $\angle B$.

 c. Suppose in $\triangle PQR$ you needed to calculate the length of $\overline{QR}$. What information would you need in order to use the Law of Cosines? Write the equation that you would use.

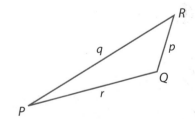

 d. Suppose in $\triangle PQR$ you knew $m\angle P$, $m\angle R$, and the lengths p and r.

  i. Could you find the length q using the Law of Cosines? Explain your reasoning.

  ii. Write an expression for calculating the length q.

 e. Using the information in Part d, write an expression for finding the length q using the Law of Sines. Which method would you prefer to use? Why?

5 Surveyors are often faced with irregular polygonal regions for which they are asked to locate and stake out boundaries, determine elevations, and estimate areas. Some of these tasks can be accomplished by using a site map and a transit as shown in the photo below. In one subdivision of property near a midsize city, a plot of land had the shape and dimensions shown.

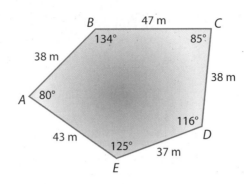

Examine the triangulation of the plot shown below.

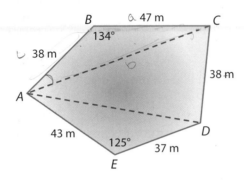

a. Find *AC* to the nearest meter.

b. Find *AD* to the nearest meter.

6 The Law of Cosines, $c^2 = a^2 + b^2 - 2ab \cos C$, states a relation among the lengths of three sides of a triangle *ABC* and the cosine of an angle of the triangle. If you know the lengths of all three sides of the triangle, you can calculate the cosine of any angle and then determine the measure of the angle itself.

a. Solve the equation $c^2 = a^2 + b^2 - 2ab \cos C$ for cos *C*.

b. Using your results from Problem 5, find the measure of ∠*BAC* to the nearest tenth of a degree. What is the measure of the third angle in △*ABC*?

c. Find the area of △*ABC*.

d. Explain how you could determine the area of the entire pentagonal plot.

7 Now that you understand how to use the Law of Cosines, examine more closely its symbolic form and the information it conveys. Consider again the linkage with arms of lengths $b = 12$ cm and $a = 16$ cm positioned at various possible angles.

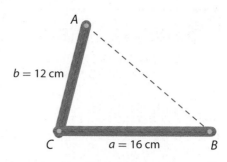

$b = 12$ cm

$a = 16$ cm

A

C

B

a. Record your data from Problem 2 in a copy of the table below. Then, using a physical linkage, interactive geometry software, or careful drawings, complete your table, showing how the distance between the ends of the linked arms changes as the angle at the link point C changes.

m∠C	30°	50°	70°	90°	110°	130°	150°
Length AB							

b. Add a row to your table from Part a showing corresponding values of $2ab \cos C$.

m∠C	30°	50°	70°	90°	110°	130°	150°
Length AB							
2ab cos C							

c. What is $\cos C$ when m∠C = 90°, and how does that simplify the equation for the Law of Cosines?

d. In what sense does the term "$2ab \cos C$" act as a *correction term,* adjusting the Pythagorean relationship for triangles in which ∠C is not a right angle?

8 As you have seen, the Law of Sines and the Law of Cosines can be used to find the measures of unknown angles as well as unknown sides of any triangle. You need to study given information about side and angle measurements to decide which law to apply. Then you work with the resulting equations to solve for the unknown angle or side measurements.

For example, suppose that two sides, $\overline{AB}$ and $\overline{BC}$, and a diagonal, $\overline{AC}$, of a parallelogram $ABCD$ measure 7 cm, 9 cm, and 11 cm, respectively.

a. Draw and label a sketch of parallelogram $ABCD$.

b. Which of the two trigonometric laws can be used to find the measure of an angle in that parallelogram?

c. Find the measure of that angle to the nearest tenth of a degree.

d. The diagonal $\overline{AC}$ splits the parallelogram $ABCD$ into two congruent triangles. Find the remaining measures of the angles in those triangles.

e. Find the length of diagonal $\overline{BD}$.

Summarize
the Mathematics

In this investigation, you explored how the Law of Cosines can be used to find unknown side lengths or angle measures in a triangle. Consider △ABC shown below.

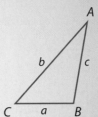

a What information would you need to know in order to use the Law of Cosines to find the length of $\overline{AC}$? What equation would you use to find that length?

b What information would you need to know in order to use the Law of Cosines to find the measure of ∠A? What equation would you use to find that angle measure?

c Suppose you know the lengths a, b, and c. What can you conclude about m∠B if $a^2 + c^2 > b^2$? If $a^2 + c^2 < b^2$? If $a^2 + c^2 = b^2$?

d What clues do you use to decide if an unknown side length of a triangle can be found using the Law of Cosines? Using the Law of Sines?

e What clues do you use to decide if an unknown angle measure of a triangle can be found using the Law of Cosines? Using the Law of Sines?

Be prepared to explain your ideas to the entire class.

✓ Check Your Understanding

A surveyor with transit at point A sights points B and C on either side of Asylum Pond. She finds the measure of the angle between the sightings to be 72°.

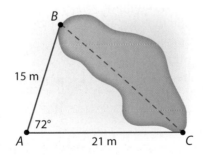

a. Find the distance BC across the pond to the nearest tenth of a meter.

b. Find m∠B and m∠C.

The Law of Sines is useful in solving problems that involve triangles when the measures of two angles and the length of a side are known. The Law of Cosines is useful when the lengths of three sides or the lengths of two sides and the measure of the angle between them are known. In this investigation, you will seek clues that may help answer the following question:

What can you conclude about triangle models for situations in which you know the lengths of two sides and the measure of an angle not included between the sides?

A cold frame is a box used to grow young plants in the spring. The top of the box is made of glass to let in light. The top can be propped open at various heights so the plants become accustomed to actual weather conditions before they are transplanted outside the box.

1 One cold frame has a rectangular top measuring 70 cm by 120 cm, hinged along the 120-cm edge. It can be held open by a prop 30 cm long. The prop is attached to the top 50 cm from the hinged end, as in the diagram at the right. Notches on the horizontal frame hold the prop in different positions, allowing the top to be opened at different angles.

a. A triangle is formed by the hinged edges of the cold frame and the prop. Make and label a diagram of the triangle showing known side lengths.

b. How do you think the measure of the angle at the hinged end changes as the prop is moved from one notch to the next? Imagine starting at the notch that is closest to the hinge and moving to the notch that is furthest from the hinge.

You will return to Problem 1 after first exploring how many differently shaped triangles can be formed given two side lengths (say, AB and BC) and the measure of an angle ($\angle A$) *not* included between them. In Problem 2, you will explore the side-side-angle (SSA) condition where $\angle A$ is acute. In Problem 3, you will explore the case where $\angle A$ is not an acute angle.

2 Use a compass and ruler or software like the interactive geometry "Explore SSA" custom tool to conduct the following experiment.

a. Draw an acute angle, $\angle A$. Fix point B on one side of $\angle A$ by marking off length AB. Suppose $\overline{BC}$ is very short, as in the figure on the right. How many triangles ABC can be formed with the three given parts: $\angle A$, $\overline{AB}$, and $\overline{BC}$? Explain.

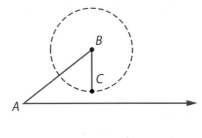

b. With the same $\angle A$ and length AB, try making $\overline{BC}$ longer. There is a minimum length of $\overline{BC}$ for which a triangle with the three given parts is formed, as shown in the next figure.

   i. In this case, what kind of triangle is ABC?

   ii. Is the triangle shown the only possible triangle determined by $\angle A$, AB, and BC?

   iii. Write the length BC in terms of length AB and a trigonometric function of $\angle A$.

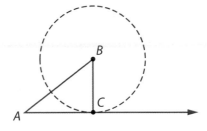

c. With the same $\angle A$ and length AB, continue to make $\overline{BC}$ longer. Try to find lengths BC for which two noncongruent triangles can be formed with the same three given parts.

   i. Draw a figure like those above that shows an example of this situation.

   ii. Over what interval of lengths BC will two noncongruent triangles be formed?

d. If BC is greater than or equal to all values in the interval you found in Part cii, how many triangles can be formed with the three given parts? Draw a figure like those above that shows an example of this situation.

e. In Parts a–d, you explored the SSA condition when $\angle A$ is acute. Summarize your findings by describing the lengths, or intervals of lengths, BC for which the given side-side-angle parts determine:

   i. no triangle.

   ii. exactly one triangle.

   iii. two noncongruent triangles.

(3) Suppose, as in Problem 2, $\overline{AB}$ and $\angle A$ are given, but now $\angle A$ is not an acute angle. Explore the two cases below.

a. Consider the case in which $\angle A$ is an obtuse angle. Explore varying lengths BC and determine conditions for which two, one, or no triangles will be determined. Make sketches to illustrate your answer.

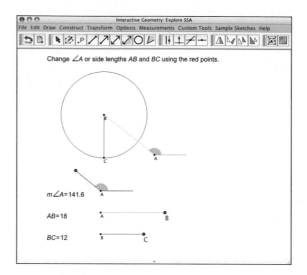

b. Suppose $\angle A$ is a right angle. Describe the triangles determined by $\overline{BC}$, $\overline{AB}$, and $\angle A$, depending on how lengths AB and BC are related.

(4) Now refer back to the cold frame in Problem 1.

a. One possible placement of the prop so that the hinged angle is $20°$ is shown below. Do you think there could be a second position for the prop along the horizontal that would satisfy the given information? If so, sketch it. If not, explain why not.

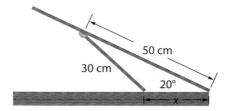

b. Use the Law of Cosines to write a quadratic equation that relates x and the three given measures. Solve the equation for x. Explain how your solution(s) relate to the diagram in Part a.

c. Suppose the gardener wants to place a notch for the prop that makes the largest hinged angle for this cold frame. In which of the settings below is the hinged angle larger? Does the larger of the two settings below maximize the hinged angle? Explain your answer.

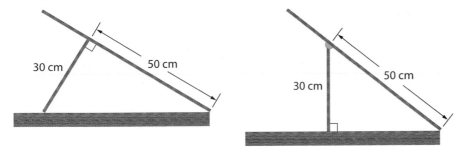

30 cm   50 cm

30 cm   50 cm

Summarize the Mathematics

In this investigation, you explored conditions under which the measure of an angle together with the lengths of two segments—one of which is opposite the angle—is sufficient to determine a triangle.

Suppose you are given side lengths *a* and *b* and a given angle measure, m∠A, of a possible triangle *ABC*.

a If at least one triangle is determined by these three parts, what is the minimum length *a* in terms of *b* and ∠A?

b Suppose ∠A is an acute angle, and there is exactly one triangle that satisfies the given condition. How are the given sides and the given angle related?

c Suppose there are two triangles that satisfy the given condition. How are the given sides and the given angle related?

d Suppose ∠A is an obtuse angle. Is it possible that two noncongruent triangles satisfy the given condition? Explain your reasoning.

Be prepared to explain your thinking to the class.

✓Check Your Understanding

The design and functioning of the automobile jack you examined in Lesson 1 and the cold frame in this lesson depend on triangle mechanisms where one side is able to vary in length. The functioning of a piston of a steam engine depends on a similar principle.

In the design below of a simple steam engine, the piston rod is connected by a rod $\overline{BC}$ to a flywheel that rotates about a point A. The piston rod moves horizontally, the connecting rod is 48 inches long, and the crank is 16 inches long. In the diagram, the connecting rod is slanted below the horizontal at an angle of 14°.

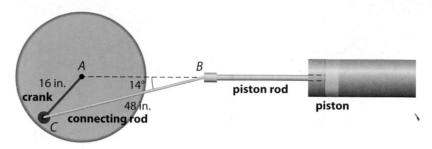

a. Are there other positions of the crank and connecting rod that set m∠ABC = 14°? If so, illustrate with labeled sketches. If not, explain why not.

b. Find the measures of the other two angles in the diagram. If you found other triangles in Part a, determine the measures of their other two angles.

c. Find the length of the third side of the triangle(s).

d. How is the *throw* of the piston (the distance between the extreme points of its motion) related to the fixed lengths AC and BC?

Applications

1 A pilot flying due east out of Denver gets word that a major thunderstorm is directly in his plane's path. The pilot turns 35° to the left of his intended course and continues on this new flight path. After avoiding the worst of the storm, he turns 45° to the right of the new course and flies until returning to his original intended line of flight. The plane reaches its original intended course at a point 80 kilometers from the start of the detour.

    a. Draw a sketch of this situation.

    b. How much farther did the aircraft travel due to the detour?

2 Refer back to the Grand Canyon mapping problem in the Think About This Situation at the beginning of this lesson (page 489). In addition to calculating the distances of particular points in the Grand Canyon from triangulation points on the rim, the surveyors wanted to map the relative depths of these points in the canyon. Using their transits, they were able to measure the **angle of depression** (that is, the angle their line of sight made with the horizontal) to the point of interest in the canyon.

    a. In Problem 4 of Investigation 1, you found that the distance from point *A* on the rim to the tip of the pointed spur at *C* was about 1.51 miles. Suppose the surveyor measured the angle of depression from point *A* to point *C* to be 28°. How far below point *A* is point *C* in vertical distance? Find the vertical distance in miles and in feet (1 mile = 5,280 feet).

    b. If point *A* is 7,392 feet above sea level, how many feet above sea level is point *C*?

    c. The Colorado River at the bottom of the Grand Canyon is 1,850 feet above sea level. How many feet above the Colorado River is point *A*? Point *C*?

3 Two lighthouses *A* and *B* are 50 km apart. At 2 A.M., a freighter moving parallel to line *AB* is sighted at point *C* as shown in the diagram below.

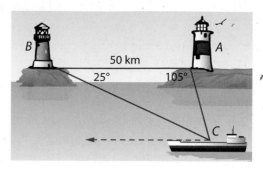

a. How far is the freighter from lighthouse *B*? From lighthouse *A*?

b. At 3 A.M., the angle at *A* is 86°. The angle at *B* is 29°. How far is the freighter from lighthouse *B*? From lighthouse *A*?

c. How far has the freighter moved in the hour between 2 A.M. and 3 A.M.?

4 The ninth hole at Duffy's Golf Club is 325 yards down a straight fairway. In his first round of golf for the season, Andy tees off and hooks the ball 20° to the left of the line from the tee to the hole. The ball stops 205 yards from the tee at point *P*, as shown in the figure.

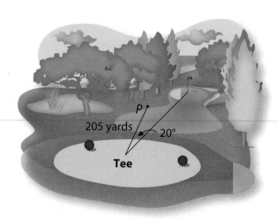

a. How far is his ball from the hole (marked by the flag)?

b. To decide which club to use on his next shot, Andy knows he hits an average of 135–145 yards with a five iron; with a four iron, he hits 145–155 yards; and with a three iron, he hits 155–165 yards. Which of these clubs would be his best choice?

5 A field is in the shape of quadrilateral *ABCD* as shown below.

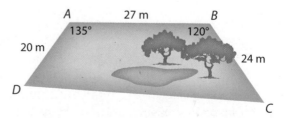

a. Find the length *CD* of its fourth side.

b. Find the measures of the remaining angles of the field.

c. Find the area of the field.

6 A side view of a reclining lawn chair is illustrated below. Its function is based on a triangle in which the lengths of two sides are fixed, and the length of the third side can vary. The key triangular component (△*ABC*) in the design of the lawn chair is shown in the diagram at the right. In △*ABC*, *AB* and *BC* are both 10 inches, and *AC* can be set to 8, 10, 12, or 14 inches. The reclining angle for someone sitting in the chair is ∠*BCD*.

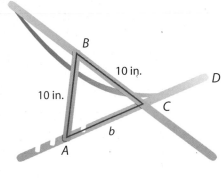

a. When the length *AC*, denoted *b*, is set, the lengths of all three sides of △*ABC* are fixed. How can you find m∠*ACB*? How is m∠*BCD*, the measure of the reclining angle, related to m∠*ACB*?

b. Write a rule that expresses m∠*BCD* in terms of *b*.

c. Use the rule in Part b to find the measure of the reclining angle for *b* = 8, 10, 12, and 14 inches. Describe how the measure of the reclining angle changes as *b* changes.

7 In the fall of 1986, three friends challenged each other to a pumpkin-throwing contest. The winner threw a pumpkin 126 feet. This event has since evolved into the annual Punkin' Chunkin' World Championship in which air compression cannons and catapult machines hurl pumpkins thousands of feet.

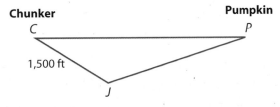

To measure the length of a throw, two judges and a spotter are involved. One judge is located at point *C* where the chunker releases the pumpkin. The second judge is on the right edge of the playing field at point *J*, a distance of 1,500 feet from point *C*. After a throw, the spotter stands at the location *P* at which the pumpkin landed. The two judges use transits to measure ∠*C* and ∠*J*, respectively. This information is then used to derive the length *CP* of the throw.

a. In 2004, Old Glory set a new world Punkin' Chunkin' record in the Adult Air Cannons category. To measure the record throw, the judges measured ∠*C* at 30.0° and ∠*J* at 135.6°. What was the distance *CP* of the record throw?

   b. Another world record was set in 2004, this by Hypertension in the Adult Catapult category. Assume for this record throw of 2,112 feet that m∠C was 28.6°.

   i. What was the distance *JP* from the position of the judge at *J* to the position of the spotter at *P* where the pumpkin landed?

   ii. What was the measure of ∠*J*?

   c. Suppose in a pumpkin throw, m∠C = 33.8° and the distance *JP* is 1,435 feet. Explain why this is not enough information to uniquely determine the length *CP* of the throw.

8 Suppose in △*ABC*, *a*, *b*, and *c* are the lengths of the sides opposites angles *A*, *B*, and *C*, respectively. Find the measures of the three unknown parts of △*ABC* to the nearest tenth, given each of the following sets of conditions.

   a. $c = 5.4$ m, m∠A = 58.0°, m∠B = 42.8°

   b. $a = 6.9$ m, m∠A = 47.2°, m∠C = 110.5°

   c. $a = 12.9$ m, $b = 16.3$ m, $c = 8.8$ m

   d. $b = 9.7$ m, $c = 10.2$ m, m∠C = 67.4°

Connections

9 To use the usual formula to find the area of a triangle, you need to know the length of one side and the altitude to that base. If you are given the lengths of two sides and the measure of the included angle, you can use the Law of Sines to find the area of the triangle.

   a. Use the three diagrams below to show that in every case, area of △*ABC* = $\frac{1}{2} bc \sin A$.

   Case I   **Case II**   **Case III**

   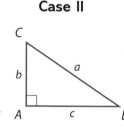   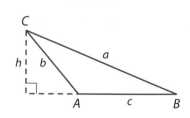

   b. Explain as precisely as you can why it is also true that area of △*ABC* = $\frac{1}{2} ac \sin B = \frac{1}{2} ab \sin C$.

   c. Determine the areas of triangles with the following side lengths and angle measures:

   i. m∠A = 73°, $b = 12$, $c = 17$

   ii. m∠A = 118°, $b = 21$, $c = 18$

10 In parallelogram *ABCD* below, information is given about one side and two angles formed by that side and the diagonals $\overline{AC}$ and $\overline{BD}$.

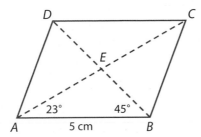

a. Verify that a parallelogram has 180° symmetry by tracing parallelogram *ABCD* on a sheet of paper and rotating the paper about point *E*.

b. Use facts about triangles and the rotational symmetry property of parallelograms to find the measures of as many of the other 7 segments and 10 angles in the given figure as you can. Do *not* use trigonometry.

c. Use the Law of Sines to find further information about the segments and angles in the figure.

11 Use your calculator or computer software to graphically examine the behavior of the sine and cosine functions for angle measures between 0° and 360°. Be sure it is set in degree mode.

a. Set your graphing window to Xmin = 0, Xmax = 360, Xscl = 45, Ymin = −1.5, Ymax = 1.5, and Yscl = 0.5. Then graph $y = \sin x$ and $y = \cos x$.

  i. Describe the pattern of change for each function.

  ii. How do these graphs compare with those of linear, exponential, and quadratic functions?

  iii. For what values of *x* does $\sin x = \cos x$?

b. Use the definitions of the trigonometric functions to explain why your descriptions of the patterns of change for the sine and cosine functions make sense.

12 Surveyors use triangulation and the methods in this lesson to determine the distance between two inaccessible points such as *P* and *Q* in the diagram. They begin by laying off a *base line* $\overline{AB}$ that can be accurately measured. Using a transit, they then measure the four angles at points *A* and *B* formed by $\overline{AB}$ and the diagonals $\overline{PB}$ and $\overline{QA}$. How can they use this information to determine *PQ*?

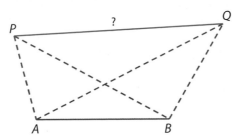

13 Find the lengths of the diagonal braces of kite *ABCD*.

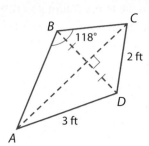

14 Make a model of a triangle *ABC* with a side whose length varies as the fixed length pieces *AB* and *BC* pivot about points *B* and *A* as shown below. Make *AB* = 10 cm and *BC* = 16 cm long (from endhole to endhole).

  a. Make strip $\overline{AD}$ with holes 2 cm apart (or draw a segment, $\overline{AD}$, on your paper and carefully mark points 2 cm apart).

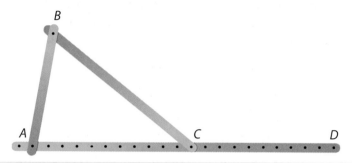

  i. What is the length needed for strip $\overline{AD}$ that would allow *C* to be extended so that *A*, *B*, and *C* are collinear?

  ii. What is the minimum length *AD* in this model?

  b. When you change the length *AC* in △*ABC*, what else changes?

  c. Adjust the length *AC* in 2-cm step sizes from its minimum to maximum possible values. At each step, use a ruler and protractor to obtain and record the length *AC* and m∠*A*.

  d. Make a scatterplot of the data pairs (*AC*, m∠*A*). Investigate whether a linear, exponential, or quadratic function would reasonably fit the pattern in the data. Do any of these appear to be a good fit for the data? Explain.

  e. Use the Law of Cosines to write an equation that gives m∠*A* as a function of *AC*. Graph the function over the interval from the minimum to the maximum length *AC*.

  f. How well does the graph of your function from Part e fit your measurement data? Is the symbolic form of your function rule that of a linear, exponential, quadratic, or some other kind of function?

15 In addition to examining the geometry of the SSA condition as you
did in Investigation 3, it is revealing to consider the condition from
trigonometric and algebraic points of view.

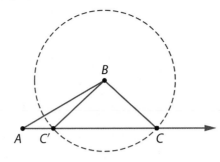

 a. Apply the Law of Sines to the
two triangles ABC and ABC' in
the diagram to the right.

  i. What equation relates $\angle A$,
AB, BC, and $\angle ACB$ in $\triangle ABC$?

  ii. What equation relates $\angle A$,
AB, BC', and $\angle AC'B$ in
$\triangle ABC'$?

  iii. How can both equations
be true when $BC = BC'$ but $m\angle ACB \neq m\angle AC'B$?

 b. Now suppose in $\triangle ABC$, $a = 1$, $c = \sqrt{3}$, and $m\angle A = 30°$.

  i. Using the Law of Cosines, write and then solve a quadratic
equation that relates the unknown side length AC and the
three given measures.

  ii. Sketch the triangle or triangles that are determined.

 c. Suppose in $\triangle ABC$, $a = 8$, $b = 10$, and $m\angle B = 120°$.

  i. Determine the number of possible triangles ABC—none, one,
or two—that can satisfy these side and angle conditions.

  ii. Find the length of the third side of any triangles that were
determined in part i.

Reflections

16 The Law of Sines for a triangle ABC can be stated in two
similar forms:

$$
\underset{\text{I}}{\frac{\sin A}{a} = \frac{\sin B}{b} = \frac{\sin C}{c}} \quad \text{and} \quad \underset{\text{II}}{\frac{a}{\sin A} = \frac{b}{\sin B} = \frac{c}{\sin C}}
$$

 a. Why are these two forms equivalent?

 b. When is form I easier to use? When is form II easier to use?

17 Bradford Washburn, the leader of a group of surveyors that remapped
the Grand Canyon in the 1970s, said that the work was "like mapping
a mountain upside down." What do you think he meant by that?

18 Models of four problem situations are shown below. For each case, describe the trigonometric method that you would use to determine the indicated length or angle measure. Would you expect one, two, or no solutions? You do not have to perform the calculations.

a.
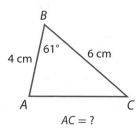
AC = ?

b.
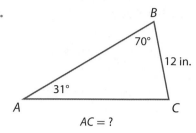
AC = ?

c.
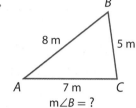
m∠B = ?

d.
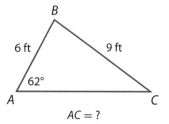
AC = ?

19 Look back at the statements of the Law of Sines (page 491) and the Law of Cosines (page 494). Write each law completely in words using no letter symbols for angles or sides of triangles.

20 The length of two sides *a* and *b* and the measure of ∠A opposite side *a* of a possible triangle *ABC* are given. Summarize in your own words the conditions under which (a) no triangle, (b) exactly one triangle, or (c) two triangles are determined.

Extensions

21 In Investigation 1, you examined a derivation of the Law of Sines for an acute triangle. In this task, you will derive the Law of Sines for the case of a right triangle. Use the fact that for right triangle *ABC* labeled as shown, sin *C* = sin 90° = 1.

a. Express sin *A* and sin *B* in terms of the sides *a*, *b*, and *c*.

b. Explain why $c = \dfrac{c}{\sin C}$.

c. Solve each equation in Part a for *c*. Use those results and your result in Part b to derive the Law of Sines.

d. What are the values of the common ratios in the formulation of the Law of Sines for △*ABC*?

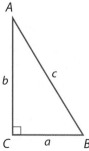

22 In this task, you will derive the Law of Sines for an obtuse triangle. Diagrams I and II below show the same triangle ABC, but the triangle is positioned differently.

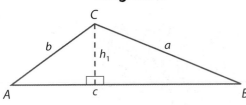

Diagram I

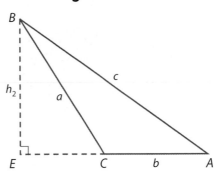

Diagram II

a. In Diagram I, the altitude from vertex C intersects side $\overline{AB}$.

   i. Write two expressions for h_1, one in terms of $\sin A$ and the other in terms of $\sin B$.

   ii. Use your expressions in part i to show that $\dfrac{\sin A}{a} = \dfrac{\sin B}{b}$.

b. In Diagram II, the altitude from vertex B to opposite side $\overline{AC}$ falls outside the triangle.

   i. With reference to right triangle ABE, write an expression for h_2 in terms of $\sin A$.

   ii. With reference to right triangle BCE, write an expression for h_2 in terms of $\sin (180 - m\angle C)°$.

   iii. Explain why the latter ratio is also equal to $\sin C$. (The figure in Connections Task 9 Part c (page 478) may be helpful.)

   iv. Use your expressions in parts i–iii to show that $\dfrac{\sin A}{a} = \dfrac{\sin C}{c}$ or $\dfrac{a}{\sin A} = \dfrac{c}{\sin C}$.

c. Combine your results in Parts a and b to complete the derivation of the Law of Sines for an obtuse triangle.

23 To find the area of a triangle using the usual formula, you need to know the length of one side and the length of the altitude to that side.

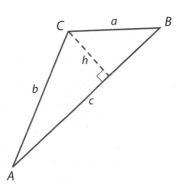

a. Suppose you know the lengths of the three sides a, b, and c. Explain how to calculate the height h.

b. How is your result in Part a related to the area formula in Connections Task 9?

24 You examined some of the geometric properties of one type of automobile jack in the Think About This Situation for Lesson 1 and in Connections Task 14 (page 480). Another kind of jack is used on racetracks and by service stations to perform quick tire changes. The jack has wheels so it can be easily moved around.

In the diagram of the automobile jack below, bar *DC* has one end *C* that can be moved between points *A* and *E* by turning a threaded rod at the back of the jack. The length *AE* is 24 inches, and *AB*, *BC*, and *BD* are each 12 inches.

a. If this jack is to be safe to use, what path must point *D* follow as point *C* moves toward point *A*?

b. What kind of triangles are $\triangle ABC$ and $\triangle ABD$?

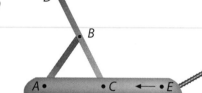

c. Use properties of isosceles triangles to explain why $m\angle CAD = 90°$ for any positive value of *CE* less than 24 inches.

d. Determine the height *AD* of the jack when *CE* = 6 inches. When *CE* = 13 inches.

25 Derivations of the Law of Cosines differ depending on whether the angle under consideration is acute or obtuse. In $\triangle ABC$ below, $\angle B$ is an obtuse angle and *h* is the length of the altitude from vertex *C*.

a. Explain why each step in the following derivation is correct. The reason for Step 5 is that $\cos(180° - \theta) = -\cos\theta$ as suggested by the figure in Connections Task 9, Part c (page 478).

(1) In right triangle *BCE*,
$h^2 = a^2 - x^2$.

(2) In right triangle *ACE*,
$h^2 = b^2 - (c + x)^2$.

(3) $b^2 - c^2 - 2cx - x^2 = a^2 - x^2$

(4) $b^2 = a^2 + c^2 + 2cx$

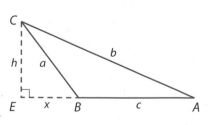

(5) In right triangle *BCE*,
$x = a\cos(m\angle CBE) = -a\cos B$.

(6) $b^2 = a^2 + c^2 - 2ac\cos B$

b. For the case of an acute angle, draw a new triangle ABC with $\angle B$ acute, and draw the altitude from vertex C to side $\overline{AB}$. Length c will then be divided into two parts of lengths x and $c - x$. The rest of the derivation of the Law of Cosines is similar to the obtuse case. Try to construct the argument.

c. Why is no additional proof needed for the case of $m\angle B = 90°$?

26 In $\triangle ABC$, ray BD bisects $\angle ABC$. Apply the Law of Sines to triangles ABD and CBD to show that the angle bisector of a triangle divides the opposite side into segments proportional in length to the adjacent sides. In symbols, show that $\dfrac{AD}{DC} = \dfrac{AB}{BC}$.

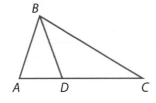

27 A carnival ride consists of six small airplanes attached to a vertical pole. As the pole rotates, the planes fly around the pole.

A rider can control the height of the plane by changing the length of the hydraulic cylinder $\overline{AC}$ attached at point C in the diagram below. In a typical design, $BD = 4$ m, $BA = 1.5$ m, $BC = 1.5$ m, $BE = 1.5$ m, and AC can vary between 1.5 and 2.2 m.

a. What is the smallest measure of $\angle ABC$? How far above the ground is the plane for that smallest angle?

b. If the hydraulic cylinder is fully extended, what is the measure of $\angle ABC$?

c. What is the lowest and what is the highest point of the plane above the ground?

d. How long should the hydraulic cylinder be so that the plane will fly 2.5 meters above the ground?

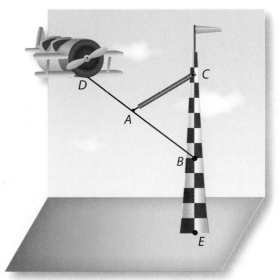

28 Suppose the length, width, and height of a rectangular prism are all different.

   a. How many different lengths of edges does the prism have?

   b. How many different shapes of faces does it have?

   c. How many different lengths of diagonals does the prism have?

29 Solve each system of equations without using the graphing or table capabilities of your calculator or computer software.

   a. $6x + 4y = 58$
      $x - 2y = 7$

   b. $y = x^2 - 3x$
      $y = 2x + 6$

   c. $y = 2x + 1$
      $y = \dfrac{-5}{x - 3}$

30 Consider matrices A and B as given below. Evaluate each expression without the use of technology.

$$A = \begin{bmatrix} -2 & 3 \\ 0 & 5 \end{bmatrix} \qquad B = \begin{bmatrix} 4 & -6 \\ 1 & -2 \end{bmatrix}$$

   a. $A + 2B$ **b.** $B - A$

   c. $A \times B$ **d.** B^2

31 On a coordinate grid, make a sketch of the triangle with vertices at $A(-2, -1)$, $B(0, 3)$, and $C(8, -1)$.

   a. Verify that $\triangle ABC$ is a right triangle.

   b. Find the perimeter of the triangle.

   c. Find the coordinates of the midpoint of the hypotenuse of $\triangle ABC$. Does the line segment that joins vertex B to the midpoint of the hypotenuse divide $\triangle ABC$ into two triangles with equal areas? Explain your reasoning.

32 Write each of the following in equivalent factored form.

   a. $12x^2 + 2x$

   b. $x^2 - 12x + 36$

   c. $x^2 - 81$

   d. $2x^2 + 20x + 48$

   e. $x^2 - 2x - 35$

   f. $2x^2 + 11x - 21$

33 Write each expression in the equivalent $ax^2 + bx + c$ form.

  a. $(x + 5)^2$

  b. $3x(x - 6) + 8(2 - 4x)$

  c. $3(2x - 10)(x + 2)$

  d. $\left(x - \frac{1}{2}\right)\left(x + \frac{1}{2}\right)$

  e. $(x + 5)(x + 4)$

34 Suppose you roll two dice and subtract the smaller number from the larger. Make a chart that shows all of the possible outcomes.

  a. What is the probability that the difference is greater than 4?

  b. What is the probability that the difference is less than or equal to 4?

  c. Explain why the probabilities you found in Parts a and b should add to 1.

  d. What is the probability that you will roll doubles and get a difference of zero?

35 Suppose that Alejandro bought a new car that cost $25,595, and the value of the car depreciates by 12% each year.

  a. What will the car be worth after two years?

  b. Write a rule that could be used to find the value V of the car for any number of years t in the future.

  c. In what year will the car be worth only half of what Alejandro originally paid for it?

36 Draw a vertex-edge graph that fits each description.

  a. Has an Euler circuit but does not have a Hamilton circuit

  b. Has a Hamilton circuit but does not have an Euler circuit

  c. Requires at least four colors to color the vertices of the graph

Looking Back

I n this unit, you extended your work with coordinates to describe angles in standard position in a coordinate plane. The coordinates of points on the terminal side of an angle in standard position were used to define a new family of functions, the trigonometric functions: sine, cosine, and tangent. These functions provide important connections between angle measure and linear measure.

When interpreted as ratios of side lengths of a right triangle, the trigonometric functions have many important applications. You also learned to use trigonometric methods, including use of the Law of Sines and the Law of Cosines for any triangle, to solve problems such as indirect measurement of inaccessible distances or angles, measurement of polygonal shapes using triangulation, and modeling mechanisms with triangles in which the measures of two parts remain fixed while the length of a side varies.

The following tasks will help you review, pull together, and apply what you have learned as you solve new problems.

1 In the 19th and early 20th centuries, water wheels were commonly used to operate grist mills that ground wheat and corn for flour. The Wayside Inn Grist Mill, Sudbury, Massachusetts, pictured below, first ground corn in 1929. Power for the grinding was provided by a water wheel driven by flowing water.

In the diagram below, the 30-foot diameter water wheel has been positioned on a coordinate system so the center of the wheel is the origin and the wheel reaches 3 feet below water level. As the wheel rotates counterclockwise, it takes 24 seconds for point *A* on the circumference of the wheel to rotate back to its starting position.

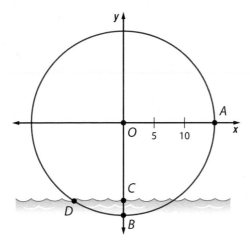

a. At its current position, how far is point *A* from the surface of the water? What are the coordinates of point *A*?

b. What are the coordinates of points *B*, *C*, and *D*?

c. Determine the distance from point *A* to the water surface as the water wheel rotates counterclockwise through each angle.

    i. 45°

    ii. 75°

   iii. 90°

   iv. 150°

d. Through what angle does point *A* rotate to first reach the water at point *D*? How many seconds does it take point *A* to first reach the water?

2 At places along the south bank of the Platte River, the river is substantially below ground level. From the edge of the water at one location, the land slopes upward with an angle of inclination of 63°. The best way to get to the water is to go directly down the slope to the water's edge, a distance of approximately 123 meters.

   a. Make a sketch of this situation showing distances and angles.

   b. How far, horizontally, is the edge of the water from the edge of the bank, just where the land begins to slope downward?

   c. How far above the surface of the Platte River is the land on the south bank?

3 In route to sea, a freighter travels 50 km due west of its home port. It then turns, making an angle of 132° with its former path. It travels 80 km before radioing its home port.

   a. Draw a diagram showing the path of the freighter.

   b. How far is the freighter from its home port?

   c. If sea conditions permitted, through what angle could the freighter have turned from its original course to go directly from its home port to the position at which it radioed the port?

   d. Describe a second way in which you could find an answer for Part c.

4 Architects sometimes use drafting tables that can be tilted forward, as shown in the photo and sketch below.

a. Explain how the structure of the table allows the top to be tilted at different angles.

b. Suppose in the sketch, $AB = 10$ inches and $AC = 7$ inches. When the tabletop is horizontal, what is the measure of $\angle ABC$? What is the length of the adjustable side BC?

c. What is the measure of $\angle ABC$ when the tabletop is tilted downward at a 30° angle with the horizontal? What is the length BC?

5 A lighthouse 30 meters high stands at the top of a cliff. The angle of elevation from a point X on the bow of a ship to point A at the top of the lighthouse is 18°. The angle of elevation from X to point B at the bottom of the lighthouse is 14°.

a. Draw a sketch of the situation and then determine to the nearest meter the height of the cliff (above the level of point X).

b. Determine to the nearest meter the horizontal distance from the lighthouse to point X.

Summarize
the Mathematics

In this unit, you extended your understanding of coordinate methods to include trigonometric methods that relate angle measures and linear measures. These methods are especially useful when solving problems involving triangulation and indirect measurement. In the process, you also extended your toolkit of basic functions to include the sine, cosine, and tangent functions for angles ranging from 0° to 360°.

a Suppose $P(a, b)$ is a point on the terminal side of an angle in standard position with measure θ.

   i. How can you find the value of $\sin \theta$? Of $\cos \theta$? Of $\tan \theta$?

   ii. How can you find the measure of θ?

b Imagine rotating a ray $\overrightarrow{OP}$ counterclockwise with initial position along the positive branch of the x-axis. As the measure θ of the angle formed increases from 0° to 360°, describe the pattern of change in:

   i. $\cos \theta$ **ii.** $\sin \theta$ **iii.** $\tan \theta$

   How are these patterns of change similar to, and different from, those of other functions you have previously studied?

c How are the definitions of the sine, cosine, and tangent of an acute angle of a right triangle related to the more general definitions of the functions in terms of an angle in standard position?

   i. How can these ratios be used to determine lengths in a right triangle that cannot be measured directly?

   ii. How can these ratios be used to determine acute angle measures that cannot be measured directly?

d Suppose you had modeled a problem with a triangle. What clues do you look for when deciding whether to use the Law of Sines? The Law of Cosines?

e Suppose for an angle A in a triangle, you know:

   i. $\sin A = p$ **ii.** $\cos A = q$ **iii.** $\tan A = r$

   In each case, can you find exactly one value for $m\angle A$? If so, how? If not, why not?

Be prepared to explain your ideas to the class.

✓Check Your Understanding

Write, in outline form, a summary of the important mathematical concepts and methods developed in this unit. Organize your summary so that it can be used as a quick reference in future units and courses.

PROBABILITY DISTRIBUTIONS

From lotteries, to weather, to disease, to your genetic make-up, to meeting the love of your life, you live in a world governed by chance. In the Course 1 unit, *Patterns in Chance*, you learned to use the Addition Rule to find the probability that event *A* happens *or* event *B* happens. In this unit, you will learn to use the Multiplication Rule to find the probability that event *A* happens *and* event *B* happens. In *Patterns in Chance*, you learned to use simulation to approximate probability distributions, such as the waiting-time distribution for the number of flips of a coin needed to get a head. In this unit, you will learn to use the Addition and Multiplication Rules to construct waiting-time distributions exactly.

Key ideas will be developed through your work on problems in three lessons.

Lessons

1 Probability Models

Use an area model to find the probability that two independent events both happen, find probabilities using the definition of conditional probability, decide whether two events are independent, use the Multiplication Rule for independent events and for dependent events.

2 Expected Value

Understand how the word "expect" is used in probability, compute the mean of a relative frequency distribution, compute the expected value of a probability distribution.

3 The Waiting-Time Distribution

Construct and analyze frequency distributions for simulations and relative frequency distributions for waiting-time situations, construct and analyze the (exact) probability distribution for a waiting-time situation, and discover the formula for the expected value of a waiting-time distribution.

Probability Models

Some physical characteristics, such as freckles, eyelash length, and the ability to roll one's tongue up from the sides, are determined in a relatively simple manner by genes inherited from one's parents. Each person has two genes that determine whether or not he or she will have freckles, one inherited from the father and one from the mother. If a child inherits a "freckles" gene from either parent or from both parents, the child has freckles. In order not to have freckles, the child must inherit a "no-freckles" gene from both parents. This explains why the gene for freckles is called *dominant,* and the gene for no freckles is called *recessive.*

A parent with two freckles genes must pass on a freckles gene to the child; a parent with two no-freckles genes must pass on a no-freckles gene. If a parent has one of each, the probability is $\frac{1}{2}$ that he or she will pass on the freckles gene, and the probability is $\frac{1}{2}$ that he or she will pass on the no-freckles gene.

Think About This Situation

Consider the chance of inheriting freckles.

a In what sense does the gene for freckles "dominate" the gene for no freckles?

b What is the probability that a child will have freckles if both parents do not have freckles?

c What is the probability that a child will not have freckles if each parent has one freckles gene and one no-freckles gene?

d Can you determine the probability that a child of freckled parents also will have freckles?

As you work on the investigations of this lesson, you will learn how to find the probability that two (or more) events, such as inheriting the freckle gene from the mother and inheriting the freckle gene from the father, both happen. In the process, you will discover the importance of distinguishing between events that are *independent* and events that are *mutually exclusive (disjoint)*.

Investigation 1 — The Multiplication Rule for Independent Events

You have found that graphical representations of data or quantitative relationships can reveal important underlying patterns and that making a "picture" of a mathematical situation can often help you understand that situation better. In Course 1 of *Core-Plus Mathematics,* you used an area model to explore patterns in chance situations. As you work on the problems of this investigation, look for answers to these questions:

How can you use an area model to find the probability that two events both happen?

How can you calculate that probability using the individual probabilities?

1 About half of all U.S. adults are female. According to a survey published in *USA Today*, three out of five adults sing in the shower.

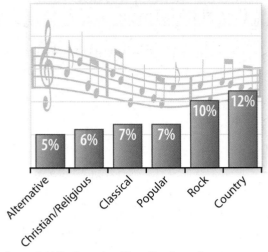

Singing in the Shower
Three out of five adults say they sing in the shower. Top types of showering music:

5% — Alternative
6% — Christian/Religious
7% — Classical
7% — Popular
10% — Rock
12% — Country

Source: Guideline Research and Consulting Corporation for Westin Hotels and Resorts

a. Suppose an adult from the United States is selected at random. From the information above, do you think that the probability that the person is a female *and* sings in the shower is equal to $\frac{3}{5}$, greater than $\frac{3}{5}$, or less than $\frac{3}{5}$?

b. Now examine the situation using the area model shown below.

   i. Explain why there are two rows labeled "No" for "Sings in Shower" and three labeled "Yes."

   ii. What assumption does this model make about singing habits of males and females?

c. On a copy of this area model, shade in the region that represents the event: *female* and *sings in the shower*.

d. What is the probability that an adult selected at random is a female and sings in the shower?

e. What is the probability that an adult selected at random is a male and does not sing in the shower?

Gender

	Male	Female
Yes		
Yes		
Yes		
No		
No		

Sings in Shower

2 Consider this problem: What is the probability that it takes exactly two rolls of a pair of dice before getting doubles for the first time?

a. Explain why it makes sense to label the rows of the area model as shown below. On a copy of this area model, label the six columns to represent the possible outcomes on the second roll of a pair of dice.

Rolling a Pair of Dice Twice

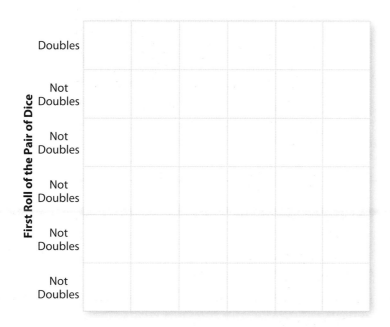

Second Roll of the Pair of Dice

First Roll of the Pair of Dice

Doubles

Not Doubles

Not Doubles

Not Doubles

Not Doubles

Not Doubles

b. On your copy of the area model, shade the squares that represent the event *not getting doubles on the first roll* and *getting doubles on the second roll.*

c. What is the probability of not getting doubles on the first roll and then getting doubles on the second roll?

d. Use your area model to find the probability that you will get doubles both times.

e. Use your area model to find the probability that you will not get doubles either time.

3 Make an area model to help you determine the probabilities that a child will or will not have freckles when each parent has one freckles gene and one no-freckles gene.

a. What is the probability that the child will not have freckles?

b. Compare your answer to Part a with your class' answer to Part c of the Think About This Situation on page 523.

4 Use area models to answer these questions:

   a. About 25% of Americans put catsup directly on their fries rather than on the plate. What is the best estimate for the probability that both your school principal and your favorite celebrity put catsup directly on their fries?

   b. About 84% of Americans pour shampoo into their hand rather than directly onto their hair. What is the best estimate of the probability that both your teacher and the President of the United States pour shampoo into their hand before putting it on their hair?

5 Look back at the situations described in Problems 1 through 4. The pairs of events in each of those problems are (or were assumed to be) **independent events**: knowing whether one of the events occurs does not change the probability that the other event occurs.

   a. For each situation, explain whether you think the assumption of independence is realistic. If independence is not a reasonable assumption, how would you change the model?

   b. Describe how you could compute the probability in each situation without making an area model (if independence is a reasonable assumption).

   c. Suppose A and B are independent events. Express your method in Part c by completing the following rule using symbols:

$$P(A \text{ and } B) = \underline{\hspace{4cm}}$$

   The notation $P(A \text{ and } B)$ is read "the probability of event A and event B."

   d. Compare your rule with those of your classmates and resolve any differences.

   e. State your agreed-upon **Multiplication Rule** in words.

Often a probability problem is easier to understand if it is written in words that are more specific than the words the original problem uses. For example, consider the problem:

What is the probability of taking exactly two tries to roll doubles?

You could express and calculate this probability in the following manner:

P(don't roll doubles on the first try and do roll doubles on the second try)
= P(don't roll doubles on the first try) · P(do roll doubles on the second try)
$$= \left(\frac{5}{6}\right)\left(\frac{1}{6}\right) = \frac{5}{36}$$

6 Suppose Shiomo is playing a game in which he needs to roll a pair of dice and get doubles and then immediately roll the dice again and get a sum of six. He wants to know the probability that this will happen.

a. Which of the following *best* describes the probability Shiomo wants to find?

   Option 1: *P(gets doubles on the first roll* or *gets a sum of six on the second roll)*

   Option 2: *P(gets doubles on the first roll* and *gets a sum of six on the second roll)*

   Option 3: *P(gets doubles* and *a sum of six)*

b. Explain why the Multiplication Rule can be used to find the probability that this sequence of two events will happen. What is the probability?

7 A modification of the game in Problem 6 involves rolling a pair of dice three times. In this modified game, Shiomo needs to roll doubles, then a sum of six, and then a sum of eleven.

a. Find the probability that this sequence of three events will happen.

b. Suppose A, B, and C are three independent events. Write a rule for calculating $P(A$ and B and $C)$ using the probabilities of each individual event.

c. Write the Multiplication Rule for calculating the probability that each of four independent events occurs.

8 For each of the following questions, explain whether it is reasonable to assume that the events are independent. Then, if it applies, use the Multiplication Rule to answer the question.

a. What is the probability that a sequence of seven flips of a fair coin turns out to be exactly HTHTTHH?

b. What is the probability that a sequence of seven flips of a fair coin turns out to be exactly TTTTTTH?

c. According to the National Center for Education Statistics, 27.9% of public school students live in a small town or rural area. (Source: nces.ed.gov/pubs2006/2006307.pdf) If you select 5 students at random, what is the probability that they all live in a small town or rural area?

d. According to the National Center for Education Statistics, the percentage of students who are homeschooled in the United States is 2.2 percent. (Source: nces.ed.gov/pubs2006/homeschool/) If you pick 10 students at random in the United States, what is the probability that none of the 10 are homeschooled?

e. Refer to Part d. You pick a family with two school-age children at random in the United States. What is the probability that both children are homeschooled?

Summarize
the Mathematics

In this investigation, you learned how to find the probability that an event A and an event B both happen, when the two events are independent.

a How can you use an area model to find the probability that an event A and an event B both happen?

b Why does it make sense to multiply the individual probabilities when you want to find the probability that two independent events both happen?

c Explain which event has the higher probability: You roll a pair of dice twice and get a sum of 7 both times. You roll a pair of dice three times and get a sum of 7 all three times.

Be prepared to share your ideas and reasoning with the class.

✓ Check Your Understanding

While playing a board game, Jenny is sent to jail. To get out of jail, she needs to roll doubles. She wants to know the probability that she will fail to roll doubles in three tries.

a. Rewrite Jenny's situation using $P(\underline{\quad})$ notation, describing the sequence of events.

b. Find the probability that Jenny fails to roll doubles in three tries.

c. Explain why you can use the Multiplication Rule for this situation.

Investigation 2 — Conditional Probability

Sometimes you are interested in the probability of an event occurring when you know another event occurs. For example, a high school athlete might be interested in knowing the probability of playing professional basketball if he or she first plays basketball at the college level. As you work on the following problems, keep in mind this basic question:

How can you find probabilities in situations with conditions?

Some boys wear sneakers and some do not. The same holds true for girls. However, in many places in the United States, boys are more likely to wear sneakers to school than are girls.

1 Count the number of students in your classroom who are wearing sneakers. Count the number of girls. Count the number of students who are wearing sneakers and are girls. Record the number of students who fall into each category in a copy of the following table.

	Wearing Sneakers	Not Wearing Sneakers	Total
Boy			
Girl			
Total			

a. Suppose you select a student at random from your class. What is the probability that the student is wearing sneakers?

b. Suppose you select a student at random from your class. What is the probability that the student is a girl?

c. Does the Multiplication Rule from Investigation 1 correctly compute the probability that the student is wearing sneakers and is a girl?

d. How is this situation different from previous situations in which the Multiplication Rule gave the correct probability?

2 The phrase *"the probability event A occurs given that event B occurs"* is written symbolically as $P(A \mid B)$. This **conditional probability** sometimes is read as "the probability of A given B." The table below categorizes the preferences of 300 students in a junior class about plans for their prom.

		Preference for Location	
		Hotel	Rec Center
Preference for Band	Hip-Hop	73	80
	Classic Rock	55	92

Suppose you pick a student at random from this class. Find each of the following probabilities.

a. *P(prefers hotel)*

b. *P(prefers hip-hop band)*

c. *P(prefers hotel and prefers hip-hop band)*

d. *P(prefers hotel or prefers hip-hop band)*

e. *P(prefers hotel | prefers hip-hop band)*

f. *P(prefers hip-hop band | prefers hotel)*

③ Recall that events A and B are **independent** if knowing whether one of the events occurs does not change the probability that the other event occurs.

  a. Using the data from Problem 1, suppose you pick a student at random. Find $P(wearing\ sneakers \mid is\ a\ girl)$. How does this compare to $P(wearing\ sneakers)$?

  b. Are the events *wearing sneakers* and *is a girl* independent? Why or why not?

  c. Consider this table from a different class.

	Wearing Sneakers	Not Wearing Sneakers
Boy	5	9
Girl	10	18

  Suppose you pick a student at random from this class.

  i. Find $P(wearing\ sneakers)$.

  ii. Find $P(wearing\ sneakers \mid is\ a\ girl)$.

  iii. Are the events *wearing sneakers* and *is a girl* independent? Why or why not?

  d. If events A and B are independent, how are $P(A)$ and $P(A \mid B)$ related?

④ Suppose that you roll a pair of dice.

  a. Which is greater? $P(doubles)$ or $P(doubles \mid sum\ is\ 2)$?

  b. Are the events *getting doubles* and *getting a sum of 2* independent? If not, how would you describe the relationship?

⑤ Refer to the table in Problem 2.

  a. If you select a junior at random, are the events *prefers hotel* and *prefers hip-hop band* independent? Explain.

  b. Recall from your work in Course 1 that two events are **mutually exclusive** if they cannot both occur on the same outcome. If you select a junior at random, are the events *prefers hotel* and *prefers hip-hop band* mutually exclusive? Explain.

  c. Are the events *prefers hotel* and *prefers rec center* mutually exclusive? Explain.

⑥ Suppose you pick a high school student at random. For each of the pairs of events in Parts a, b, and c, write the mathematical equality or inequality that applies:

$$P(A) = P(A \mid B), \quad P(A) > P(A \mid B), \quad \text{or} \quad P(A) < P(A \mid B).$$

  a. A is the event that the student is male, and B is the event that the student is over six feet tall.

  b. A is the event that the student is female, and B is the event that the student has brown eyes.

c. *A* is the event that the student is a member of the French club, and *B* is the event that the student is taking a French class.

d. Which of the pairs of events is it safe to assume are independent? Explain your reasoning.

Summarize
the Mathematics

In this investigation, you learned how to find conditional probabilities.

a What is the difference between $P(A)$ and $P(A|B)$?

b Use the two symbolic expressions in Part a to write a definition (in if-and-only-if form) that tells when events *A* and *B* are independent.

c Suppose you ask 30 juniors and 70 seniors whether they studied the previous night. Make three possible tables where the events *being a junior* and *studied last night* are independent events in this group of 100 students.

Be prepared to share your ideas and examples with the class.

✓ Check Your Understanding

A survey of 505 teens by the American Academy of Dermatology included 254 boys and 251 girls. Thirty-three percent of the boys said they wear sunscreen, and 53% of the girls said they wear sunscreen.

a. Fill in a copy of the following table, showing the number of teenagers who fell into each category.

	Boy	Girl	Total
Wear Sunscreen			
Don't Wear Sunscreen			
Total			505

Source: www.aad.org/public/News/NewsReleases/Press+Release+Archives/Skin+Cancer+and+Sun+Safety/Teen+Survey+Results.htm

b. Suppose you pick one student at random from these 505 teens. Find the probability of each of the following events.

  i. $P(\textit{wears sunscreen})$

  ii. $P(\textit{is a boy})$

  iii. $P(\textit{wears sunscreen and is a boy})$

  iv. $P(\textit{wears sunscreen or is a boy})$

  v. $P(\textit{wears sunscreen} \mid \textit{is a boy})$

  vi. $P(\textit{is a boy} \mid \textit{wears sunscreen})$

c. Are being a boy and wearing sunscreen independent events? How could you tell this from the information given about the survey? How could you tell this from your table?

The Multiplication Rule When Events Are Not Independent

In Investigation 1, you discovered a rule for calculating the probability that two *independent* events, A and B, both happen. As you saw in Investigation 2, sometimes the occurrence of one event influences the probability that another event occurs—that is, the events are *not* independent. As you work on the problems of this investigation, look for an answer to this question:

> *How do you find P(A and B) when A and B are not independent?*

1 About half of all U.S. adults are male. The *USA Today*-reported data in Investigation 1 indicate that three out of five adults sing in the shower. Some people think that males are more likely to sing in the shower than are females. Suppose that they are right and 80% of males sing in the shower, but only 40% of females sing in the shower.

   a. Make an area model that represents this situation.

   b. Suppose you pick an adult at random. What is the probability that you get a female who sings in the shower?

   c. Complete the following equation (in words) that describes how you found the probability in Part b.

   $$P(\textit{female and sings in shower}) =$$

   d. Suppose you pick an adult at random. What is the probability that you get a male who sings in the shower?

   e. Write an equation in words that describes how you found the probability in Part d.

2 As you have seen in Problem 1, if events A and B are not independent, you can find P(A and B) by using either of the following rules.

   $$P(A \text{ and } B) = P(A) \cdot P(B \mid A)$$
   $$P(A \text{ and } B) = P(B) \cdot P(A \mid B)$$

a. For the situation of rolling a pair of dice once, let event A be rolling doubles and event B be getting a sum of 8.

   i. Using the sample space below, find each of the following probabilities.

   - $P(A)$
   - $P(B)$
   - $P(A \mid B)$
   - $P(B \mid A)$
   - $P(A$ and $B)$

Number on Second Die

Number on First Die	1	2	3	4	5	6
1	1, 1	1, 2	1, 3	1, 4	1, 5	1, 6
2	2, 1	2, 2	2, 3	2, 4	2, 5	2, 6
3	3, 1	3, 2	3, 3	3, 4	3, 5	3, 6
4	4, 1	4, 2	4, 3	4, 4	4, 5	4, 6
5	5, 1	5, 2	5, 3	5, 4	5, 5	5, 6
6	6, 1	6, 2	6, 3	6, 4	6, 5	6, 6

   ii. Verify that both rules for $P(A$ and $B)$ hold for the probabilities that you found in part i.

b. Show that both rules also work for each of the following situations.

   i. You roll a pair of dice once. Event A is rolling doubles. Event B is getting a sum of 7.

   ii. You roll a pair of dice once. Event A is getting 1 on the first die. Event B is getting a sum of 7.

3 A Web site at Central Michigan University collects data from statistics students. In one activity, students were asked whether they were right-handed or left-handed. Students were also asked which thumb is on top when they fold their hands (intertwining their fingers).

The following table shows the results for the first 80 students who submitted their information.

	Left-Handed	Right-Handed	Total
Left Thumb on Top	2	46	48
Right Thumb on Top	4	28	32
Total	6	74	80

Source: stat.cst.cmich.edu/statact/index.php

Suppose you pick one of these 80 students at random.

a. Find each probability.

   i. *P(left-handed)*

   ii. *P(left thumb on top)*

   iii. *P(left thumb on top | left-handed)*

   iv. *P(left-handed | left thumb on top)*

b. Are being left-handed and having the left thumb on top independent events? Are they mutually exclusive events?

c. Use your results from Part a and the formula to find *P(left-handed and left thumb on top)*. Check your answer by using the table directly.

4 Think about a single roll of two dice. For each of the situations below, tell whether the two events are mutually exclusive. Then tell whether they are independent.

a. Event *A* is rolling doubles. Event *B* is getting a sum of 8.

b. Event *A* is rolling doubles. Event *B* is getting a sum of 7.

c. Event *A* is getting 1 on the first die. Event *B* is getting a sum of 7.

d. Event *A* is getting 1 on the first die. Event *B* is getting doubles.

5 Recall that the rule for finding the probability that event *A or* event *B* occurs is

$$P(A \text{ or } B) = P(A) + P(B) - P(A \text{ and } B).$$

When *A* and *B* are mutually exclusive, the rule simplifies to

$$P(A \text{ or } B) = P(A) + P(B).$$

a. Use the appropriate formula and your results from Problem 3 to find *P(left-handed* or *left thumb on top)*. Check your answer by using the table directly.

b. Suppose you know *P(A)* and *P(B)*. When you find *P(A or B)* when *A* and *B* are mutually exclusive, why is there no need to subtract *P(A and B)*?

c. Suppose you know *P(A)* and *P(B)*. How can you find *P(A or B)* when *A* and *B* are independent?

d. Suppose you know *P(A)* and *P(B)*. Can you find *P(A or B)* when *A* and *B* are neither mutually exclusive nor independent? Explain your thinking.

In this investigation, you learned how to compute probabilities when two events are not independent.

a How do you find the probability that event *A* and event *B* both happen when the two events are not independent?

b Does the method in Part a work when events *A* and *B* are independent? Explain.

c What is the difference between mutually exclusive and independent events?

Be prepared to share your ideas with the class.

✓ Check Your Understanding

Rebekka has 75 books in her library. She categorized them in the following way.

	Fiction	Non-Fiction	Total
Book for Teens	20	30	50
Book for Adults	10	15	25
Total	30	45	75

Suppose Rebekka picks a book at random.

a. Find *P(book for teens)*, *P(fiction)*, *P(book for teens | fiction)*, and *P(fiction | book for teens)*.

b. Are *book for teens* and *fiction* independent events?

c. Are *book for teens* and *fiction* mutually exclusive events?

d. Use your results from Part a and the formula to find *P(book for teens and fiction)*. Check your answer by using the table directly.

e. Use your results from the previous parts of this task and the formula to find *P(book for teens or fiction)*. Check your answer by using the table directly.

Applications

1 In a famous trial in Sweden, a parking officer had noted the position to the nearest "hour" of the valve stems on the tires on one side of a car. For example, in the following picture, the valve stems are at 3:00 and at 10:00. The officer issued a ticket for overtime parking. Upon returning later, the officer noted that the valve stems were still in the same position. However, the owner of the car claimed he had moved the car and returned to the same parking place.

a. Use an area model to estimate the probability that if a vehicle is moved, both valve stems return to the same position (to the nearest hour) that they were in before the car was moved.

b. Use the Multiplication Rule to estimate the probability that if a vehicle is moved, the valve stems return to the same position (to the nearest hour) that they were in before the car was moved.

c. What assumption are you making? How can you find out if it is reasonable?

d. Do you think that the judge ruled that the owner was guilty or not guilty?

2 Refer to your area model for the probabilities of whether a child will have freckles if both parents have one freckles gene and one no-freckles gene. (See Problem 3 on page 525.)

a. Compare the following.

P(freckles gene inherited from father)

P(freckles gene inherited from father and *freckles gene inherited from mother)*

b. How is the independence of the events *freckles gene inherited from father* and *freckles gene inherited from mother* shown in your area model?

3 Suppose you draw one card from a shuffled standard deck of cards.

  a. Find the following probabilities.

- *P(card is an ace)*
- *P(card is a heart | card is an ace)*
- *P(card is a heart)*
- *P(card is an ace and card is a heart)*
- *P(card is an ace | card is a heart)*
- *P(card is an ace or card is a heart)*

  b. Are the events *card is an ace* and *card is a heart* independent? Explain.

  c. Are the events *card is an ace* and *card is a heart* mutually exclusive? Explain.

  d. Give an example of two events that would be mutually exclusive.

4 For each situation below,

- find $P(A)$ and $P(A \mid B)$.
- say whether the pair of events A and B are independent or dependent.
- say whether events A and B are mutually exclusive or not mutually exclusive.

  a. You roll a pair of tetrahedral dice once. Event A is getting doubles, and event B is getting a sum of 3.

  b. You flip a coin twice. Event A is getting a head on the second flip, and event B is getting a head on the first flip.

  c. You pick a day of the week at random. Event A is getting a Monday, and event B is getting a school day.

5 Suppose you are trying to draw a heart from a regular deck of 52 cards.

  a. After each draw, you do not replace that card before you draw again.

   i. What is the smallest number of cards you might have to draw in order to get a heart?

   ii. What is the largest number of cards you might have to draw in order to get a heart?

   iii. Are the draws independent? Explain.

  b. After each draw, you do replace that card (and reshuffle) before you draw again.

   i. What is the smallest number of cards you might have to draw in order to get a heart?

   ii. What is the largest number of cards you might have to draw in order to get a heart?

   iii. Are the draws independent? Explain.

  c. Should you replace the card or not if you want to get a heart in the fewest number of draws? Why does this make sense?

6 Consider the table below, which shows how many juniors and seniors at a small high school have a driver's license.

	Juniors	Seniors	Total
Have Driver's License	60	55	115
Do Not Have License	20	15	35
Total	80	70	150

Suppose you pick a student at random.

a. Find *P(junior)*, *P(has driver's license)*, *P(junior | has driver's license)*, and *P(has driver's license | junior)*.

b. Are being a junior and having a driver's license independent events? Are they mutually exclusive events?

c. Use your results from Part a and the formula to find *P(junior and has driver's license)*. Check your answer by using the table directly.

d. Use your results from the previous parts of this problem and the formula to find *P(junior or has driver's license)*. Check your answer by using the table directly.

7 There were about 300 million people in the United States in 2006. About 14.3 million of them watched Game 5 of the 2006 National Basketball Association championship between the Miami Heat and the Dallas Mavericks. About 15.7 million people watched the sixth and final game. (Source: en.wikipedia.org/wiki/National_Basketball_Association_Nielsen_ratings)

a. What proportion of people watched the fifth game? The sixth game?

b. Why is it unreasonable to multiply your two numbers in Part a to find the proportion of people who watched both Game 5 and Game 6?

c. What would be a better estimate?

Connections

8 Refer to Applications Task 1 about the Swedish overtime parking ticket. Design a simulation to estimate the probability that if a car is moved, the valve stems return to the same position (to the nearest hour) that they were in before the car was moved. You must make an assumption about how tires rotate. Will your simulation give approximately the same answer you found in Applications Task 1?

9 *Tree graphs* are a way of organizing all possible sequences of outcomes. For example, the tree graph below shows all possible families of exactly three children (with no twins or triplets). Each *G* means a girl was born, and each *B* means a boy was born. In the United States, the probability that a girl is born is approximately 49%.

First Child   **Second Child**   **Third Child**

G(0.49)

B(0.51)

G

B

G

B

G

B

G

B

G

B

G

B

a. Use the graph to find the probability that a family of three children will consist of two girls and a boy (not necessarily born in that order).

b. Make a tree graph that shows all possible outcomes if you roll a die twice and each time read the number on top. What is the probability you will get the same number twice?

c. Make a tree graph that shows all possible outcomes if you flip a coin four times. What is the probability you will get exactly two heads?

d. Make a tree graph that shows all possible outcomes if you buy three boxes of cereal, each equally likely to contain one of the following stickers: bird of paradise, tiger, elephant, or crocodile. What is the probability that you will get three different stickers?

10 A board game has several PAY DAY spaces throughout the board. On each turn, the player spins a spinner similar to the one below and moves the indicated number of spaces around the board.

Suppose you are 5 spaces away from the next PAY DAY on the game board. You could land on the space by spinning a 5 on your next turn. Another way to land on the space is by spinning a 1 on your next turn, a 3 on the following turn, and a 1 on the turn after that.

a. Make a tree graph that shows all possible sequences of spins that would get you to this PAY DAY.

b. Make sure that you have 16 sequences in your tree graph and then compute the probability of each sequence.

c. What is the probability that you will land on this PAY DAY space on this trip around the board?

11 Consider the situation of rolling two dice. Give an example of events A and B in which the following is true:

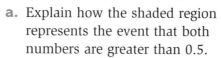

12 If you select two random numbers that are both between 0 and 1, what is the probability that they are both greater than 0.5? You can think geometrically about this kind of problem, as shown at the right.

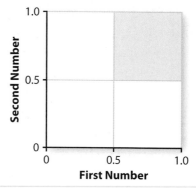

a. Explain how the shaded region represents the event that both numbers are greater than 0.5.

b. What is the probability that both numbers are greater than 0.5?

c. What is the probability that at least one of the numbers is greater than 0.5?

13 Use a copy of the coordinate grid below to help answer the following questions.

a. If you select two random numbers that are both between 0 and 1, what is the probability that they are both less than 0.25?

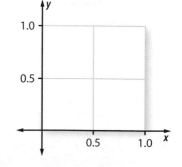

b. Suppose you select two random numbers that are both between 0 and 1.

   i. Make an area model and shade the region where their sum is less than 0.4.

   ii. What are the equations of the lines that border the region?

c. Suppose you select two random numbers that are both between 0 and 1.

   i. Make an area model and shade the region where their sum is greater than 1.5.

   ii. What are the equations of the lines that border the region that represents this event?

Reflections

14 If the probability that an event *A* will occur is *p* (that is, $P(A) = p$), what is the probability that the event will *not* occur, $P(not\ A)$? Explain why your conclusion makes sense.

15 Suppose you want to compute the proportion of students in your school who are sophomores and play a varsity sport. You know the proportion of students who are sophomores, and you know the proportion of students who play a varsity sport. To find the proportion of students who fit into both categories, you multiply the two proportions. Later, you discover that sophomores are less likely to play a varsity sport than are students in general. Is the proportion you computed correct, too large, too small, or could it be either too large or too small? Give an example to illustrate your answer.

16 In which of the following situations do you think it is reasonable to assume the events are independent?

 a. Rolling a pair of dice twice in a row: the first event is not getting doubles on the first roll. The second event is getting doubles on the second roll.

 b. Selecting two people at random: the first event is the first person pouring shampoo directly onto his or her hair. The second event is the second person pouring shampoo directly onto his or her hair.

 c. Selecting a person at random: the first event is getting a person who puts catsup directly on his or her fries. The second event is getting a person who puts shampoo directly on his or her hair.

 d. Selecting one person at random: the first event is getting a person with voice (singing) training. The second event is getting a person who can play a musical instrument.

 e. Waiting for the results of next year's sports championships: the first event is the Celtics winning the NBA championship. The second event is the Red Sox winning the World Series.

 f. Selecting a pair of best friends at random from a high school: the first event is the first friend attending the last football game. The second event is the second friend attending the last football game.

17 Can two events with nonzero probabilities be both independent and mutually exclusive? Explain.

18 The idea of independent events can be somewhat difficult to understand. Suppose that someone in your class has asked you to explain it. Write an explanation of the difference between independent events and dependent events. Include examples that would interest students in your high school.

Extensions

19 Some genetic diseases result only when a baby inherits a "disease" gene from both parents. Suppose that 1 in 30 people carry the gene for a certain disease and that the probability that a person who carries the gene passes it on to a baby is $\frac{1}{2}$. What proportion of babies will have the disease? What assumption must you make to do this computation?

20 Jesse read a report that said that 90% of carpool riders said they would go into the same carpool again, and 72% of carpool drivers said they would go into the same carpool again. He computed the probability that, for a random carpool, both the driver and the rider would say that they would go into the same carpool again as follows:

$$P(\textit{rider would} \text{ and } \textit{driver would}) = P(\textit{rider would}) \cdot P(\textit{driver would})$$
$$= (0.90)(0.72)$$
$$= 0.648$$

Is Jesse correct? Explain your reasoning.

21 Suppose you are sorting 100 items that have been donated for a charity auction. You classify each as antique (A) or not antique (*not A*) and as broken (B) or not broken (*not B*). You make a two-way table after you finish sorting and get the row and column totals in the table below. Keep these totals unchanged throughout this task. You will select one of these 100 items at random.

	B	Not B	Total
A			40
Not A			60
Total	10	90	100

a. Finish filling in the table so that events A and B are independent. Is there more than one way to do this?

b. Using your table from Part a:
   i. are event *not A* and event B independent?
   ii. are A and *not B* independent?
   iii. are *not A* and *not B* independent?

c. Can you fill in the table so that events A and B are independent, but none of the other three pairs of events are independent?

d. Next, fill in the table so that events A and B are not independent. Is there more than one way to do this?

e. Now fill in the table so that events A and B are mutually exclusive.

Review

22 Use algebraic reasoning to solve each equation.

　a. $3(t + 5) + 5(t - 3) = 6t + 12$

　b. $3(2d^2) - 4d = (2d)^2$

　c. $10^{x-3} = 592$

　d. $(3x - 7)^2 = 121$

　e. $2n^2 + 28n = 30$

23 When Daniel stands 20 feet from the base of the Veterans' Monument, he needs to look up at an angle of 68° to see the top of the monument. Daniel is 6 feet tall. About how tall is the monument?

24 Write each product in equivalent $ax^2 + bx + c$ form.

　a. $(x + 5)(x + 8)$　　　　**b.** $(3x + 1)^2$

　c. $(6 - 2x)(5x + 3)$　　　**d.** $(10x - 4)(10x + 4)$

25 Write each quadratic expression in equivalent form as a product of linear factors.

　a. $x^2 - 4x - 32$　　　　**b.** $x^2 + 7x - 18$

　c. $x^2 - 36$　　　　　　**d.** $3x^2 + 30x + 75$

26 In late July and August of 2002, the Division of Visitor Services and Communications, National Wildlife Refuge System, U.S. Fish and Wildlife Service conducted a survey of visitors to 45 "high visitation" National Wildlife Refuges in order to gather information about visitor satisfaction. The table below gives the responses of 2,811 visitors to the statement: "Considering my overall experiences with this National Wildlife Refuge, I am satisfied with the quality of the recreational/educational experience."

Response	Frequency	Percent
Strongly disagree	60	2.1
Disagree	26	0.9
Neither agree nor disagree	71	2.5
Agree	1,065	37.9
Strongly agree	1,589	56.5

Source: *U.S. Fish and Wildlife Service National Wildlife Refuge Visitor Satisfaction Survey: Data Analysis and Report*, page 23.
www.fws.gov/refuges/generalInterest/pdfs/VSS_part1.pdf

　a. So that you can quickly compare responses from year to year, you plan to get a mean number for the visitors' responses. To do this, you must first turn the "Response" column into a numerical response. How could you do this? Make a new column with these numerical responses.

b. Using your numerical responses and the frequency column, compute the mean response. On average, are people generally satisfied with their experience at the national wildlife refuges?

c. Make a new column, converting "Percent" to "Relative Frequency." Using your numerical responses and the relative frequency column, compute the mean response. Is your answer the same as that in Part b? Should it be the same?

27 Glendale and Arcadia High Schools both need to order new supplies for their math classrooms. Price information for ordering from School Central (SC) and Discount Educational Supplies (DES) is in the matrix below.

$$\begin{array}{c c} & \begin{array}{cc} \text{SC} & \text{DES} \end{array} \\ \begin{array}{c} \text{Rulers} \\ \text{Graph Paper Pads} \\ \text{Compasses} \end{array} & \begin{bmatrix} 1.35 & 0.95 \\ 3.50 & 4.15 \\ 1.75 & 1.25 \end{bmatrix} \end{array}$$

a. Glendale High needs 120 rulers, 15 graph paper pads, and 60 compasses. Arcadia High needs 60 rulers, 7 graph paper pads, and 90 compasses. Organize this information into a 2 × 3 matrix. Be sure to label the rows and columns of your matrix.

b. Use matrix multiplication to help you determine which supplier is less expensive for each school. What is the size of your product matrix?

28 The amount of time it takes to fill an empty tanker truck varies inversely with the number of gallons pumped per minute. When the pump is set at 8 gallons per minute, it takes 85 minutes to fill an empty tanker truck.

a. Write a general rule that describes the relationship between the amount of time it takes to fill the empty tanker and the pump rate in gallons per minute.

b. What is the constant of variation for your rule? What does this constant indicate about the tanker?

c. If the pump rate is doubled, how does the amount of time it takes to fill the empty tanker change? Explain your reasoning.

29 Determine to the nearest degree the measure of ∠A in each triangle.

a.   **b.**

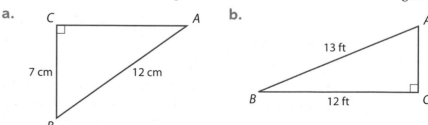

c.

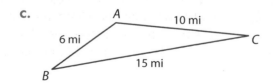

$	1,000
$	5,000
$	10,000
$	25,000
$	50,000
$	75,000
$	100,000
$	200,000
$	300,000
$	400,000
$	500,000
$	750,000
$	1,000,000

Choose one of the suitcases
This will be your suitcase

Expected Value

If you are like many people, you are probably intrigued by television game shows that provide chances of winning large sums of money. On one popular game show, the contestant chooses one of 26 suitcases which remains unopened. The 26 suitcases contain these amounts of money in dollars: 0.01, 1, 5, 10, 25, 50, 75, 100, 200, 300, 400, 500, 750, 1,000, 5,000, 10,000, 25,000, 50,000, 75,000, 100,000, 200,000, 300,000, 400,000, 500,000, 750,000, 1,000,000. The contestant then chooses six other suitcases, which are opened.

At this stage, the player can either keep the suitcase she selected or take the amount of money offered by the "bank." If she does not take the money offered by the bank, more suitcases are opened, and the bank gives new offers. The amount of each offer depends on which amounts are left in the unopened suitcases. Ultimately, the player either accepts one of the bank's offers or keeps her suitcase.

Consider the mathematics behind this television game.

a) What is the probability that the player picks a suitcase that contains at least half a million dollars? At the beginning of the game, what is the average amount the player can expect to win if she keeps the suitcase she selected?

b) After several rounds of this game one evening, a player got to the point where there were only four suitcases left unopened. The remaining suitcases contained $75, $5,000, $500,000, and $750,000. The bank offered her $212,050 for her suitcase. Was this a fair offer?

c) She refused the offer and opened one additional suitcase. It contained $750,000. Only three suitcases were left containing $75, $5,000, and $500,000. The bank's offer went down to $155,038. If you were the player, would you take the offer? Explain your thinking.

In this lesson, you will learn how to compute the average **(expected value)** of a probability distribution.

Investigation 1 — What's a Fair Price?

In mathematics, the **fair price** for a game is defined as the price that should be charged so that in the long run, the players come out even. In other words, in the long run, the amount won by the players is equal to the amount the players were charged to play. Of course, people who run carnival games or who sell insurance want to cover their expenses and make a profit so they charge more than the "fair price." To decide how much to charge, they must first compute the fair price. As you work on the problems of this investigation, look for answers to the following question:

How can you compute the fair price for a game or insurance policy?

1. Suppose your school decides to hold a raffle. The prizes in the raffle will be a bicycle that costs $400, an audio player that costs $175, and a video game that costs $100. Exactly 2,000 tickets will be sold. Three tickets will be drawn at random. The person who holds the first ticket drawn gets the bicycle, the second the audio player, and the third the video game.

   a. What is a fair price for a ticket?

   b. Write a procedure for finding the fair price of a raffle ticket.

2 At a fund-raising carnival for a service organization, Renee is trying to get Leroy to play a game she has invented. Leroy would spin the spinner shown below and get a gift certificate worth the amount indicated. The organization charges $5 to play this game.

a. Leroy thinks he has the advantage because he has two chances of coming out ahead, one chance of coming out even, and only one chance of winning less than the game costs to play. What would you say to him?

b. How much would Leroy expect to win if he played 100 games? What is Leroy's expected net earnings?

c. Is $5 the fair price to play this game? If not, what would be the fair price?

d. Design a spinner that would make $5 a fair price to charge to play.

3 In a carnival game, a wheel is spun. The wheel has an equal chance of ending up in any one of 38 positions. The wheel has 18 red positions, 18 blue positions, and 2 white positions. Suppose a player picks red. If the wheel ends up on red, the player wins $2. The price to play this game is $1.

a. Complete the table below, which shows the outcomes and their probabilities for this game.

Outcome	Probability
Win $2	
Win $0	

b. Is $1 a fair price for playing this game? Why or why not?

4 The National Center for Health Statistics reports that the death rate for males aged 15–24 in the United States is 113 deaths for every 100,000 males in this age group. There are 43 deaths per year for every 100,000 females in that age group. (Source: www.cdc.gov/nchs/fastats/pdf/mortality/nvsr54_19t01.pdf) Use these statistics to answer the following.

a. Ignoring all factors other than gender, what would be the fair price for an insurance company to charge to insure the life of a male in that age group for one year for $50,000?

b. Ignoring all factors other than gender, what would be the fair price for an insurance company to charge to insure the life of a female in that age group for one year for $50,000?

c. If the insurance company is not allowed to have different rates for each gender, what would be the fair price for a $50,000 policy for one year? Assume that the same number of insurance policies are sold to males as to females.

d. Compare the procedure you used to answer Parts a and b with your procedure in Part b of Problem 1.

e. In what ways is insurance mathematically similar to a raffle? In what ways is it different?

Summarize
the Mathematics

In this investigation, you explored how to compute the fair price for games and insurance policies.

a What is the relationship between the fair price of a game and the average winnings of a player in the long run?

b Describe a general procedure for finding the fair price of raffle or lottery tickets.

c Describe a general procedure for finding the fair price of an insurance policy.

Be prepared to explain your ideas to the class.

✔Check Your Understanding

Apply your method of finding a fair price to the three situations below.

a. After several rounds of the game show described at the beginning of this lesson, the contestant had opened all but five suitcases. The remaining suitcases contained $75, $500, $5,000, $500,000, and $750,000. The bank offered her $129,519 for her suitcase. Is this a fair offer?

b. The prizes in a raffle are ten $15 CDs and one $500 stereo. If 1,000 raffle tickets will be sold, what is the fair price for a ticket?

c. According to the Youth Risk Behavior Survey, about 30% of high school students reported that they had some property stolen or deliberately damaged at school within the previous year. Suppose the average value of the stolen or damaged property in that year was $60. Assuming these statistics stay the same each year, what would be the fair price to charge a student who wanted to be insured against theft or damage for one year of high school?
(Source: www.cdc.gov/mmwr/preview/mmwrhtml/ss5302a1.htm)

**Expected Value of a
Probability Distribution**

You have learned how to find the fair price for games of chance and for insurance. In this investigation, you will see how organizing your work in a table leads to a formula for *expected value*. Keep in mind the following questions as you work on the problems of this investigation:

> *How can you compute the fair price of a game if you are given the probability distribution of the prizes?*

> *In general, how can you find the expected value of a probability distribution?*

1 A game of chance has the probability distribution given in the table below.

Prize Value, x	Probability, P(x)
$1	$\frac{1}{6}$
$2	$\frac{1}{6}$
$3	$\frac{1}{6}$
$4	$\frac{1}{6}$
$5	$\frac{1}{6}$
$6	$\frac{1}{6}$
Total	$\frac{6}{6}$

a. What is the fair price for this game?

b. Make a histogram of this probability distribution. Locate the fair price on the histogram. How does this compare to the mean of the distribution?

c. Complete a copy of this table, including the total.

Prize Value, x	Probability, P(x)	x · P(x)
$1	$\frac{1}{6}$	
$2	$\frac{1}{6}$	
$3	$\frac{1}{6}$	$\frac{3}{6}$
$4	$\frac{1}{6}$	
$5	$\frac{1}{6}$	
$6	$\frac{1}{6}$	
Total	$\frac{6}{6}$	

d. Compare your total from Part c to the fair price from Part a. What do you notice?

2 The chart below gives the possible outcomes and their probabilities for a version of a scratch-off game played at a fast-food restaurant. What is the fair price of one scratch-off card?

Prize	Probability
Win free fries worth 90¢	$\frac{1}{6}$
Win nothing	$\frac{5}{6}$
Total	$\frac{6}{6}$

3 Here is the table for another scratch-off game.

Prize Value	Probability
$1	$\frac{4}{10}$
$2	$\frac{2}{10}$
$3	$\frac{2}{10}$
$5	$\frac{1}{10}$
Win nothing	$\frac{1}{10}$
Total	$\frac{10}{10}$

a. What is the fair price of one scratch-off card?

b. Make a histogram of this probability distribution.

c. Estimate the balance point of the histogram. Compare this answer to the fair price of the card that you computed in Part a. What do you notice?

4 The mean of a probability distribution is called the **expected value** (*EV*).

a. Suppose that your probability distribution represents the chances of winning the various prizes in a game. Explain why the expected value gives the fair price for a game.

b. Describe how to implement a procedure for finding the expected value of a probability distribution in an efficient way on your calculator or computer.

c. Use your method to find the expected value of the probability distribution table for the sum of two dice.

5 The table and histogram at the top of page 551 give the proportion of families in the United States that are a given size based on a census that tried to count all families in the United States.

a. If you were to pick a family at random from the United States, what is the probability that it would have four people in it? What is the probability it would have four or fewer people in it?

b. What is the mean number of people per family? (Count 7 or more as 7 people.)

Size of Family	Relative Frequency
2	0.43
3	0.23
4	0.20
5	0.09
6	0.03
7 or more	0.02
Total	1.00

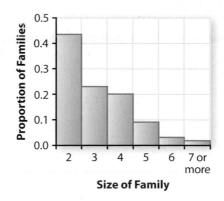

Source: *U.S. Bureau of the Census, Statistical Abstract of the United States: 2000 (120th edition).* Washington, D.C., 2000.

c. Is the standard deviation of this distribution closest to 1, 2, or 3? (A method for computing the standard deviation is outlined in Extensions Task 18.)

Summarize
the Mathematics

Look back at your method for finding the expected value of a probability distribution.

a How is the method for finding the expected value of a probability distribution similar to the method for finding the mean of a frequency table? How is it different?

b What property of a probability distribution explains the difference in the methods?

Be prepared to share your ideas with the class.

✓ Check Your Understanding

Find the expected value (fair price) of a ticket from the scratch-off game described in the following table.

Prize/Value	Probability
Small soft drink (89¢)	$\frac{15}{100}$
Small hamburger ($1.29)	$\frac{8}{100}$
T-shirt with restaurant logo ($7.50)	$\frac{3}{100}$
Movie passes ($15.00)	$\frac{1}{100}$
You lose! ($0.00)	$\frac{73}{100}$
Total	$\frac{100}{100}$

Applications

1. A charity raffle has five prizes: a car worth $25,000, a vacation worth $2,800, and three MP3 players, each worth $400.

   a. If 10,000 tickets are to be sold, what is the fair price of a ticket?

   b. If you buy 1 ticket, what is the probability that you win a prize?

   c. If you buy 10 tickets, what is the probability that you win the car?

2. The average claim for collision damage to a fairly new car involved in a collision is about $3,910. For every 100 fairly new cars that are insured, each year, there are about 7.8 collisions in which a claim is filed. (Source: Highway Loss Data Institute, www.iihs.org/research/hldi_facts/collision_coverage_trends.pdf)

   a. For every 100 fairly new cars that are insured, how much money would you expect to be paid out to insurance claims for collision damage?

   b. What is the fair price to charge for collision insurance for one year for a fairly new car?

3. If there is a 40% chance of rain today, it means that it rained on 40% of the days in the past that had weather conditions similar to those today.

   a. On 14 different days, the weather report says there is a 40% chance of rain. On how many of these days do you expect it to rain? On how many of these days do you expect it not to rain?

   b. On 20 different days, the weather report says there is a 50% chance of rain. It actually rained on 9 of those days. Do you think the meteorologist did a good job of predicting rain? Explain.

4. The table below is copied from the back of a ticket in a scratch-off state lottery game.

Prize	Probability of Winning
$0.75	$\frac{1}{10}$
$2.00	$\frac{1}{14.71}$
$4.00	$\frac{1}{50}$
$10.00	$\frac{1}{71}$
$20.00	$\frac{1}{417}$
$250.00	$\frac{1}{1,000}$

   a. What is the probability of winning nothing with one ticket? Add a line to a copy of the table that shows this outcome and its probability.

   b. What is a fair price for a ticket?

   c. The tickets in this lottery sell for $1.00 each. How much money does the state expect to make if 1,000,000 tickets are sold?

(5) The player with the highest field goal percentage in the history of the National Basketball Association (NBA) is Artis Gilmore. In his career in the NBA, Gilmore attempted 9,570 field goals and made 5,732 of them.

   a. What was Gilmore's field goal percentage?

   b. During a typical game, Gilmore might attempt 25 field goals. In a typical game, how many field goals would you expect Gilmore to make?

The NBA player with the highest lifetime free throw percentage is Mark Price. Price had a free throw "percentage" of 0.904. He made a total of 2,135 free throws.

   c. Why do you think the word percentage is in quotation marks above?

   d. How many free throws did Price attempt?

   e. How many free throws would you expect Price to make in 50 attempts?

   f. Write an equation that relates the number of free throws T expected for a player who makes A attempts and whose free throw percentage is p.

(6) A fast-food restaurant once had a scratch-off game in which a player picked just one of the four games on the card to play. In each game, the player stepped along a path, scratching off one of two adjacent boxes at each step. To win, the player had to get from start to finish without scratching off a "lose" box. Here's how the games on one card would have looked. (Of course, the words were covered until the player scratched off the covering.)

Game	Prize
A	free food worth 55¢
B	free food worth 69¢
C	free food worth $1.44
D	free food worth $1.99

Game A
Start

GO	GO
LOSE	GO

→ Finish

Game B
Start

LOSE	GO	LOSE
GO	GO	GO

→ Finish

Game C
Start

LOSE	LOSE	LOSE
GO	GO	GO

→ Finish

Game D
Start

LOSE	GO	GO	GO
GO	LOSE	LOSE	LOSE

→ Finish

   a. What is the probability of winning each game?

   b. For each game, make a probability distribution table and find the expected value.

c. Which is the best game to pick if you just want to win something?

d. Which is the best game to pick if you want to have the largest expected value?

7 While a prisoner of war during World War II, J. Kerrich conducted an experiment in which he flipped a coin 10,000 times and kept a record of the outcomes. A portion of the results is given in the table below.

Number of Flips	Number of Heads
10	4
50	25
100	44
500	255
1,000	502
5,000	2,533
10,000	5,067

Source: J. Kerrich. *An Experimental Introduction to the Theory of Probability.* Copenhagen: J. Jorgenson and Co., 1964.

a. How many heads would you expect if you flip a coin 100 times? 32 times? 15 times?

b. After how many flips is the number of heads in Kerrich's table closest to the expected number of heads? Furthest?

c. Was the percentage of heads closer to the expected percentage of 50% after flipping 10 times or 10,000 times? Is this what you would expect? Explain.

Connections

8 Imagine an amateur archer shooting at the target below. The square board has a side length of six feet. Suppose the archer can always hit the board, but the spot on the board is random.

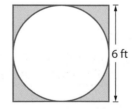

6 ft

a. Describe how to use a coordinate system and the **rand** function of your calculator to simulate the point where an arrow will land.

b. How can you tell whether the point is inside the circle?

c. Describe a simulation to estimate the probability that an arrow shot will land in the circle.

d. Simulate three arrow shots and tell whether each lands in the circle or not.

e. Describe another way to estimate the probability in Part c.

9 In this task, you will explore some of the geometry and algebra connected with the probability distribution for the sum of two dice.

  a. Make a probability distribution table for the sum of two octahedral dice (which have the numbers 1 through 8 on them).

  b. Plot the points given in your table $(x, P(x))$, where x is the sum and $P(x)$ is its probability.

  c. Write a single equation whose graph contains the points for $x = 2, 3, 4, 5, 6, 7, 8, 9$.

  d. Write another single equation whose graph contains the points for $x = 9, 10, 11, 12, 13, 14, 15, 16$.

  e. How are the slopes of these two graphs related?

  f. Use absolute value to write one equation whose graph contains all 15 points.

  g. Write an equation whose graph contains the points that represent the probability distribution for the sum of two regular polyhedral dice, each with n sides.

10 A summation sign, Σ, is useful in writing expressions involving expected value.

  a. Write an expression that uses a summation sign and gives the expected value of the probability distribution of a single roll of a die. Use x to represent the numbers in the faces of the die.

  b. Using the probability distribution table from Problem 1, Investigation 2, page 549, find each of the following sums.

   i. Σx **ii.** $\Sigma P(x)$ **iii.** $\Sigma[x^2 \cdot P(x)]$

  c. Using the probability distribution table from Problem 3, Investigation 2, page 550, find each of the sums above.

  d. Complete this sentence: In any probability distribution table, $\Sigma P(x) = $ _____ because _____.

Reflections

11 In this lesson, you calculated the expected value for various probability distributions. Expected value is another name for theoretical average.

  a. If you flip a fair coin 10 times, on average, how many heads would you expect to get?

  b. If you flip a coin 5 times, on average, you would expect to get 2.5 heads. Explain mathematically why that makes sense.

  c. If you roll a pair of dice 42 times, how many doubles do you expect? How many doubles do you expect if you roll 41 times?

d. About 45% of the U.S. population has type O blood. If 200 random people walk into a blood bank, how many do you expect to have type O blood? If 50 random people walk into a blood bank, how many do you expect to have type O blood?

e. If the probability of a success on each trial of a chance situation is p, how many successes would you expect to get in n trials?

12 Select one of the following projects and write a brief report summarizing your findings.

a. If your state has a lottery, investigate the amount of money income from sales, the amount of money paid in prizes, the operating costs of the lottery, the profit your state makes, and what your state does with the profits of its lottery.

b. Find the information about a scratch-off lottery ticket. Compute the expected value of a ticket.

13 Why don't gambling games charge the fair price for playing? Why do people gamble when the price of playing a game is always more than the expected value of the play?

14 According to the National Center for Health Statistics, in 2003, a newborn male in the United States could expect to live 74.8 years. A 20-year-old male could expect to live to the age of 75.9. A newborn female could expect to live to the age of 80.1 and a 20-year-old female to the age of 80.9.
(Source: www.cdc.gov/nchs/data/nvsr/nvsr53/nvsr53_15.pdf)

a. In this case, what is meant by the words "expect to live to the age of"?

b. Why is the life expectancy for a 20-year-old greater than for a newborn?

15 This probability distribution table gives the probability of getting a specified number of heads if a coin is tossed five times.

Number of Heads	Probability
0	$\frac{1}{32}$
1	$\frac{5}{32}$
2	$\frac{10}{32}$
3	$\frac{10}{32}$
4	$\frac{5}{32}$
5	$\frac{1}{32}$
Total	$\frac{32}{32}$

a. What is the expected number of heads if a coin is tossed five times? Find the answer to this question in at least two different ways.

b. José says that the answer to Part a cannot involve half a head. How would you help him understand why it can?

Extensions

16 In an episode of a television show, a man receives an anonymous letter that correctly predicts the winner of a sports event. In the next four weeks, similar letters arrive, each making a prediction that turns out to be correct. The fifth and final letter asks the man for money before he receives another prediction. The whole thing turns out to be a scam.

Two versions of the first letter had been sent out, each to a large number of people. Half of the people received letters that predicted Team A would win, and half of the people received letters that predicted Team B would win. Those people who received letters with the correct prediction were sent letters the second week. Again, half of the letters predicted Team C and half predicted Team D.

How many letters should have been sent out the first week so that exactly one person would be guaranteed to have all correct predictions at the end of the five weeks?

17 To help raise money for charity, a college service organization decided to run a carnival. In addition to rides, the college students planned games of skill and games of chance. For one of the booths, a member suggested using a large spinner wheel with the numbers from 1 to 10 on it. The group considered three different games that could be played. The games are described below. For each game, do the following:

a. Make a table showing all possible outcomes and their probabilities.

b. Calculate the expected value of the game, and compare that value to the price of playing the game.

c. Determine if the game will raise money, lose money, or break even in the long run. If the game will not make money, recommend a change that the group might consider.

Double Dare To play this game, the player must pay the booth attendant $1. The player then chooses two numbers from one to ten and gives the wheel a spin. If the wheel stops on either of the two numbers, the attendant gives the player a prize worth $15.

Anything Goes For this game, the player can choose either one or two numbers. The player again must pay the attendant $1. After paying the attendant and choosing numbers, the player spins the wheel. If the wheel stops on a number the player chose, the attendant gives the player a prize. If the player selected only one number, the prize is worth $10. If the player selected two numbers, the prize is worth $5.

Triple Threat This game is a little more expensive to play. The player must pay $3 to spin the wheel. If the wheel stops on 1, 2, or 3, the player loses, receiving no prize. If the wheel stops on 4, 5, 6, or 7, the attendant gives the player a prize worth $2. If the wheel stops on 8, 9, or 10, the player gets a prize worth $6.

18 To find the standard deviation of a probability distribution given in a table, first compute the expected value, *EV*. Then complete columns as in the following table, or use the List feature of your calculator for the calculations. Finally, sum the numbers in the last column (or list) and take the square root.

Value, x	Probability, $P(x)$	$(x - EV)^2$	$(x - EV)^2 \cdot P(x)$
2	0.43		
3	0.23		
4	0.20		
5	0.09		
6	0.03		
7	0.02		

a. Find the standard deviation for the probability distribution above.

b. Check your answer to Problem 5 Part c (page 551) of Investigation 2.

c. Refer to Applications Task 2. Suppose that you insure just one car. Complete the probability distribution table of possible insurance claims below. Then find the expected value and standard deviation. Use these two numbers in a sentence about insurance.

Claim, x	Probability, $P(x)$
3,910	
0	

d. Write a formula for the standard deviation of a probability distribution.

Review

19 Javier bought a used car for $5,600. The value of his car is depreciating at the rate of 1.25% per month.

  a. How much is Javier's car worth after 6 months?

  b. Write a *NOW-NEXT* rule that indicates how the value of Javier's car changes from one month to the next.

  c. Write a function rule that can be used to find the value of Javier's car after any number of months.

  d. Javier took out a three-year loan to pay for his car. How much is his car worth when he makes his final payment?

  e. When is Javier's car first worth less than $2,500?

20 Suppose that you roll two dice and record the numbers showing on the top faces of the two dice. Find each of the following probabilities.

  a. *P(sum is 5)*   **b.** *P(sum is less than 4)*

  c. *P(sum is odd)*   **d.** *P(roll doubles)*

21 Write each expression in the simplest equivalent form that uses only positive exponents.

  a. $2x^3y(-3xy^{-2})$ **b.** $\dfrac{8a^4b^{10}}{2a^{-1}b^5c}$

  c. $2(-3x^2y^3)^3$ **d.** $3a^3b - 2a(a^2b + a)$

22 Consider the triangle shown below.

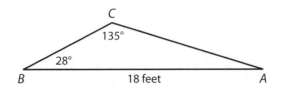

  a. Find *BC*.   **b.** Find *AC*.

23 Play the game below several times.

  In a game of matching, two people each flip a coin. If both coins match (both heads or both tails), Player B gets a point. If the coins don't match, Player A gets two points.

  a. What is the probability that Player A wins a round? That Player B wins?

  b. Explain why this is or is not a fair game.

24 Write rules for quadratic functions with graphs that have the following properties.

  a. opens down and has *y*-intercept at (0, 6)

  b. *x*-intercepts at (2, 0) and (8, 0) and opens up

  c. minimum point at (3, 0) and opens up

  d. *x*-intercepts at (−2, 0) and (4, 0) and maximum point at (1, 5)

The Waiting-Time Distribution

Anita and her friends are playing a board game. Movement around the board is determined by rolling a pair of dice. Winning is based on a combination of chance and a sense for making smart real estate deals. While playing this game, Anita draws the card shown above. She must go directly to the "jail" space on the board.

Anita may get out of jail by rolling doubles on one of her next three turns. If she does not roll doubles on any of the three turns, Anita must pay a $50 fine to get out of jail. Anita takes her first turn and does not roll doubles. On her second turn, she does not roll doubles again. On her third and final try, Anita does not roll doubles yet again. She grudgingly pays the $50 to get out of jail. Anita is feeling very unlucky.

Anita's situation suggests several questions.

a How likely is it that a player who is sent to jail (and does not have a "Get Out of Jail Free" card) will have to pay $50 to leave? As a class, think of as many ways to find the answer to this question as you can.

b In games and in real life, people are occasionally in the position of waiting for an event to happen. In some cases, the event becomes more and more likely to happen with each opportunity. In some cases, the event becomes less and less likely to happen with each opportunity. Does the chance of rolling doubles change each time Anita rolls the dice?

c On average, how many rolls will it take to roll doubles? Do you think Anita should feel unlucky? Explain your reasoning.

In this lesson, you will use the rules of probability that you have learned so far to analyze **waiting-time situations**, where you perform repeated, identical *trials* and are waiting for a specific event to happen.

Investigation 1 — Waiting for Doubles

In this investigation, you will explore several aspects of Anita's situation of waiting for doubles to get out of jail while playing the board game. For this investigation, assume that a player must stay in jail until he or she rolls doubles. That is, a player cannot pay $50 to get out of jail in this version of the game, and there is no "Get Out of Jail Free" card. As you work on the following problems, make notes of answers to the following questions:

How can you roll dice to estimate the probability that it will take a specified number of rolls to get doubles?

How can you calculate that probability exactly?

What is the shape of a waiting-time distribution?

1 Suppose you are playing Anita's game under this new rule and have just been sent to jail. Take your first turn and roll a pair of dice. Did you roll doubles and get out of jail? If so, stop. If not, roll again. Did you roll doubles and get out of jail on your second turn? If so, stop. If not, roll again. Did you roll doubles and get out of jail on your third turn? If so, stop. If not, keep rolling until you get doubles.

a. Copy the frequency table below and put a tally mark in the frequency column next to the event that happened to you. Add rows as needed.

Rolling Doubles

When First Rolled Doubles	Number of Rolls	Frequency
First try	1	
Second try	2	
Third try	3	
Fourth try	4	
⋮	⋮	
	Total	100

b. With other members of your class, perform this experiment a total of 100 times. Record the results in your frequency table.

c. Do the events in the frequency table appear to be *equally likely*? That is, does each of the events have the same chance of happening?

d. Use your frequency table to estimate the probability that Anita will have to pay $50, or use a "Get Out of Jail Free" card, to get out of jail when playing a standard version of her game. Compare this estimate with your original estimate in Part a of the Think About This Situation.

e. Make a histogram of the data in your frequency table. Describe the shape of this histogram.

f. Explain why the frequencies in your table are decreasing even though the probability of rolling doubles on each attempt does not change.

2 Suppose you compared your class' histogram of the waiting time for rolling doubles with another class' histogram.

a. Explain why the histograms should or should not be exactly the same.

b. What characteristics do you think the histograms will have in common?

In Problem 1, you constructed an approximate *waiting-time distribution*. As you noted in Problem 2, two different groups could get quite different histograms when they constructed them by rolling dice. The two groups would then have different estimates of a probability. In the next problem, you will construct waiting-time distributions exactly, so everyone should get the same answers to the probability questions related to these distributions.

3 Imagine 36 students are playing a modification of Anita's game in a class tournament. All are sent to jail. A student cannot pay $50 to get out of jail but must roll doubles. (There is no other way out.)

a. How many of the 36 students do you expect to get out of jail by rolling doubles on the first try? (Remember that the word "expect" has a mathematical meaning.) How many students do you expect to remain in jail?

b. How many of the remaining students do you expect to get out of jail on the second try? How many students do you expect to remain in jail then?

c. How many of the remaining students do you expect to get out of jail on the third try? How many students do you expect to remain in jail then?

d. Complete a table like the one below. Round numbers to the nearest hundredth. The first three lines should agree with your answers to Parts a–c.

Rolling Dice to Get Doubles

Number of Rolls to Get Doubles	Expected Number of Students Released on the Given Number of Rolls	Expected Number of Students Still in Jail
1		
2		
3		
4		
5		
6		
7		
8		
9		
10		
11		
12		

e. What patterns of change do you see in this table? If possible, describe each pattern using the idea of *NOW* and *NEXT*.

(4) The table you started in Problem 3 never really can be completed because there are infinitely many possible outcomes. That is, the rows could be continued indefinitely.

   a. How many of the 36 students do you expect to be in jail after 12 tries to roll doubles?

   b. Add a "13 or more" row to your table and write the expected number of students in the appropriate place.

(5) Make a histogram of the "expected number of students released on the given number of rolls" from the frequency table you constructed in Problem 3.

   a. Compare this histogram to the one you constructed in Problem 1 of this investigation.

   b. Examine your histogram of the theoretical distribution. The height of each bar is what proportion of the height of the bar to its left?

   c. Using the balance point of your histogram, estimate the mean of the distribution.

   d. Using the frequency table from Problem 3, calculate the mean number of rolls of the dice it takes to get doubles. Compare your calculated mean to your estimate in Part c.

(6) If there were 1,000 people (rather than 36) who had been sent to jail in Problem 3, how would the histogram change? How would the mean change? How many of these people would you expect to need 13 or more rolls to get out of jail?

Summarize
the Mathematics

In this investigation, you explored the distribution of the waiting time for rolling doubles.

a Suppose the probability of an event is p. How many times would you expect this event to happen in a series of n independent trials?

b Suppose the probability of an event is p and you have made a histogram of the waiting-time distribution for the event. How is the height of each bar of the histogram related to the height of the bar to its left?

Be prepared to share your ideas and reasoning with the class.

✓ Check Your Understanding

Change the rules of Anita's game so that a player must flip a coin and get heads in order to get out of jail.

a. Is it harder or easier to get out of jail with this new rule instead of by rolling doubles? Explain your reasoning.

b. Play this version 24 times, either with a coin or by simulating the situation. Put your results in a table like the one in Problem 1. Then make a histogram of your results.

c. What is your estimate of the probability that a player will get out of jail in three flips or fewer?

d. Now, suppose you start with 1,000 people in jail. Make a table that shows the theoretical distribution of exactly how many you would expect to get out of jail on each flip. Continue until your table has four rows.

Investigation 2 — The Waiting-Time Formula

In Investigation 1, you created waiting-time distributions using simulation and theoretical probabilities. The theoretical distributions had a well-defined pattern. As you complete this investigation, keep in mind this basic question:

What general formula can be used to calculate the probability that it will take exactly x trials to get the first success in a waiting-time distribution?

1 Imagine that all students in your class are playing Anita's game. All are sent to jail. To get out of jail, a student must roll doubles. (You cannot pay $50 or use a Get Out of Jail Free card).

a. What is the probability of getting doubles on the first roll? Enter your answer as a fraction on the first line of a copy of the table at the right.

Number of Rolls to Get Doubles for the First Time	Probability
1	
2	
3	
4	
5	
6	
7	
8	
9	
10	
11	
12	
13 or more	

b. To get out of jail on the second roll, two events must happen. You must not roll doubles on the first roll, and you must roll doubles on the second roll.

    i. What rule of probability can you use to compute $P(A \text{ and } B)$ when A and B are independent events?

    ii. What is the probability of not getting doubles on the first roll and getting doubles on the second roll? Write your answer as a fraction on the second line of the table.

c. To get out of jail on the third roll, three events must happen.

    i. What are these three events?

    ii. What is the probability that all three events will happen? Write your answer as a fraction on the third row of the table.

d. To get out of jail on the fourth roll, four events must happen.

    i. What are these four events?

    ii. What is the probability that all four events will happen? Write your answer as a fraction on the fourth row of the table.

2 Explain why your work for Problem 1 can be summarized in the following manner.

$$P(1) = \frac{1}{6}$$

$$P(2) = \left(\frac{5}{6}\right)\left(\frac{1}{6}\right)$$

$$P(3) = \left(\frac{5}{6}\right)^2\left(\frac{1}{6}\right)$$

$$P(4) = \left(\frac{5}{6}\right)^3\left(\frac{1}{6}\right)$$

3 Examine the equations in Problem 2.

a. What patterns do you see in the equations in Problem 2?

b. What is $P(5)$? $P(6)$? Explain the meaning of each of these probabilities.

c. What is $P(x)$? That is, what is the probability the first doubles will appear on the xth roll of the dice?

d. Use your formula in Part c to complete the rest of the probability distribution table. Leave the answers in the form of those in Problem 2. For the row "13 or more," include the probability that makes the total probability equal to 1.

e. Make another column, multiplying out the probabilities and expressing them as decimals to the nearest thousandth.

f. Make a graph from the probability distribution table in Part e. How does the height of each bar compare to the one on its left? Explain why this is the case.

g. When they have to wait longer than almost everyone else, lots of people begin to feel that something very unusual has happened. Thus, a **rare event** for a waiting-time distribution is defined as an event that falls in the upper 5% of a waiting-time distribution. Would it be a rare event to require 12 rolls to get doubles for the first time?

4. Use your completed probability distribution table to find the following probabilities. "Rolls" stands for the number of rolls to get doubles for the first time.

a. $P(rolls \leq 3)$

b. $P(rolls > 3)$

c. $P(rolls \geq 5)$

5. In this problem, you will develop a formula for waiting-time distributions. In this general case, use the letter p to represent the probability of getting the waited-for event. (In the situation of waiting for doubles on a pair of dice, $p \approx 0.167$ on each trial. In the situation of flipping a coin until a head appears, $p = 0.5$ on each trial.)

a. The first row of the table below gives the probability that the waited-for event will occur on the first trial. On a copy of this table, fill in the first row.

Number of Trials to Get First Success, x	Probability, P(x)
1	
2	
3	
4	
5	
x	

b. What is the probability of not getting the waited-for event on the first trial and then getting it on the second trial? Fill in the second row of the table.

c. What is the probability of not getting the waited-for event on the first trial, not getting it on the second trial, and then getting it on the third trial? Fill in the third row of the table.

d. Finish filling in the rows of the table. In the last row, write a general formula for the probability it will take x trials to get the first success. Compare your formula with those of your classmates. Resolve any differences.

e. How is your general formula like an exponential function rule? How is it different?

6 Use your formula from Problem 5 to answer these questions.

a. The Current Population Survey of the U.S. Census Bureau recently found that 27% of adults over the age of 25 in the United States have at least a bachelor's degree. If a line of randomly-selected U.S. adults over the age of 25 is walking past you, what is the probability that the fifth person to pass will be the first with four or more years of college?

b. What is the probability that parents would have seven boys in a row and then have a girl? Use $P(boy) = 0.51$ and assume births are independent.

Summarize
the Mathematics

In this investigation, you developed a formula for constructing a waiting-time distribution.

a Describe what each term in the formula represents.

b Explain how you know your formula gives the correct probability.

c How can you determine if a specific event in a waiting-time distribution is a rare event?

Be prepared to share your ideas and reasoning with the class.

✓ Check Your Understanding

Complete a probability distribution table like the one below for the experiment of flipping a coin until a head appears. Complete the "Probability, $P(x)$" column using decimals rounded to the nearest thousandth.

Number of Flips to Get a Head, x	Probability, P(x)
1	
2	
3	
4	
5	
6	
7	
8	
9	
10	

a. Make a graph of your distribution.

b. What is the probability that the first head occurs on the second flip?

c. What is the formula for $P(x)$, the probability of getting the first head on flip number x? Simplify this formula.

d. What is the probability that Scott will require at most 5 flips to get a head?

e. What is the probability that Michele will require at least 8 flips to get a head?

f. What numbers of flips are considered rare events?

g. Suppose that the parents pictured had planned to have children until they had a boy. Did a rare event occur? Assume that the probability of getting a boy on each birth is 0.5.

Expected Waiting Time

According to the company that makes them, M&M's® Milk Chocolate Candies are put randomly into bags from a large vat in which all the colors have been mixed. The percentage of different color coatings varies—20% are orange.

The probability distribution table below shows the waiting time to draw an orange one from a very large bag of the candies. In this situation, the probability of success on a single trial is 0.2. As you work on the problems in this investigation, look for an answer to this question:

How can you find the expected value of a waiting-time distribution?

Drawing Candies

Number of Draws to Get First Orange Candy	Probability
1	0.2
2	(0.8)(0.2)
3	$(0.8)^2(0.2)$
4	$(0.8)^3(0.2)$
5	$(0.8)^4(0.2)$
6	$(0.8)^5(0.2)$
7	$(0.8)^6(0.2)$
8	$(0.8)^7(0.2)$
9	$(0.8)^8(0.2)$
10	$(0.8)^9(0.2)$
11	$(0.8)^{10}(0.2)$
12	$(0.8)^{11}(0.2)$
⋮	

1 In this problem and the next one, you will explore how to find the expected number of draws until an orange candy appears. In other words, if a large number of people each draws candies until an orange one appears, what is the average number they draw?

a. What is the difficulty in trying to find the expected value of this waiting-time distribution?

b. Make an estimate of the expected value of the distribution by using just the first 25 rows of the table to compute the expected value. Keep at least 6 decimal places in all calculations.

c. How much would adding the 26th row change your expected value?

d. Is the real expected value larger or smaller than your estimate in Part b?

2 Place your estimated expected value from Problem 1, Part b in the appropriate space in a copy of the table below.

Expected Waiting Time

Probability of a Success, p, on Each Trial	Estimated Average Waiting Time (Expected Value)
0.10	10
0.20	
0.30	
0.40	
0.50	
0.60	
0.70	
0.80	
0.90	

a. The students in your class should regroup, if necessary, into seven groups, one for each of the remaining values of p in the table above. Each group should construct a waiting-time probability distribution table for its value of p. Tables should have at least 25 rows. Then estimate the expected waiting time for your value of p.

b. Get the expected waiting time from each of the groups and fill in the rest of the table.

c. Using a copy of this coordinate grid, make a scatterplot of the values from your table.

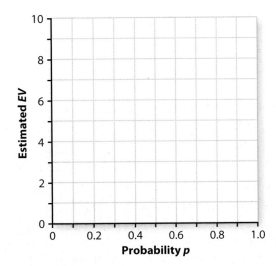

d. What is the expected value when $p = 1.00$? Plot the corresponding point on the scatterplot.

e. Describe the overall pattern relating probability p and expected value EV. Find an equation whose graph fits these points well.

f. According to your equation, how many candies would you expect to draw (from a large bag that contains 20% orange candies) until you get an orange one?

(3) Use your equation from Problem 2 to answer the following questions.

a. Suppose you want a sum of 6. How many times do you expect to have to roll the dice?

b. About 25% of adults bite their fingernails. How many adults do you expect to have to choose at random until you find a fingernail biter?

c. There is about 1 chance in 14,000,000 of winning the California lottery with a single ticket. If a person buys one ticket a week, how many weeks do you expect to pass until he or she wins the lottery? How many years is this? If an average lifetime is 75 years, how many lifetimes is this?

Summarize
the Mathematics

In this investigation, you discovered the formula for the expected value of a waiting-time distribution. Suppose the probability of success on each trial of a waiting-time distribution is *p*.

a Write a formula for the expected value of this distribution. What information do you need to know about a distribution in order to use this formula?

b Explain what happens to the expected waiting time as *p* gets larger.

c How can you estimate the expected waiting time just using the histogram?

Be prepared to share your ideas and examples with the class.

✓Check Your Understanding

Suppose 30% of the beads in a very large bag are red.

a. How many beads would you expect to draw from the bag until you got a red one?

b. Can you conclude from Part a that about half of the people trying this experiment would need that many draws or even more to get a red bead? Explain.

Applications

1. Cereal manufacturers often place small prizes in their cereal boxes as a marketing scheme. Boxes of cereal once contained one of four endangered animal stickers: bird of paradise, tiger, African elephant, and crocodile. Suppose that these stickers were placed randomly into the boxes and that there were an equal number of each kind of animal sticker.

a. Polly likes birds and wanted the bird of paradise sticker. Describe a simulation to estimate the average number of boxes of cereal that Polly would have had to buy before she got a bird sticker.

b. Do five runs of your simulation. Add your results to those in the frequency table below so that there is a total of 100 runs. If needed, add additional rows.

Getting a Bird Sticker

Number of Boxes	Frequency	Number of Boxes	Frequency
1	19	15	0
2	14	16	2
3	15	17	1
4	13	18	0
5	11	19	1
6	6	20	0
7	4	21	0
8	2	22	0
9	1	23	0
10	1	24	0
11	1	25	0
12	3	26	1
13	0	Total	
14	0		

c. Make a histogram from the completed frequency table and describe its shape.

d. Use the completed frequency table to estimate the mean number of boxes a person would have to buy to get a bird of paradise sticker.

e. Estimate the chance that a person would need to buy more than 10 boxes.

2 Two statisticians have estimated that a penny has about a 6% chance of going out of circulation each year. About 10,000,000,000 pennies are minted each year in the United States. Assume that was the number minted the year you were born.

a. Complete this theoretical probability distribution table for the number of those pennies that go out of circulation each year. Add as many rows as you need to get to this year.

Circulation of Pennies from Your Birth Year

Years Since Your Birth	Number of Pennies That Go out of Circulation	Number of Pennies Still Left in Circulation
0	—	10,000,000,000
1	600,000,000	9,400,000,000
2		
3		
4		
5		

b. Write a *NOW-NEXT* equation describing the pattern of change in each of the last two columns.

c. Approximately what percentage of the pennies minted the year you were born are still in circulation?

d. Estimate the mean length of time until a penny goes out of circulation.

e. About how long will it (or did it) take for half of the pennies minted in the year you were born to go out of circulation? (This length of time is called the half-life of a penny.)

3 Suppose that you are drawing one card at a time from a deck of cards and want to get a heart.

a. Describe a simulation using random digits to estimate the mean number of draws from a regular deck of cards needed to get a heart if the card is replaced after each draw.

b. How would you modify your simulation model if the card is not replaced after each draw?

c. What is the probability that it will take you exactly 4 draws to get the first heart if the card is replaced after each draw?

d. What is the probability that it will take you no more than 4 draws to get the first heart if the card is replaced after each draw?

e. What is the probability that it will take you at least 4 draws to get the first heart if the card is replaced after each draw?

f. Would it be a rare event if it took you 5 draws to get the first heart?

4 Suppose there are 1,000 people who, like Polly in Applications Task 1, are buying boxes of cereal in order to get a bird of paradise sticker.

a. Make a chart with six rows that shows how many people you would expect to get a bird of paradise sticker with the first box, with the second box, etc.

b. Write a formula for $P(x)$, the probability of getting a tiger sticker on the xth purchase.

5 Sixteen percent of "M&M's®" Plain Chocolate Candies are green. Suppose that you are drawing candies one at a time from a very large bag of M&M's® in which all the colors have been mixed. (The bag has so many candies in it that you can assume that the probability of drawing a green candy is still about 0.16, no matter how many you have drawn before.)

a. Find the first five entries of the table for the waiting-time probability distribution for drawing a green candy.

b. What is the probability it takes you at least four draws to get a green candy?

6 Consider a situation in which a blood bank is testing people at random until it finds a person with type O blood. About 45% of the U.S. population has type O blood.

a. Make a probability distribution table with five rows for this situation.

b. Write a formula for $P(x)$, the probability that the xth person tested is the first with type O blood.

c. What is the probability that the 5th person tested is the first with type O blood?

d. What is the probability that it takes 5 or more people before getting one with type O blood?

e. Would it be a rare event if it takes 5 people before getting one with type O blood?

7 Painkillers are often given as shots to people who have sustained injuries. The time that it takes for a person's body to get the medicine out of his or her system varies from person to person. Suppose one person is given 400 mg of a medicine, and her body metabolizes the medicine so 35% is removed from her bloodstream each hour.

a. Complete the following chart.

Hours	Milligrams of Medicine Leaving the Blood	Milligrams of Medicine Left in the Blood
0	0	400
1		
2		
3		
4		
5		
6		

b. How is this situation similar to and different from a waiting-time distribution?

c. On average, how long does a milligram of medicine stay in the blood?

d. What is the approximate half-life of medicine in the blood? That is, how long does it take for half of the medicine to be gone?

8 In the 1986 nuclear reactor disaster at Chernobyl in the former Soviet Union, radioactive atoms of strontium-90 were released. Strontium-90 decays at the rate of 2.5% a year.

a. What is the average time it takes for a strontium-90 atom to decay?

b. Supposedly, it will be safe again for people to live in the area after 100 years. What percentage of the strontium-90 released will still be present after 100 years?

Connections

(9) Refer to the table for waiting for doubles in Problem 3 of Investigation 1 on page 563.

    a. Make a scatterplot of the points (*number of rolls to get doubles, expected number of students who get out of jail on this roll*).

    b. Find an algebraic model of the form "$y = ...$" that is a good fit for these points.

    c. Let *NOW* be the number of people who are expected to get doubles on a roll, and let *NEXT* be the number of people who are expected to get doubles on the next roll. Write a rule relating *NOW* and *NEXT*.

(10) Let *A* be the event of requiring exactly 5 rolls of a pair of dice to get doubles for the first time. Let *B* be the event that the first four rolls are not doubles. Find the following probabilities.

    a. $P(A)$ **b.** $P(B)$

    c. $P(A \mid B)$ **d.** $P(B \mid A)$

    e. $P(A \text{ and } B)$ **f.** $P(A \text{ or } B)$

(11) Consider the median as a measure of center of a probability distribution.

    a. Find the median of the waiting-time-for-doubles probability distribution produced in Problems 1–3 of Investigation 2. In this situation, what does the median tell you?

    b. Describe how to find the median of any probability distribution.

    c. Make a rough sketch of the box plot of a waiting-time distribution.

(12) You have studied many different models for expressing relationships between quantitative variables, including linear, exponential, power, and quadratic.

    a. When examining the pattern in your scatterplot from Problem 2 of Investigation 3, which of these models could you immediately rule out as not reasonable? Explain your reasoning.

    b. Refer to Applications Task 7. Write an equation that represents the number of milligrams of medicine left in the blood at any hour *x*.

    c. Refer to Applications Task 2. Write an equation that represents the number of pennies minted in your birth year that are still in circulation *x* years later.

13 When given some information about a waiting-time distribution, you can use connections to reason to other information.

 a. If the expected value of a waiting-time distribution is 6.5, what is the probability *p* of success on each trial?

 b. James draws marbles one at a time from a bag of green and white marbles. He replaces each marble before drawing the next. If the probability that he gets his first green marble on the second draw is 0.24, what percentage of the marbles in the bag are green?

14 Most of the waiting-time distributions that you investigated in this unit could be represented by formulas of the form $P(x) = pq^{x-1}$.

 a. For a given situation, what would the symbols $P(x)$, p, and q represent? How are p and q related?

 b. What is the expected value of this distribution?

15 The waiting-time distribution below was constructed by a computer simulation.

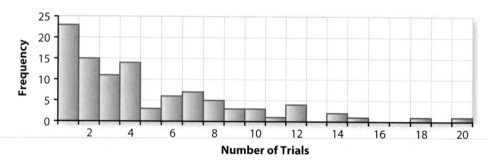

Think of at least two methods that you could use to estimate *p*, the probability of the event occurring on each trial. What is your estimate of *p* with each method?

Reflections

16 April and May are playing Anita's game (See page 561.) and are both in jail. April has tried twice to roll doubles and failed both times. May has tried only once, and she was also unsuccessful. Who has the better chance of rolling doubles on her next turn? Explain your reasoning.

17 Suppose that two dice are rolled repeatedly.

  a. What is the probability of getting doubles on the first roll? On the fourth roll?

  b. What is the probability of getting doubles for the first time on the first roll? What is the probability of getting doubles for the first time on the fourth roll?

  c. Compare your answers to Parts a and b. Why is the probability of getting doubles for the first time on the fourth roll less than the probability of getting doubles on the fourth roll?

18 The table and histogram below give the mileage at which each of 191 buses had its first major motor failure. Is the shape of the distribution the same as that of other waiting-time distributions you have seen? Explain why this makes sense.

Mileage before Failure	Number of Buses
0 to 19,999	6
20,000 to 39,999	11
40,000 to 59,999	16
60,000 to 79,999	25
80,000 to 99,999	34
100,000 to 119,999	46
120,000 to 139,999	33
140,000 to 159,999	16
160,000 and up	4
Total	**191**

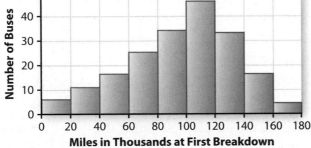

Source: Mudholkar, G.S., D.K. Srivastava, and M. Freimer. "The Exponential Weibull Family: A Reanalysis of the Bus-Motor-Failure Data." *Technometrics* 37 (Nov. 1995): 436–445.

19 Examine each of the following probability distributions.

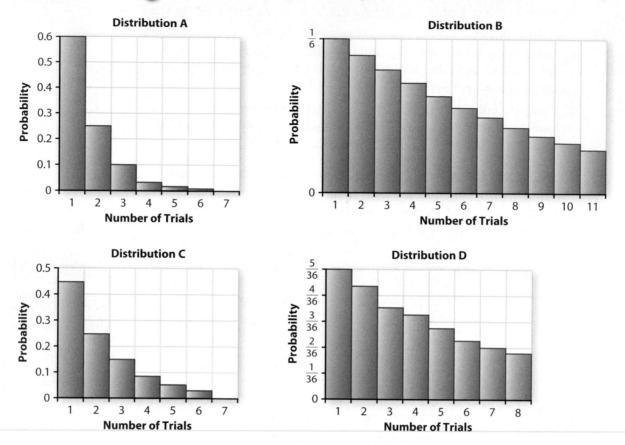

a. Match the following descriptions to the probability distributions.

- rolling a die until a 6 appears
- rolling two dice until a sum of 6 appears
- counting the days on which the weather report states there is a 60% chance of rain until there is a rainy day
- selecting a person at random until one with type O blood appears (About 45% of the U.S. population has type O blood.)

b. What is the height of the second bar (representing 2 trials) in each case? The third bar?

20 Which of the following, if any, are correct interpretations of the expected value (*EV*) of a waiting-time distribution?

- Half of the people wait longer than the EV and half shorter.
- The EV is the most likely time to wait.
- More than half of the people will wait longer than the *EV*.

Extensions

21 Now that you can describe waiting-time probability distributions with an algebraic formula, you can use a calculator to help analyze situations modeled by these distributions. A helpful calculator procedure to produce a waiting-time probability distribution table makes use of a sequence command found on some graphing calculators. The command is usually of the form **seq(*formula, variable, begin, end, increment*)**. The following keystroke procedure uses *A* for the variable and 1 for both the beginning value and the increment. (The exact structure of this command and how to access it vary among calculator models. You may need to refer to the manual for your calculator.)

First, you need to access the sequence command. For example, from the home screen on a TI-84, press 2nd STAT ▶ **5**. Then, complete the command by entering the following example:

Store the resulting values in a list for easier access. The following are sample display screens for this procedure.

 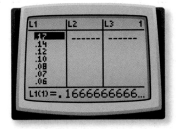

The list now holds the first 12 probabilities of the waiting-time probability distribution for rolling doubles. (Note that the decimal display was set to show only two digits.)

a. Compare the probabilities produced by this procedure with those you calculated in Problem 1 of Investigation 2.

b. Compare the sequence command with the general formula you wrote in Problem 5, Part d of Investigation 2. How are they similar and how are they different?

c. Modify the sequence command to produce the first ten probabilities of the waiting-time probability distribution for drawing a green M&M® as in Applications Task 5. If you completed that task, compare the first five entries.

d. Write a summary of what you need to know in order to use this procedure for producing a waiting-time probability distribution.

e. How could this calculator procedure help you analyze questions about rare events associated with waiting-time probability distributions?

You have seen two ways to compute the expected value of the waiting-time distribution for rolling doubles. The first is to use the formula $\Sigma x \cdot P(x)$, which gives an *infinite series*:

$$1\left(\frac{1}{6}\right) + 2\left(\frac{5}{6}\right)\left(\frac{1}{6}\right) + 3\left(\frac{5}{6}\right)^2\left(\frac{1}{6}\right) + 4\left(\frac{5}{6}\right)^3\left(\frac{1}{6}\right) + 5\left(\frac{5}{6}\right)^4\left(\frac{1}{6}\right) + \cdots$$

The second way is to use the formula you discovered in this investigation:

$$EV = \frac{1}{\frac{1}{6}} = 6$$

Because these two methods give the same expected value, you can set them equal. So,

$$6 = 1\left(\frac{1}{6}\right) + 2\left(\frac{5}{6}\right)\left(\frac{1}{6}\right) + 3\left(\frac{5}{6}\right)^2\left(\frac{1}{6}\right) + 4\left(\frac{5}{6}\right)^3\left(\frac{1}{6}\right) + 5\left(\frac{5}{6}\right)^4\left(\frac{1}{6}\right) + \cdots$$

It is rather amazing that a series can keep going forever and still add up to 6.

a. Write the next three terms of the infinite series above.

b. Here is another example of an infinite series:

   i. Show by dividing 3 into 1 that $\frac{1}{3} = 0.333333333\ldots$.

   ii. Why can you then write
$$\frac{1}{3} = \frac{3}{10} + \frac{3}{100} + \frac{3}{1{,}000} + \frac{3}{10{,}000} + \frac{3}{100{,}000} + \cdots\,?$$

   iii. Multiply both sides of $\frac{1}{3} = 0.3333333\ldots$ by 3.

      What do you conclude?

c. Write $\frac{2}{3}$ as an infinite series.

d. The following square is 1 unit on each side.

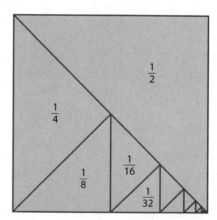

   i. What is the area of the square?

   ii. Describe its area by adding the areas of the (infinitely many) individual triangles.

   iii. What can you conclude?

23 Study the following method to find the sum of one kind of an *infinite geometric series*:

To find the sum of a series like $\frac{1}{2} + \frac{1}{4} + \frac{1}{8} + \frac{1}{16} + \frac{1}{32} + \cdots$ first, set the sum of the series equal to S:

$$S = \frac{1}{2} + \frac{1}{4} + \frac{1}{8} + \frac{1}{16} + \frac{1}{32} + \cdots \qquad \textbf{Equation I}$$

Multiply both sides by 2:

$$2S = 2\left(\frac{1}{2} + \frac{1}{4} + \frac{1}{8} + \frac{1}{16} + \frac{1}{32} + \cdots\right)$$

$$2S = \frac{2}{2} + \frac{2}{4} + \frac{2}{8} + \frac{2}{16} + \frac{2}{32} + \cdots$$

$$2S = 1 + \frac{1}{2} + \frac{1}{4} + \frac{1}{8} + \frac{1}{16} + \cdots \qquad \textbf{Equation II}$$

Finally, subtract each side of Equation I from the corresponding side of Equation II:

$$2S = 1 + \frac{1}{2} + \frac{1}{4} + \frac{1}{8} + \frac{1}{16} + \cdots$$
$$\underline{-S = -\left(\frac{1}{2} + \frac{1}{4} + \frac{1}{8} + \frac{1}{16} + \cdots\right)}$$
$$S = 1$$

In this example, both sides of Equation I were multiplied by 2. With other infinite sums of this type, other numbers must be used.

a. Use the method above to find the sum of each of the infinite series below. You must first find the number to use for the multiplication.

  i. $S = \frac{1}{3} + \frac{1}{9} + \frac{1}{27} + \frac{1}{81} + \cdots$

  ii. $S = \frac{7}{10} + \frac{7}{100} + \frac{7}{1,000} + \cdots$

b. Why doesn't the method work with the following infinite series?

$$\frac{1}{2} + \frac{1}{3} + \frac{1}{4} + \frac{1}{5} + \frac{1}{6} + \frac{1}{7} + \cdots$$

c. Show that the sum of the probabilities in the waiting-time distribution for rolling doubles is equal to 1.

Review

24 Consider the function $f(x) = \frac{64}{x^2}$.

a. Sketch a graph of $f(x)$.

b. Evaluate $f(4)$.

c. Evaluate $f\left(-\frac{1}{2}\right)$.

d. Solve the equation $f(x) = 4$.

25 Shown below is a scatterplot matrix of information about the number of cell phones, main line telephones, and personal computers per 100 persons in 13 different countries. (Source: *Statistical Abstract of the United States: 2007*. U.S. Census Bureau)

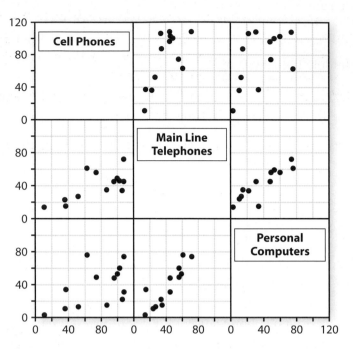

a. Which pair of variables has the strongest association? The weakest association? Explain your reasoning.

b. Which of the following is the correlation for the (*main line telephones, personal computers*) data pairs?

$$r = 0.43 \qquad r = -0.85 \qquad r = 0.88 \qquad r = 1.15$$

c. Saudi Arabia has 15 telephones and 34 personal computers per 100 persons. If Saudi Arabia were deleted from the data set, would the correlation for the (*main line telephones, personal computers*) data pairs increase or decrease?

d. If you convert the data so that it gives the number of cell phones, main line telephones, and personal computers per person rather than per 100 persons will the correlation for the (*main line telephones, personal computers*) data pairs be the same, larger, or smaller? Explain your reasoning.

26 Systems of equations can have no solutions, exactly one solution, or an infinitely many solutions. Consider the following system of equations.

$$4x + 18y = 32$$
$$2x + ky = c$$

a. Find values for k and c so that the system has exactly one solution.

b. Find values for k and c so that the system has no solution.

c. Find values for k and c so that the system has infinitely many solutions.

27 Namid is considering using regular pentagons with sides of length 10 cm as part of the pattern on her quilt.

   a. How could she find the measure of each angle of the regular pentagon? What is the angle measure?

   b. How could she find the area of each pentagon? What is the area?

   c. Namid has lots of squares of material left over from a previous project. Each square has sides of length 20 cm. Could she use these scraps to make her pentagons?

   d. Could she make her entire quilt using just regular pentagons sewn side-to-side? Explain your reasoning.

28 Solve each equation by reasoning with the symbols themselves.

   a. $(x - 5)(x + 3) = 9$

   b. $-3(2x + 7) + 2(12 - x) = 21$

   c. $\dfrac{3}{x} = x - 2$

   d. $2x^2 + 6x = 4$

29 Consider $\triangle ABC$ with vertices at $A(-2, 5)$, $B(2, 8)$ and $C(7, 8)$.

   a. Is $\triangle ABC$ an isosceles triangle?

   b. Find the measure of all three angles of $\triangle ABC$.

   c. Find the coordinates of the vertices of the final image of $\triangle ABC$ if it is first reflected across the line $y = x$ and then the image is rotated clockwise 90° about the origin.

   d. Write a coordinate rule for the composite transformation described in Part c.

   e. Represent $\triangle ABC$ with a matrix. What matrices could you use to find the image of $\triangle ABC$ in Part c?

30 Write each radical expression in equivalent form so that the number under the radical sign is the smallest possible whole number.

   a. $\sqrt{49}$

   b. $\sqrt{50}$

   c. $\sqrt{52}$

   d. $5\sqrt{24}$

Looking Back

In this unit, you learned about probability distributions and the rules used to construct them. Among the concepts you investigated were the Multiplication Rule, independent events, conditional probability, fair price, and expected value. You used these concepts to construct the probability distribution table for a waiting-time situation. In this lesson, you will review and apply many of these ideas in new contexts.

1. North Carolina collects data on all bicycle crashes with motor vehicles that are reported to the police. The table below gives the injury status and whether the bicyclist was wearing a helmet for all crashes in a recent year. Children and young adults are the most frequent victims.

	Killed/ Disabling Injury	Less Serious or No Injury	Total
Helmet Not Used	68	598	666
Helmet Used	2	27	29
Total	70	625	695

Source: www.pedbikeinfo.org/pbcat/bike_main.htm

Suppose you pick a crash at random.

a. Compare P(*killed/disabled* | *helmet not used*) with P(*killed/disabled* | *helmet used*). What can you conclude about helmet use?

b. Are being killed or getting a disabling injury and using a helmet independent events? Are they mutually exclusive events?

c. Use the Multiplication Rule to find P(*killed/disabled* and *helmet used*). Check your answer by using the table directly.

d. Use the Addition Rule to find P(*killed/disabled* or *helmet used*). Check your answer by using the table directly.

2. Throughout the basketball season, Delsin has maintained a 60% free-throw shooting average.

   a. Suppose in the first game of the post-season tournament, Delsin is in a two-shot foul situation. That is, he gets two attempts to make a free throw, which are worth one point each. Use an area model to determine:

   i. P(*Delsin scores 2 points*)

   ii. P(*Delsin scores 1 point*)

   iii. P(*Delsin scores 0 points*)

   b. Show how to use a formula to determine the probability that Delsin will score 2 points.

   c. What is the expected number of points Delsin will score?

   d. Suppose that later in the game, Delsin is in a one-and-one free throw situation. That is, if he makes the free throw on his first try, he gets a second free throw. If he misses the first free throw, he does not get a second attempt. Determine the expected number of points Delsin will score in this situation.

3. The Bonus Lotto game described below is similar to those played in many states. The jackpot starts at $4,000,000. On Saturday, 6 numbers from 1 through 47 are drawn. A seventh number, called the Bonus Ball, is then drawn from the remaining numbers. A player wins if the numbers he or she selects match the Bonus Ball and at least two of the numbers drawn.

   The probabilities of winning various prizes are given in the following table.

Match	Winnings	Probability
6 of 6	$4,000,000	$\frac{1}{10,737,573}$
5 of 6 + bonus ball	$50,000	$\frac{1}{1,789,595}$
4 of 6 + bonus ball	$1,000	$\frac{1}{17,896}$
3 of 6 + bonus ball	$100	$\frac{1}{688}$
2 of 6 + bonus ball	$4	$\frac{1}{72}$
Other	0	

   a. What is the probability of winning nothing? Write your answer in decimal form.

   b. What would be a fair price to pay for a ticket?

   c. Bonus Lotto costs $2 to play. How much does the state expect to earn on every 1,000,000 tickets sold?

d. If the jackpot is not won on the Saturday drawing, it grows by $4,000,000 for the next week. Other prize winnings remain the same. If you buy one ticket that second week, what is the probability of winning the jackpot?

e. What is a fair price for a ticket the second week?

f. What is the probability that a person who plays Bonus Lotto once this week and once next week will not win anything either week?

g. Suppose a person buys one ticket a week. What is the expected number of weeks he or she will have to wait before winning the jackpot? How many years is this? How much money will have been spent on ticket purchases?

h. The table above actually simplified the situation. In fact, if there is more than one winner, the $4,000,000 jackpot is shared. Explain why this fact makes the answer to Part b even smaller.

4 In the *Patterns in Chance* unit in Course 1, one investigation focused on the population issues in China. In 2000, the population of China was more than 1,200,000,000. To control population growth, the government of China has attempted to limit parents to one child each. This decision has been unpopular in the areas of rural China where the culture is such that many parents desire a son.

Suppose that a new policy has been suggested by which parents are allowed to continue having children until they have a boy. For the following problems, assume that the probability that a child born will be a boy is 0.51 and that births are independent.

a. Describe a method using random digits to simulate the situation of parents having children until they get a boy.

b. Out of every 100 sets of parents, how many would you expect to get the first boy with the first baby? With the second baby? With the third baby?

c. Construct a theoretical probability distribution table for this situation, using a copy of the table below.

Number of Children to Get First Boy	Probability
1	
2	
3	
4	
5	
6	
7	
8 or more	

d. From your table, estimate the mean number of children two parents will have.

e. From the formula for expected value, what is the expected number of children?

f. Explain whether the population will increase, decrease, or stay the same under this plan.

g. If this new policy were adhered to, what percentage of the population would be boys? Explain your reasoning.

Summarize
the Mathematics

In this unit, you explored the multiplication rules of probability and used those rules to construct a theoretical waiting-time distribution.

a Give an example that illustrates the difference between $P(A)$ and $P(A|B)$.

b How can you find $P(A \text{ and } B)$ when events A and B are independent? When they are not?

c What is the expected value of a probability distribution? How do you compute it?

d Describe a waiting-time distribution. Include:

- how to construct the probability distribution table.
- how to develop the formula.
- what the shape of the distribution looks like.
- ways to find the average waiting time.

Be prepared to share your ideas and examples with the class.

✓Check Your Understanding

Write, in outline form, a summary of the important mathematical concepts and methods developed in this unit. Organize your summary so that it can be used as a quick reference in future units and courses.

Glossary/Glosario

Math Online A mathematics multilingual glossary is available at www.math.glencoe.com/multilingual_glossary. The Glossary includes the following languages:

Arabic   English   Korean   Tagalog
Bengali   Hatian Creole   Russian   Urdu
Cantonese   Hmong   Spanish   Vietnamese

English	Español

(A)

Addition Rule Formula for the probability that event A occurs, or event B occurs, or both events occur: $P(A \text{ or } B) = P(A) + P(B) - P(A \text{ and } B)$.

Regla de adición Fórmula para encontrar la probabilidad que el evento A ocurra o que el evento B ocurra, o que ambos eventos ocurran: $P(A \text{ ó } B) = P(A) + P(B) - P(A \text{ y } B)$.

Addition Rule for Mutually Exclusive Events Simplified form of the Addition Rule that can be used only when event A and event B are mutually exclusive: $P(A \text{ or } B) = P(A) + P(B)$.

Regla de adición para eventos mutuamente excluyentes Forma simplificada de la Regla de adición que puede usarse solamente cuando el evento A y el evento B son mutuamente excluyentes: $P(A \text{ ó } B) = P(A) + P(B)$.

Additive identity matrix (p. 134) An $m \times n$ matrix E such that $B + E = B = E + B$, for all $m \times n$ matrices B. (Also called a *zero matrix*, since every entry in an additive identity matrix is 0.)

Matriz identidad aditiva (p. 134) Una E matriz $m \times n$ de modo que $B + E = B = E + B$, para todas las B matrices $m \times n$. (También se le llama *matriz cero*, debido a que cada entrada en una matriz de identidad aditiva es 0.)

Additive inverse matrix (p. 134) The matrix which when added to a given matrix yields the zero matrix.

Matriz inversa aditiva (p. 134) Matriz que al sumarla a una matriz dada da como resultado la matriz cero.

Adjacency matrix for a digraph (p. 93) Matrix representation of a digraph in which the vertices are used as labels for the rows and columns of a matrix and each entry of the matrix is a "1" or a "0" depending on whether or not there is a directed edge from the row vertex to the column vertex. (Sometimes an adjacency matrix is constructed such that each entry is the *number* of directed edges from the row vertex to the column vertex, thus an entry could be larger than 1.)

Matriz de adyacencia para un dígrafo (p. 93) La representación de una matriz de un dígrafo en el cual los vértices se usan como rótulos para las filas y columnas de una matriz y cada entrada de la matriz es un "1" o un "0" dependiendo de si hay o no una arista dirigida de la fila del vértice hacia la columna del vértice. (Algunas veces una matriz de adyacencia se construye de tal manera que cada entrada es el *número* de aristas dirigidas de la fila del vértice a la columna del vértice, por lo tanto una entrada puede ser mayor que 1.)

Algorithm (p. 167) A list of step-by-step instructions, or a systematic step-by-step procedure.

Algoritmo (p. 167) Lista de instrucciones detalladas o procedimiento detallado.

Angle in standard position (p. 461) A directed angle with vertex at the origin of a rectangular coordinate system and initial side the positive x-axis.

Ángulo en posición estándar (p. 461) Un ángulo dirigido con vértice en el origen de un sistema rectangular de coordenadas y lado inicial del eje x positivo.

Angle of depression (p. 476) The acute angle between the line of sight and the horizontal when sighting from one point downward to a second point.

Ángulo de depresión (p. 476) Ángulo agudo entre la línea de visión y la horizontal cuando se aprecia de un punto hacia abajo a un segundo punto.

Angle of elevation (also called *angle of inclination*) (p. 481) The acute angle between the line of sight and the horizontal when sighting from one point upward to a second point.

Ángulo de elevación (también llamado *ángulo de inclinación*) (p. 481) Ángulo agudo entre la línea de visión y la horizontal cuando se aprecia de un punto hacia arriba de un segundo punto.

Glossary/Glosario

English	Español
Association (p. 299) A relationship between two variables. Association can be positive or negative, weak or strong, curved or linear. Compare with *correlation*.	**Asociación** (p. 299) Una relación entre dos variables. La asociación puede ser positiva o negativa, débil o fuerte, curva o lineal. Comparar con *correlación*.

••••••••••••••••••••••••••••••• **B** •••••••••••••••••••••••••••••••

Bivariate data (p. 257) Data consisting of ordered pairs that are responses for two variables for each person or object of study, such as age and height for each tree in a park.	**Datos covariantes** (p. 257) Datos que constan de pares ordenados que son las respuestas para dos variables para cada persona u objeto de estudio, tales como edad y altura para cada árbol de un parque.
Brute-force method (p. 409) A problem-solving method that involves finding and checking all possibilities.	**Método de la fuerza bruta** (p. 409) Método de resolución de problemas relacionado con hallar y comprobar todas las posibilidades.

••••••••••••••••••••••••••••••• **C** •••••••••••••••••••••••••••••••

Cause-and-effect relationship (p. 299) A change in the value of one variable (called the *explanatory* or *independent variable*) tends to cause a change in the value of a second variable (called the *response* or *dependent variable*).	**Relación de causa y efecto** (p. 299) Un cambio en el valor de una variable (llamada la *variable explicativa* o *independiente*) tiende a causar un cambio en el valor de una segunda variable (llamada *variable de respuesta* o *dependiente*).
Centroid (p. 286) On a scatterplot of points (x, y), the point $(\overline{x}, \overline{y})$, where $\overline{x}$ is the mean of the values of x and $\overline{y}$ is the mean of the values of y. (In a polygon, the point that is the "center of gravity.")	**Centroide** (p. 286) En una gráfica de dispersión, de los puntos (x, y), el punto $(\overline{x}, \overline{y})$, donde $\overline{x}$ es la media de los valores de x, y $\overline{y}$ es la media de los valores de y. (En un polígono, el punto que es el "centro de gravedad.")
Circle The set of all points in a plane that are a fixed distance r, called the *radius*, from a given point O, called the *center* of the circle.	**Círculo** Conjunto de todos los puntos de un plano que están a una distancia dada r, llamada *radio*, de un punto O dado, denominado *centro* del círculo.
Circuit (p. 403) A route through a vertex-edge graph that starts and ends at the same vertex and does not repeat any edges.	**Circuito** (p. 403) Una ruta a través de un grafo que empieza y termina en el mismo vértice y no repite ningún vértice.
Closed interval A continuous interval of real numbers that includes the endpoints of the interval; the interval from a to b, including a and b, is denoted $[a, b]$ or $\{x: a \le x \le b\}$.	**Intervalo cerrado** Intervalo continuo de números reales que incluye todos sus puntos límite; el intervalo de a a b, incluyendo a y b, se denota como $[a, b]$ ó $\{x: a \le x \le b\}$.
Column matrix (p. 119) A matrix consisting of one column. (Also called a *column vector* or a *one-column matrix*.)	**Matriz columna** (p. 119) Matriz formada por una columna. (También llamada *vector de columna* o *matriz de una columna*.)
Column of an $m \times n$ matrix (p. 76) A vertical array of m numbers in the matrix.	**Columna de una matriz $m \times n$** (p. 76) Matriz o conjunto vertical de m números en la matriz.
Column sum of a matrix (p. 81) The sum of all numbers in a column of a matrix.	**Suma de columna de una matriz** (p. 81) La suma de todos los números en la columna de una matriz.

Glossary/Glosario

|

Complementary angles (p. 481) Two angles whose measures sum to 90°.

Components of a translation (p. 199) The horizontal and vertical directed distances (left or right, up or down) through which all points in the plane are moved by the translation.

Composition of transformations (p. 213) The result of applying two transformations in succession. The transformation that maps the *original preimage* to the *final image* is called the *composite transformation*.

Conditional probability (p. 528) The probability that an event *A* occurs given that another event *B* occurs, written $P(A \mid B)$. When $P(B) > 0$,
$$P(A \mid B) = \frac{P(A \text{ and } B)}{P(B)}.$$

Congruent figures Figures that have the same shape and size, regardless of position or orientation. (For angles: having the same measure; for segments: having the same length.)

Connected graph (p. 406) A vertex-edge graph that is all in one piece; that is, from each vertex there is at least one path to every other vertex.

Correlation (p. 257) A measure, usually Pearson's *r*, of the linear association between two variables. A number between −1 and 1 that tells how closely the points on a scatterplot cluster about the regression line.

Cosine function (p. 457) If $P(x, y)$ is a point (not the origin) on the terminal side of an angle θ in standard position and $r = \sqrt{x^2 + y^2}$, then $\cos \theta = \frac{x}{r}$. If *A* is an acute angle in a right triangle, then $\cos A = \dfrac{\text{length of side } adjacent \text{ to } \angle A}{\text{length of } hypotenuse}$.

Critical path (p. 434) A path through a *project digraph* that corresponds to the earliest finish time for the project.

Ángulos complementarios (p. 481) Dos ángulos cuyas medidas suman 90°.

Componentes de una traslación (p. 199) Distancias dirigidas horizontal y vertical (izquierda o derecha, arriba o abajo) por las cuales se mueven todos los puntos en un plano por una traslación.

Composición de transformaciones (p. 213) Resultado de aplicar dos transformaciones en sucesión. La transformación que traza la *figura geométrica original* (o preimagen) a la *imagen final* se llama *transformación compuesta*.

Probabilidad condicional (p. 528) La probabilidad que un suceso *A* ocurra a condición de que otro suceso *B* ocurra, de manera escrita es $P(A \mid B)$. Cuando $P(B) > 0$, $P(A \mid B) = \dfrac{P(A \text{ y } B)}{P(B)}$.

Figuras congruentes Figuras de la misma forma y tamaño, sin importar su posición u orientación. (Para los ángulos: tener la misma medida; para los segmentos: tener la misma longitud.)

Grafo conexo (p. 406) Un grafo que es de una sola pieza, o sea, de cada vértice hay por lo menos un camino a cada uno de los otros vértices.

Correlación (p. 257) Una medida, usualmente la *r* de Pearson, de la asociación lineal entre dos variables. Un número entre −1 y 1 que indica qué tan cerca los puntos en una gráfica de dispersión se agrupan acerca de la línea de regresión.

Función de coseno (p. 457) Si $P(x, y)$ es un punto (no el origen) en el lado terminal de un ángulo θ en posición estándar y $r = \sqrt{x^2 + y^2}$, entonces coseno θ $= \frac{x}{r}$. Si *A* es un ángulo agudo en un triángulo rectángulo, entonces $\cos A = \dfrac{\text{medida del cateto } adyacente \text{ a } \angle A}{\text{medida de la } hipotenusa}$.

Trayectoria crítica (p. 434) Una trayectoria a través de un *dígrafo de proyecto* que corresponde al tiempo más temprano del final para el proyecto.

Glossary/Glosario

English	Español

Critical Path Method (CPM) (p. 437) A method using critical path analysis to optimally schedule large projects consisting of many subtasks; developed at about the same time as and similar to the Program Evaluation and Review Technique (PERT).

Método de Trayectoria Crítica (CPM por sus siglas en inglés) (p. 437) Método que emplea un análisis crítico de trayectoria para programar de manera óptima proyectos grandes compuestos de muchos subproyectos; desarrolladas aproximadamente al mismo tiempo y similares a la Evaluación de Programa y Técnica de Revisión (PERT por sus siglas en inglés).

Critical task (p. 440) A task on a *critical path*.

Tarea crítica (p. 440) Tarea o trabajo en una *trayectoria crítica*.

Ⓓ

Dart A nonconvex quadrilateral with two pairs of congruent consecutive sides.

Dardo Cuadrilátero no convexo con dos pares de lados consecutivos congruentes.

Degree of a vertex The number of edges touching the vertex. If an edge loops back to the same vertex, that counts as two edge-touchings.

Grado de un vértice Número de aristas que concurren en el vértice. Si un extremo se regresa al mismo vértice, eso cuenta como dos veces.

Dependent variable (p. 3) A variable whose value changes in response to change in one or more related independent variables. (Also called a *response variable*.)

Variable dependiente (p. 3) Variable cuyo valor cambia en respuesta a cambios en una o más variables independientes relacionadas. (También llamada *variable de respuesta*.)

Diagonal of a polygon A line segment connecting two vertices that are not adjacent.

Diagonal de un polígono Un segmento de recta que conecta dos vértices que no son adyacentes.

Digraph (p. 93) A vertex-edge graph in which all the edges are directed, that is, the edges have arrows indicating a direction. (Also called a *directed graph*.)

Dígrafo (p. 93) Grafo en el cual todas las aristas están dirigidas, es decir, las aristas tienen flechas que indican una dirección. (También se le llama *gráfico dirigido*.)

Direct variation (p. 1) If variables x and y are related by an equation in the form $y = kx$ or $\frac{y}{x} = k$, then y is said to vary directly with x, or be directly proportional to x.

Variación directa (p. 1) Si las variables x y y están relacionadas por una ecuación en la forma $y = kx$ o $\frac{y}{x} = k$, entonces se dice que y varía directamente con x, o que es directamente proporcional a x.

Directed edge (p. 93) An edge in a vertex-edge graph with a direction indicated.

Arista dirigida (p. 93) Arista de un grafo en que se indica la dirección de la misma.

Directed graph See *digraph*.

Grafo dirigido Véase *dígrafo*.

Distance Formula (p. 171) Formula for calculating the distance between two points in the coordinate plane.

Fórmula de distancia (p. 171) Fórmula para calcular la distancia entre dos puntos en un plano de coordenadas.

Distance matrix (p. 414) A matrix representation of a weighted graph in which the vertices are labels for the rows and columns and each entry is the length of a shortest path between the corresponding vertices.

Matriz distancia (p. 414) Representación de una matriz de un gráfico cargado en el cual los vértices son rótulos para las filas y columnas y cada entrada es el largo de la trayectoria más corta entre los vértices correspondientes.

Glossary/Glosario

English	Español

Domain of a function (p. 330) For a function f, all values of the independent variable x that have corresponding $f(x)$ values. (Also called *input values* for the function.)

Dominio de una función (p. 330) Para una función f, todos los valores de la variable independiente x que tienen valores $f(x)$ correspondientes. (También llamados *valores de entrada* para la función.)

····················· E ·····················

Earliest Finish Time (EFT) (p. 438) The minimum amount of time needed to complete a large project that consists of numerous subtasks.

Tiempo mínimo de resolución (EFT por sus siglas en inglés) (p. 438) La cantidad mínima de tiempo necesario para terminar un proyecto grande compuesto por varios subproyectos.

Elimination method (p. 55) A method used to solve a system of linear equations. One or both of the equations may be multiplied by a nonzero constant so that the coefficient of one of the variables is the same in both equations, subtracting the equations eliminates that variable.

Método de eliminación (p. 55) Método que se utiliza para resolver un sistema de ecuaciones lineales. Una o ambas ecuaciones se pueden multiplicar por una constante no-cero para que el coeficiente de una de las variables sea el mismo en ambas ecuaciones, restar las ecuaciones elimina esa variable.

Equally likely outcomes (p. 562) Outcomes that all have the same probability of occurring.

Resultados equiprobables (p. 562) Resultados que tienen la misma oportunidad de ocurrir.

Equation of a line An equation that can be expressed as $ax + by = c$ where a and b are not both 0.

Ecuación de una línea Una ecuación que puede expresarse como $ax + by = c$ en donde a y b no son ambas 0.

Error in prediction (p. 283) For points not used to calculate the regression equation, the difference between the observed value of y and the value of y predicted by the regression equation.

Error en la predicción (p. 283) Para aquellos puntos no utilizados para calcular la ecuación de regresión, la diferencia entre el valor observado de y y el valor de y predicho por la ecuación de regresión.

Euler circuit A route through a connected graph such that (1) each edge is used exactly once, and (2) the route starts and ends at the same vertex.

Circuito de Euler Camino en un grafo conexo de modo que (1) cada arista se recorre sólo una vez y (2) el camino empieza y termina en el mismo vértice.

Expected value or expectation (p. 525) The mean, or average, of a probability distribution.

Valor previsto o de expectativa (p. 525) Media, o promedio, de una distribución de probabilidad.

Experiment (p. 4) A research study in which subjects are randomly assigned to two or more different treatments in order to compare how the responses to the treatments differ.

Experimento (p. 4) Estudio de investigación en el cual se asignan sujetos al azar a dos o más tratamientos diferentes para comparar cómo se diferencian las respuestas a los tratamientos.

Explanatory variable See *independent variable*.

Variable explicativa Véase *variable independiente*.

Exponential function A function with rule of correspondence that can be expressed in the algebraic form $f(x) = a(b^x)$ ($a, b > 0$).

Función exponencial Función con regla de correspondencia que se puede expresar en forma algebraica $f(x) = a(b^x)$ ($a, b > 0$).

Glossary/Glosario

English	Español

Fair price (p. 546) The price that should be charged to play a game so that, in the long run, the player wins the same amount that he or she pays to play.

Precio justo (p. 546) Precio que se debe cobrar para jugar un juego de modo que, a la larga, el jugador gane la misma cantidad que éste paga por jugar.

Function (p. 161) A relationship between two variables in which each value of the independent variable x corresponds to exactly one value of the dependent variable y. The notation $y = f(x)$ is often used to denote that y is a function of x.

Función (p. 161) Una relación entre dos variables en la cual cada valor de la variable independiente x corresponde exactamente a un valor de la variable dependiente y. La notación $y = f(x)$ comúnmente se usa para denotar que y es una función de x.

Glide-reflection (p. 224) A rigid transformation that is the composition of a reflection across a line and a translation in a direction parallel to the line.

Reflexión del deslizamiento (p. 224) Transformación rígida que es la composición de una reflexión en una recta y una traslación en dirección paralela a esa recta.

Graph See *vertex-edge graph*.

Gráfica Véase *grafo*.

Half-turn (p. 229) A 180° rotation about a point.

Media vuelta (p. 229) Rotación de 180° con relación de un punto.

Hamilton circuit (p. 408) A route through a vertex-edge graph that starts at one vertex, visits all the other vertices exactly once, and ends at the starting vertex.

Circuito de Hamilton (p. 408) Camino a través de un grafo que empieza en una vértice, visita todas las demás vértices exactamente una vez y termina en el vértice inicial.

Identity matrix (p. 135) An $n \times n$ square matrix I such that $A \times I = I \times A = A$ for all $n \times n$ matrices A. (Also sometimes called a *multiplicative identity matrix*.)

Matriz identidad (p. 135) Una matriz I cuadrada $n \times n$ de modo que $A \times I = I \times A = A$ para todas las matrices A $n \times n$. (También llamada en ocasiones una *matriz multiplicativa de identidad*.)

Identity transformation A rigid transformation that maps each point of the plane onto itself.

Transformación de identidad Transformación rígida que traza cada punto de un plano en sí mismo.

Immediate prerequisite table (p. 436) A table showing the immediate prerequisites for each task within a large project.

Tabla de prerrequisitos inmediatos (p. 436) Tabla que muestra los prerrequisitos inmediatos para cada tarea dentro de un proyecto grande.

Independent events (p. 523) Two events A and B are independent if the occurrence of one of the events does not change the probability that the other event occurs. That is, $P(A \mid B) = P(A)$. Alternatively, events A and B with nonzero probabilities are independent if $P(A \text{ and } B) = P(A) \cdot P(B)$.

Eventos independientes (p. 523) Dos eventos A y B son independientes si el que uno ocurra no afecta la probabilidad de que el otro ocurra. Es decir, $P(A \mid B) = P(A)$. Igualmente, los eventos A y B con probabilidad no-cero son independientes si $P(A \text{ y } B) = P(A) \cdot P(B)$.

Glossary/Glosario

English	Español

Independent variable (p. 3) Variables whose values are restricted only by the context of the problem or by mathematical restrictions on allowed values. These variables influence the values of other variables called *dependent variables*. (Also called an *explanatory variable*.)

Variable Independiente (p. 3) Variable cuyos valores están restringidos solamente por el contexto de un problema o por restricciones matemáticas sobre los valores permitidos. Estas variables influyen en los valores de otras variables, las llamadas *variables dependientes*. (También llamada *variable explicativa*.)

Influential point (p. 287) On a scatterplot, an outlier such that when it is removed from the data set, the slope or y-intercept of the regression line changes quite a bit, where "quite a bit" must be determined by the real-life situation.

Punto influyente (p. 287) En una gráfica de dispersión, un valor atípico dado que cuando se elimina del conjunto de datos, la pendiente o intersección y de la línea de regresión cambia un poco, situación en la cual "un poco" debe determinarse de acuerdo con la situación real.

Initial side of an angle (p. 461) The position of a ray that is one side of the angle before it rotates about the angle's vertex to the terminal side.

Lado inicial de un ángulo (p. 461) Posición de una semirrecta que está en un lado del ángulo antes de rotar en el vértice de un ángulo al lado terminal.

Inscribed angle (p. 178) An angle whose vertex is on the circumference of a circle and whose sides are segments connecting the vertex to two other points on the circumference.

Ángulo inscrito (p. 178) Ángulo cuyo vértice está en la circunferencia de un círculo y cuyos lados son segmentos que conectan el vértice con otros dos puntos en la circunferencia.

Inverse matrix (p. 137) For a given square matrix A, the matrix denoted A^{-1} (if it exists) that satisfies $A \times A^{-1} = A^{-1} \times A = I$, where I is the identity matrix. (Also called *multiplicative inverse matrix*.)

Matriz inversa (p. 137) Para una matriz cuadrada dada A, la matriz denominada A^{-1} (si existe) que satisface $A \times A^{-1} = A^{-1} \times A = I$, donde I es la matriz identidad. (También llamada *matriz inversa multiplicativa*.)

Inverse variation (p. 42) If variables x and y are related by an equation in the form $y = \frac{k}{x}$ or $yx = k$, then y is said to vary inversely with x, or be inversely proportional to x.

Variación inversa (p. 42) Si las variables x y y están relacionadas por una ecuación en la forma de $y = \frac{k}{x}$ o $yx = k$, entonces se dice que y está inversamente con x, o que es inversamente proporcional a x.

Isosceles triangle A triangle with at least two congruent sides. The side that joins the congruent sides is called the *base*, and the angles that lie opposite the congruent sides are called the *base angles*.

Triángulo isósceles Triángulo con por lo menos dos lados congruentes. El lado que une a los lados congruentes se llama *base* y los ángulos opuestos a los lados congruentes se llaman *ángulos basales*.

· (K) ·

Kite A convex quadrilateral with two distinct pairs of congruent consecutive sides.

Deltoide Cuadrilátero convexo con exactamente dos pares de lados congruentes consecutivos.

Kruskal's Algorithm (p. 405) An algorithm for finding a minimum spanning tree in a connected graph.

Algoritmo de Kruskal (p. 405) Algoritmo para hallar un árbol de expansión en un grafo conectado.

Glossary/Glosario

····················· (L) **·····················**

Law of Cosines (p. 457) In any triangle ABC with sides of lengths a, b, and c opposite $\angle A$, $\angle B$, and $\angle C$, respectively: $c^2 = a^2 + b^2 - 2ab \cos C$.

Law of Large Numbers In a simulation, the more runs there are, the closer the probability determined by the simulation tends to the theoretical probability.

Law of Sines (p. 457) In any triangle ABC with sides of lengths a, b, and c opposite $\angle A$, $\angle B$, and $\angle C$, respectively: $\frac{\sin A}{a} = \frac{\sin B}{b} = \frac{\sin C}{c}$.

Least squares regression line (p. 257) The line on a scatterplot that has the smallest sum of squared residuals (SSE). (Also called the *regression line*.)

Line reflection (p. 201) A rigid transformation which associates with each point P in a plane an image point P' such that the "mirror line" (or line of reflection) is the perpendicular bisector of the segment $\overline{PP'}$ if P is not on the line of reflection. A point on the line of reflection is its own image.

Linear data (p. 264) The points on a scatterplot are called "linear" if they form an elliptical cluster so that a line is an appropriate summary.

Linear function A function with rule of correspondence that can be expressed in the algebraic form $f(x) = mx + b$.

Linear scale (p. 388) A scale for which the difference between equally spaced scale points is a constant.

Logarithm (p. 325) If $y = 10^x$ then x is the common or base-10 logarithm of y. This relationship is often indicated by the notation $y = \log x$ or $y = \log_{10} x$.

Logarithmic function (p. 381) A function with rule of correspondence that can be expressed in the form $f(x) = \log x$.

Ley de cosenos (p. 457) En cualquier triángulo ABC con longitudes de sus lados a, b, y c opuestos $\angle A$, $\angle B$, y $\angle C$, respectivamente: $c^2 = a^2 + b^2 - 2ab \cos C$.

Ley de números grandes En una simulación, si el número de repeticiones es mayor, más cercana a la probabilidad teórica tiende a ser la probabilidad determinada por la situación

Ley de senos (p. 457) En un triángulo ABC con longitudes de sus lados a, b, y c opuesto $\angle A$, $\angle B$, y $\angle C$, respectivamente: $\frac{\sin A}{a} = \frac{\sin B}{b} = \frac{\sin C}{c}$.

Línea de regresión de cuadrados mínimos (p. 257) La línea en una gráfica de dispersión que tiene la menor suma de residuos cuadrados (SSE por sus siglas en inglés). (También llamada *línea de regresión*.)

Línea reflexión (p. 201) Una transformación rígida que se relaciona con cada punto P en un plano con un punto de imagen P' de modo que la "línea de espejo" (o línea de reflexión) es el bisector perpendicular del segmento $\overline{PP'}$ si P no está en la línea de reflexión. Un punto en la línea de reflexión es su propia imagen.

Datos lineales (p. 264) Los puntos en una gráfica de dispersión se llaman "lineales" si forman un grupo elíptico para que una recta sea un resumen apropiado.

Función lineal Función con regla de correspondencia que puede expresarse en la forma algebraica $f(x) = mx + b$.

Escala lineal (p. 388) Escala para la cual la diferencia entre los puntos de la misma están separados por igual.

Logaritmo (p. 325) Si $y = 10^x$ entonces x es el logaritmo común o de base 10 de y. La notación $y = \log x$ ó $y = \log_{10} x$ usualmente indica esta relación.

Función logarítmica (p. 381) Función con regla de correspondencia que puede expresarse en la forma $f(x) = \log x$.

Glossary/Glosario

English	Español

Logarithmic scale (p. 388) A scale for which the ratio between consecutive scale points is a constant.

Lurking variable (p. 300) When explaining the association between two variables, a third variable that affects both of the original variables.

Escala logarítmica (p. 388) Escala para la cual el radio entre puntos consecutivos de la escala es constante.

Variable latente (p. 300) Al explicar la asociación entre dos variables, una tercera variable que afecta a ambas variables originales.

· (M) ·

Main diagonal of a square matrix (p. 80) The entries in the matrix that run from the top-left corner of the matrix to the bottom-right corner.

Matrix (p. 73) A rectangular array of numbers (plural: *matrices*).

Matrix addition (p. 134) Two matrices A and B, having the same size, are combined by adding their corresponding entries to produce the sum matrix, $A + B$.

Matrix multiplication (p. 105) An $m \times k$ matrix A and a $k \times n$ matrix B are multiplied to produce the $m \times n$ product matrix, $A \times B$, in which the entries of $A \times B$ are computed by a specific method of combining rows of A with columns of B.

Matrix of coefficients of a system of linear equations (p. 133) A matrix whose entries are the coefficients of the variables in the system of linear equations.

Midpoint (p. 163) The point on a segment that is equidistant from the endpoints of the segment.

Midpoint Formula (p. 169) Formula for calculating the coordinates of the midpoint of the segment connecting two points in the coordinate plane.

Minimum spanning tree (p. 403) A spanning tree in a vertex-edge graph that has minimum total weight.

Multiplication rule (p. 532) If A and B are two events, $P(A \text{ and } B) = P(A)P(B \mid A)$.

Multiplication Rule for Independent Events (p. 523) When events A and B are independent, the multiplication rule simplifies to $P(A \text{ and } B) = P(A) \cdot P(B)$.

Diagonal principal de una matriz cuadrada (p. 80) Entradas de una matriz que van de la esquina superior izquierda de la matriz a la esquina inferior derecha.

Matriz (p. 73) Arreglo rectangular de números (plural: *matrices*).

Adición de la matriz (p. 134) Dos matrices A y B, que tienen el mismo tamaño, se combinan al sumar sus entradas correspondientes para producir la suma de la matriz, $A + B$.

Multiplicación de la matriz (p. 105) Una matriz A $m \times k$ y una matriz B $k \times n$ se multiplican para producir el producto $m \times n$ de matriz, $A \times B$, en el cual las entradas de $A \times B$ se computan con un método específico de combinación de filas de A con columnas de B.

Matriz de coeficientes de un sistema de ecuaciones lineales (p. 133) Matriz cuyas entradas son los coeficientes de las variables en el sistema de ecuaciones lineales.

Punto medio (p. 163) El punto en un segmento que está a la misma distancia de los extremos del segmento.

Fórmula de punto medio (p. 169) Fórmula para calcular las coordenadas de un punto medio de los segmentos que conectan dos puntos en el plano de coordenadas.

Árbol de expansión mínima (p. 403) Árbol que atraviesa en un grafo que tiene peso total mínimo.

Regla de multiplicación (p. 532) Si A y B son dos eventos, $P(A \text{ y } B) = P(A)P(B \mid A)$.

Regla de multiplicación para eventos independientes (p. 523) Cuando los eventos A y B son independientes, la regla de multiplicación se simplifica a $P(A \text{ y } B) = P(A) \cdot P(B)$.

Glossary/Glosario

English	Español

Multiplicative inverse matrix See *inverse matrix.*

Matriz inversa multiplicativa Véase *matriz inversa.*

Multiply a matrix by a number (p. 85) Multiply each entry in a matrix, *A*, by the same number, *k*, to generate the entries in a new matrix, *kA*. (Also called *scalar multiplication.*)

Multiplicar una matriz por un número (p. 85) Multiplicar cada entrada en una matriz, *A*, por el mismo número, *k*, para generar las entradas en una nueva matriz, *kA*. (También conocida como *multiplicación escalar.*)

Mutually exclusive events (or *disjoint events*) (p. 523) Events that cannot occur on the same outcome.

Eventos mutuamente excluyentes (disjuntos) (p. 523) Eventos que no pueden ocurrir en el mismo resultado.

N

Negative correlation (p. 264) The points on a scatterplot have a downwards trend from left to right and so the slope of the regression line is negative.

Correlación negativa (p. 264) Los puntos en un diagrama de dispersión tienden a bajar de izquierda a derecha y por lo tanto la pendiente de la línea de regresión es negativa.

O

One-to-one function (p. 354) A function *f* for which each value of *f*(*x*) in the range of *f* corresponds to exactly one value of *x* in the domain of *f*.

Función biunívoca (p. 354) Función *f* para la cual cada valor de *f*(*x*) en el rango de *f* corresponde exactamente al valor de *x* en el dominio de *f*.

Orientation of a figure (p. 212) Can be determined by clockwise or counterclockwise labeling of consecutive vertices of a figure.

Orientación de una figura (p. 212) Se puede determinar por los rótulos de los vértices consecutivos de una figura en dirección de las manecillas del reloj o contrario a éstas.

Outlier on a scatterplot (p. 77) A point that does not follow the trend of the other points and so lies outside the main cluster of points.

Valor atípico (o Dato aberrante) en una gráfica de dispersión (p. 77) Punto que no sigue la tendencia de los demás puntos y queda fuera del grupo principal de puntos.

P

Parallel lines Lines that are coplanar and do not intersect.

Rectas paralelas Rectas coplanarias que no se intersecan.

Parallelogram (p. 154) A quadrilateral with opposite sides congruent.

Paralelogramo (p. 154) Cuadrilátero de lados opuestos congruentes.

Perfect correlation (p. 260) All points on a scatterplot fall on the regression line so that the correlation is 1 or −1.

Correlación perfecta (p. 260) Todos los puntos en una gráfica de dispersión recaen en la recta de regresión de modo que la correlación es biunívoca.

Perpendicular lines Lines that intersect at right angles.

Rectas perpendiculares Rectas que se intersecan en ángulos rectos.

Point matrix (p. 233) A one-column matrix whose entries are coordinates of a point in the plane. (See also *column matrix.*)

Matriz punto (p. 233) Matriz de una columna cuyas entradas son coordenadas un punto en el plano. (Véase también *columna de matriz.*)

Glossary/Glosario

English	Español

Positive correlation (p. 264) The points on a scatterplot have an upwards trend from left to right and so the slope of the regression line is positive.

Correlación positiva (p. 264) Los puntos en un diagrama de dispersión muestran una tendencia hacia arriba de izquierda a derecha y por lo tanto la inclinación de la línea de regresión es positiva.

Power function (p. 10) A function with rule of correspondence that can be expressed in the algebraic form $f(x) = ax^r$ $(r \neq 0)$.

Función exponencial (p. 10) Función con regla de correspondencia que puede expresarse con la forma algebraica $f(x) = ax^r$ $(r \neq 0)$.

Preimage (p. 199) If point A' is the image of a point A under a transformation, then point A is the preimage of point A'.

Preimagen (p. 199) Si el punto A' es la imagen de un punto A después de una transformación, entonces el punto A es la preimagen del punto A'.

Probability distribution A description of all possible quantitative (numerical) outcomes of a chance situation, along with the probability of each outcome; the distribution may be in table, formula, or graphical form.

Distribución probabilística Descripción de todos los resultados posibles de una situación aleatoria, junto con la probabilidad de cada uno; la distribución puede estar en forma de tabla, fórmula o gráfica.

Program Evaluation and Review Technique (PERT) (p. 435) A technique using critical path analysis to optimally schedule large projects consisting of many subtasks; developed in the 1950s to help create military defense systems. (See also *Critical Path Method*.)

Técnica de Evaluación y Revisión de Programa (PERT, por sus siglas en inglés) (p. 435) Técnica que utiliza el análisis de trayectoria crítico para programar de manera óptima proyectos grandes compuestos de muchos subproyectos; desarrollado en la década de 1950 como ayuda para crear sistemas de defensa militar. (Véase también *Método de trayectoria crítica*.)

Project digraph (p. 437) A digraph representing a large project, in which the vertices represent the subtasks of the project and the directed edges show the immediate prerequisite(s) for each task.

Dígrafo del proyecto (p. 437) Grafo que representa un proyecto grande, en el cual los vértices representan los sub-proyectos del proyecto y las aristas dirigidas muestran los prerrequisito(s) para cada trabajo.

· (Q) ·

Quadratic equation An equation that can be expressed in the form $ax^2 + bx + c = 0$ $(a \neq 0)$.

Ecuación cuadrática Ecuación que puede expresarse en la forma $ax^2 + bx + c = 0$ $(a \neq 0)$.

Quadratic formula (p. 340) A formula for the solutions of a quadratic equation in the form $ax^2 + bx + c = 0$: $x = \frac{-b}{2a} \pm \frac{\sqrt{b^2 - 4ac}}{2a}$.

Formula cuadrática (p. 340) Fórmula para las soluciones a una ecuación cuadrática que puede expresarse en la forma $ax^2 + bx + c = 0$: $x = \frac{-b}{2a} \pm \frac{\sqrt{b^2 - 4ac}}{2a}$.

Quadratic function A function with rule of correspondence that can be expressed in the algebraic form $f(x) = ax^2 + bx + c$ $(a \neq 0)$.

Función cuadrática Función con regla de correspondencia que puede expresarse en la forma algebraica $f(x) = ax^2 + bx + c$ $(a \neq 0)$.

Glossary/Glosario

English	Español

Random digit A digit selected from 0, 1, 2, 3, 4, 5, 6, 7, 8, 9 in a way that makes each of the digits equally likely to be chosen (has probability $\frac{1}{10}$); successive random digits should be independent, which means that if you know what random digits have already been selected, each digit from 0 through 9 still has probability $\frac{1}{10}$ of being the next digit.

Range of a function (p. 330) For a function f, the values of the dependent variable y corresponding to values of x in the domain of f. (Also called *output values* of the function.)

Rank correlation (p. 259) A correlation based on two different rankings of the same items. Two types of rank correlation are Spearman's and Kendall's.

Rare event (p. 566) In a waiting-time distribution, an event that falls in the upper 5% of the distribution.

Rectangle A quadrilateral with opposite sides congruent and four right angles.

Regression equation (p. 282) The equation of the least squares regression line for the points on a scatterplot.

Regression line See *least squares regression line.*

Regular polygon A polygon in which all sides are congruent and all angles are congruent.

Residual (error) (p. 283) For points used to calculate the regression equation, the difference between the observed value of y and the value of y predicted by the regression equation, $y - \hat{y}$.

Response variable See *dependent variable.*

Rhombus A quadrilateral with all four sides congruent.

Rigid transformation (p. 196) A transformation of points in the plane that repositions figures without changing their shape or size.

Dígito aleatorio Dígito escogido de 0, 1, 2, 3, 4, 5, 6, 7, 8, 9 de modo que cada uno tenga la misma probabilidad de elegirse que cualquier otro (tiene probabilidad $\frac{1}{10}$); los dígitos aleatorios consecutivos deben ser independientes, o sea, se conocen los dígitos aleatorios ya escogidos, cada dígito de 0 a 9 aún tiene $\frac{1}{10}$ de escogerse cono el dígito siguiente.

Rango de una función (p. 330) Para una función f, los valores de las variables dependientes y que corresponden con los valores de x en el dominio de f. (También llamada *valores de salida* de la función.)

Correlación de rango (p. 259) Correlación basada en dos diferentes rangos del mismo objeto. Dos tipos de rangos son los de Spearman y de Kendall.

Evento raro (p. 566) En una distribución de espera de tiempo, un evento que está en el 5% superior de la distribución.

Rectángulo Cuadrilátero con lados opuestos congruentes y cuatro ángulos rectos.

Ecuación de regresión (p. 282) Ecuación de la recta de regresión de los mínimos cuadrados para los puntos en una gráfica de dispersión.

Recta de regresión Véase *línea de regresión de mínimos cuadrados.*

Polígono regular Polígono cuyos lados y ángulos son todos congruentes.

Residuo (error) (p. 283) Para puntos que se usan para calcular la ecuación de la regresión, la diferencia entre el valor observado de y y el valor y predicho por la ecuación de la regresión, $y - \hat{y}$.

Variable respuesta Véase *variable dependiente.*

Rombo Cuadrilátero con cuatro lados congruentes.

Transformación rígida (p. 196) Una transformación de puntos en un plano que vuelve a colocar figuras sin cambiar su forma o tamaño.

Glossary/Glosario

English	Español

English

Rotation (p. 161) A rigid transformation of points in the plane that rotates (or turns) figures about a specified point, called the *center of rotation*, through a specified angle, called the *directed angle of rotation*.

Rotation matrix (p. 232) A matrix, which when multiplied on the right by a point matrix, has the effect of rotating the point about the origin through a specified angle.

Row matrix (p. 105) A matrix consisting of one row. (Also called a *one-row matrix* or a *row vector*.)

Row of an $m \times n$ matrix (p. 76) A horizontal array of n numbers in the matrix.

Row sum of a matrix (p. 81) The sum of all the numbers in a row of a matrix.

(S)

Sample space A list of all possible outcomes of a chance situation.

Scalar multiplication See *multiply a matrix by a number*.

Scale factor of a size (or similarity) transformation (p. 215) The ratio of the distance between any two image points and the distance between their preimages under the transformation.

Scatterplot matrix (p. 266) A matrix where each entry is a scatterplot formed using a pair of variables from a set of multivariate data.

Similar figures (p. 215) Figures that are related by a similarity transformation. These figures have the same shape, regardless of position or orientation, but may be of different scales.

Similarity transformation (p. 196) Composition of a size transformation and a rigid transformation (possibly the *identity transformation*). Such a transformation resizes a figure in the plane without changing its shape.

Simulation Imitating a real-life situation by creating a mathematical model that captures the situation's essential characteristics.

Español

Rotación (p. 161) Transformación rígida de puntos en un plano en la que una figura gira en torno a un punto fijo, llamado *centro de rotación*, y a través de un ángulo especificado, llamado *ángulo dirigido de rotación*.

Matriz rotación (p. 232) Matriz que al multiplicarse a la derecha por un punto de matriz tiene el efecto de rotar el punto del origen a través de un ángulo dado.

Matriz fila (p. 105) Matriz que consta de una fila. (También llamada *matriz de una fila* o *vector fila*.)

Fila de una matriz $m \times n$ (p. 76) Arreglo horizontal de n números en la matriz.

Suma de fila de una matriz (p. 81) Suma de todos los números de la fila de una matriz.

Espacio muestral Lista de todos los resultados posibles de un suceso.

Multiplicación escalar Véase *multiplicar una matriz por un número*.

Factor de escala de una transformación de tamaño (o similitud) (p. 215) Radio de distancia entre cualquiera dos puntos de una imagen y la distancia entre sus preimagenes bajo la transformación.

Matriz de gráfica de dispersión (p. 266) Matriz en la cual cada entrada es una gráfica de dispersión formada usando un par de variables de un conjunto de datos multivariados.

Figuras semejantes (p. 215) Figuras que se relacionan por una transformación de similitud. Estas figuras tienen la misma forma, sin importar la posición u orientación pero pueden ser de diferentes escalas.

Transformación de similitud (p. 196) Composición de una transformación de tamaño y una transformación rígida (posiblemente la *transformación de identidad*). Tal transformación cambia de tamaño en el plano sin cambiar su forma.

Simulación Imitación de una situación real al crear un modelo matemático que captura las características esenciales de la situación.

Glossary/Glosario

English	Español

Sine function (p. 457) If $P(x, y)$ is a point (not the origin) on the terminal side of an angle θ in standard position and $r = \sqrt{x^2 + y^2}$, then $\sin \theta = \frac{y}{r}$. If A is an acute angle in a right triangle, then $\sin A = \frac{\text{length of side } opposite \angle A}{\text{length of } hypotenuse}$.

Función de seno (p. 457) Si $P(x, y)$ es un punto (no el origen) en el lado terminal de un ángulo θ en posición estándar y $r = \sqrt{x^2 + y^2}$, entonces $\sin \theta = \frac{y}{r}$. Si A es un ángulo agudo en un triángulo rectángulo, entonces $\sin A = \frac{\text{medida del cateto } opuesto\ a \angle A}{\text{medida de la } hipotenusa}$.

Size of a matrix (p. 76) The number of rows and columns in a matrix, denoted by (*number of rows*) × (*number of columns*).

Tamaño de una matriz (p. 76) Número de filas y columnas en una matriz, indicado por (*número de filas*) × (*número de columnas*).

Size transformation (or *dilation*) (p. 205) A transformation that moves each point P in the plane along a ray through P from a specified point O, called the *center of the transformation*, according to the rule $OP' = kOP$, where P' is the image of P and $k \neq 0$ (k is called the *scale factor* or *magnitude* of the transformation).

Transformación de tamaño (o *dilatación*) (p. 205) Transformación que mueve cada punto P en el plano a lo largo de una semirrecta a través de P del punto dado O, llamado *centro de la transformación*, según la regla $OP' = kOP$, donde P' es la imagen de P y $k \neq 0$ (k se llama el *factor de posicionamiento* o *magnitud* de la transformación.

Slope of a line Ratio of change in y-coordinates to change in x-coordinates between any two points on a nonvertical line; $\frac{change\ in\ y}{change\ in\ x}$ or $\frac{\Delta y}{\Delta x}$; indicates the direction and steepness of a line.

Pendiente de una recta Radio de cambio en las coordenadas y para cambiar en las coordenadas x entre dos puntos cualesquiera en una recta no vertical; $\frac{cambio\ en\ y}{cambio\ en\ x}$ o $\frac{\Delta y}{\Delta x}$; indica la dirección e inclinación de una recta.

Square matrix (p. 138) A matrix with the same number of rows and columns.

Matriz cuadrada (p. 138) Matriz con el mismo número de filas y columnas.

Spanning tree (p. 403) A tree in a vertex-edge graph that includes all the vertices of the graph.

Árbol de expansión (p. 403) Árbol en un grafo que incluye todos los vértices de la gráfica.

Strength of a correlation (p. 260) The association between two variables is strong if the points cluster closely to the regression line and weak if the distances from the regression line to the points tend to be large.

Fuerza de una correlación (p. 260) La asociación entre dos variables es fuerte si los puntos se agrupan cerca de la recta de regresión y débil si las distancias de la recta de regresión a los puntos tiende a ser grande.

Subscript notation (p. 189) A letter or number displayed slightly below a variable. Subscripts are used to discriminate between variables that have the same letter symbol.

Notación de subíndice (p. 189) Una letra o número que aparece ligeramente debajo de una variable. Las notaciones de subíndice se usan para discriminar entre variables que tienen el mismo símbolo de la letra.

Substitution method (p. 51) A method used to solve a system of linear equations. Two equations with two variables are combined into a single equation with only one variable by *substituting* an expression for a variable from one equation into the other.

Método de substitución (p. 51) Método que se utiliza para resolver un sistema de ecuaciones lineales. Dos ecuaciones con dos variables se combinan en una sola ecuación con únicamente una variable al *sustituir* una expresión por una variable de una ecuación a la otra.

Glossary/Glosario

English	Español

Sum of squared errors (SSE or sum of squared residuals) (p. 285) The sum of the squared residuals, $SSE = \Sigma(y - \hat{y})^2$.

Suma de cuadrados debido al error (SSE, por sus siglas en inglés, o suma de residuales cuadrados) (p. 285) La suma de residuos cuadrados, $SSE = \Sigma(y - \hat{y})^2$.

Symmetric matrix (p. 92) A square matrix that exhibits reflection symmetry about its main diagonal.

Matriz simétrica (p. 92) Matriz cuadrada que muestra simetría de reflexión de su diagonal principal.

System of equations A set of two or more equations with two or more variables for which common solution(s) are sought.

Sistema de ecuaciones Conjunto de dos o más ecuaciones con dos o más variables para las cuales se buscan solución(es) en común.

- - - - - - - - - - - - - - - - - - - **T** - - - - - - - - - - - - - - - - - - -

Tangent (p. 180) A line is *tangent* to a given curve at a point, called the *point of tangency*, if the line touches the curve at that point, but does not cross the curve at that point.

Tangente (p. 180) Una recta es *tangente* de una curva dada en un punto, llamado *punto de tangencia*, si la recta pasa a través de la curva en sólo un punto pero no cruza la curva en ese punto.

Tangent function (p. 462) If $P(x, y)$ is a point (not the origin) on the terminal side of an angle θ in standard position and $r = \sqrt{x^2 + y^2}$, then $\tan \theta = \frac{y}{x}$ ($x \neq 0$). If A is an acute angle in a right triangle, then $\tan A = \frac{\text{length of side } opposite \angle A}{\text{length of side } adjacent \text{ to } \angle A}$.

Función de tangente (p. 462) Si $P(x, y)$ es un punto (no el origen) en el lado terminal de un ángulo θ en posición estándar y $r = \sqrt{x^2 + y^2}$, entonces $\tan \theta = \frac{y}{x}$ ($x \neq 0$). Si A es un ángulo agudo en un triángulo rectángulo, entonces, $\tan A = \frac{\text{la medida del cateto } opuesto \text{ a } \angle A}{\text{la medida del cateto } adyacente \text{ a } \angle A}$.

Taxi-distance (p. 191) The shortest distance between two locations following a path along the edges of a square grid (or parallel to those edges).

Distancia taxi (p. 191) La menor distancia entre dos posiciones seguida de un camino a lo largo de las aristas de una cuadrícula cuadrada (o paralela a esas aristas).

Terminal side of an angle (p. 461) The position of a ray that is the side of an angle after rotating about the angle's vertex from the initial position.

Lado terminal de un ángulo (p. 461) La posición de una semirrecta que es el lado de un ángulo después de rotar en torno al vértice del ángulo de una posición inicial.

Transformation (p. 161) A one-to-one correspondence (function) between points of a plane.

Transformación (p. 161) Correspondencia (función) biunívoca entre dos puntos de un plano.

Translation (p. 197) A rigid transformation that shifts all points in the plane a specified distance and direction, determined by the *translation vector*.

Traslación (p. 197) Transformación rígida que cambia todos los puntos en el plano a una distancia y dirección dadas, determinadas por el *vector de traslación*.

Transpose of a matrix (p. 126) The matrix obtained from a given matrix A by interchanging the rows and columns of A. Denoted by A^T.

Trasposición de una matriz (p. 126) Matriz que se obtiene de una matriz dada A al intercambiar las filas y columnas de A, la cual se denomina A^T.

Glossary/Glosario

English	Español

Traveling Salesperson Problem (p. 407) A problem related to vertex-edge graphs stated informally as follows: A sales representative wants to visit several different cities, each exactly once, and then return home. Among the possible routes, which will minimize the total distance traveled? (Also called the *Traveling Salesman Problem* or the *TSP*.)

Tree (p. 403) A connected graph that has no circuits.

Trial (or sometimes *run*) One repetition of a simulation.

Trigonometric functions (p. 457) The sine, cosine, and tangent functions and (to be defined in a later course) their reciprocals.

TSP See *Traveling Salesperson Problem*.

Problema del vendedor viajero (p. 407) Problema relacionado informalmente con grafos de la siguiente manera: un vendedor quiere visitar diferentes ciudades, solamente una vez cada una, y después regresar a casa. De entre los resultados posibles, ¿cuál minimizará la distancia total que se viaja? (También conocido como *TSP*, por sus siglas en inglés.)

Árbol (p. 403) Gráfica conectada que no tiene circuitos.

Prueba Repetición de un simulacro.

Funciones trigonométricas (p. 457) Funciones de seno, coseno y tangente y (se definirán en un curso posterior) sus recíprocos.

TSP Véase *Problema del vendedor viajero*.

Venn diagram A diagram where mutually exclusive events are represented by non-overlapping circles and events that are not mutually exclusive are represented by overlapping circles.

Vertex-edge graph (p. 68) A diagram consisting of points (called *vertices*) along with segments or arcs (called *edges*) joining some of the points. (Also simply called a *graph*.)

Diagrama de Venn Diagrama en el cual los eventos mutuamente excluyentes se presentan por círculos no sobrepuestos y los eventos que no son mutuamente excluyentes se presentan por círculos sobrepuestos.

Grafo (p. 68) Diagrama que consta de un conjunto de puntos (los *vértices*) junto con segmentos o arcos (las *artistas*) que unen algunos de los puntos. (También llamada simplemente una *gráfica*.)

Waiting-time (geometric) distribution (p. 525) A probability distribution of the number of independent trials required to get a specified outcome called a "success." The probability that a success will occur must be the same on every trial.

Weighted graph (p. 403) A vertex-edge graph with numbers (*weights*) on its edges.

Weights (p. 403) Numbers that are placed on the edges (or vertices) of a vertex-edge graph.

With replacement Selecting a sample from a set so that each selection is replaced before selecting the next; thus, a member of the set can be selected more than once.

Distribución (geométrica) del tiempo de espera (p. 525) Distribución probabilística del número de pruebas independientes requeridas para obtener un resultado especificado llamado "éxito." La probabilidad de que un éxito suceda debe ser la misma en cada prueba.

Gráfico cargado (p. 403) Grafo con números (*pesos*) en sus aristas.

Pesos (p. 403) Números que se colocan en las artistas (o vértices) de un grafo.

Con devolución Selección de una muestra de un conjunto de modo que cada selección se devuelve antes de elegir la siguiente; así cada miembro del conjunto puede escogerse más de una vez.

Glossary/Glosario

English	Español
Without replacement Selecting a sample from a set so that each selection is not replaced before selecting the next; no member of the set can be selected more than once.	**Sin devolución** Selección de una muestra de un conjunto de modo que cada selección no se devuelve antes de elegir la siguiente; así cada miembro del conjunto no puede escogerse más de una vez.

· (X) ·

English	Español
x-intercept(s) of a graph The point(s) where the graph intersects the x-axis.	**Intersección(es) x de una gráfica** El punto o los puntos en que una gráfica interseca el eje x.

· (Y) ·

English	Español
y-intercept(s) of a graph The point(s) where the graph intersects the y-axis.	**Intersección(es) y de una gráfica** El punto o los puntos en que una gráfica interseca el eje y.

· (Z) ·

English	Español
Zero matrix (p. 134) A matrix in which every entry is 0. See *additive identity matrix*.	**Matriz cero** (p. 134) Matriz en la cual cada entrada es cero. Véase *matriz identidad aditiva*.

Index of Mathematical Topics

Index of Mathematical Topics (continued)

Index of Mathematical Topics (continued)

Index of Mathematical Topics (continued)

Index of Mathematical Topics (continued)

Index of Contexts

Index of Contexts (continued)

Index of Contexts (continued)

Index of Contexts (continued)

Index of Contexts (continued)

Index of Contexts (continued)

Photo Credits